THE ART OF DICING MICROELECTRONIC SUBSTRATES

Gideon Levinson

ISBN 979-8-35094-226-2

I would like to thank my dear wife Haya, who should be endorsed for her patience during the many days and hours that I had to leave her while searching, writing, and preparing the many detailed sketches.

TABLE OF CONTENTS

CHAPTER 1

INTRODUCTION

I have been in the dicing business for almost forty years, from the times when all dicing saws in the microelectronic industry were bench-type saws, most of them manual operated. During those days, in most dicing processes, mainly in the semiconductor industry, dicing silicon was done in a manual mode, dicing only partial cuts and then manually breaking the substrates to separate into individual dies. I remember customers using a bottle of wine or a round wooden roller warped with a towel to roll over the partly diced silicon wafers to separate the dies. (see fig. 1 & fig. 2)

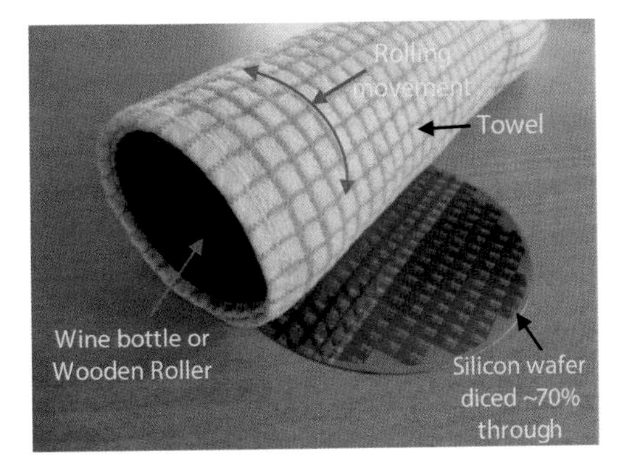

Figure 1. – Rolling over a silicon wafer

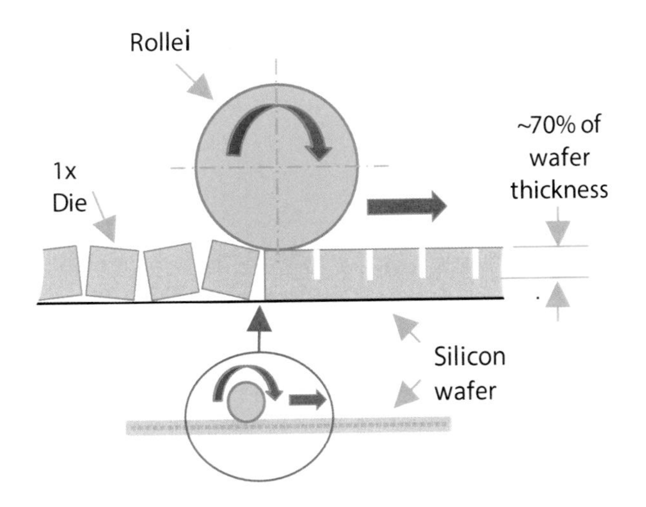

Figure 2. – Individual dies breaking during rolling

When the industry went into automation in the manufacturing processes, new mounting techniques were developed in order to ease the handling from station to station in the manufacturing process. At this point, the cut-through was introduced with new requirements to dice fast with better cut location accuracy and better cut quality specs. Those new requirements required newly developed and improved dicing saws with superior new accuracy and new features to handle the new, demanding requirements. In addition, new applications were introduced in the market using much harder and more brittle material substrates. This led to the development of new blade matrices/binders in order to meet the new quality specs and new UPH requirements. During all those years, only a few technical articles were published, but no real technical book that covered the blade technology and the application side was published. This present technical review will cover the blade and the application sides in a more detailed manner to better support and help dicing process engineers and managers to get a better understanding of the dicing process and a bit of its history.

I would like to take this opportunity to thank Mr. Dan Vilenski, who was the VP R&D manager of Kulicke and Soffa in Philadelphia and later the General Manager of Kulicke and Soffa Israel – (Kulso). Dan chose me to be the development manager for the dicing blade development project in the US. Dan was later the initiator and General Manager of KLA in Israel, The US Israeli BIRD Foundation Manager and later the CEO of Applied Materials in Israel.

I would like to thank two dear gentlemen for working with me in the K&S US at the initial development of the diamond nickel blade process. Dave Mynick was my partner engineer and was deeply involved in all the initial diamond plating process. My dear friend Jack Lyons was the K&S logistic Manager for all our development activities and later helped me in the marketplace as the East Coast Marketing and Sale Manager.

I would like to thank Mr. Moshe Jacobi the K&S Israel General Manager of supporting me in my technical assignment in North America as the dicing blade Technical Marketing Manager. This assignment got me to learn and understand better the Microelectronic industry with their many different dicing applications.

I would like to thank Mr. Avi Ben-Har, who was the Bonding Tools Sales and Marketing Manager during our time in Kulicke and Soffa. Avi pushed me into writing technical articles and was probably the main initiator that got me into the world of technical writing.

I would also like to thank Mr. Paul Amberg from Minitron in Germany, who was a sales and technical mentor in the marketplace. Paul is a very knowledgeable person with a lot of experience and know-how in the marketplace. Paul recommended that I add an important chapter to the book and gave me a few hints to better organize the book and make it easier for future readers. Needless to say, Paul has been a dear friend for many years.

Last, I would like to thank my Sons Oren Levinson and Tomer Levinson. Oren has a few engineering degrees and is managing a large high tech engineering group running a large-scale project. Oren helped me to forward my initial word document prior to the printing process. Tomer is a Senior Engineering Manager in a Semiconductor high tech unique product development projects. Tomer's help came at the last critical step process of preparing and organizing the text and sketches to meet the printing requirements for the book.

CHAPTER 2: DICING PRINCIPLES

2.1 - WHY DICING?

The high demand for mass production in the microelectronic industry, mainly of small components, led the industry to develop unique manufacturing processes. The idea was to manufacture the devices on round wafers and other flat substrate geometries (see fig. 3).

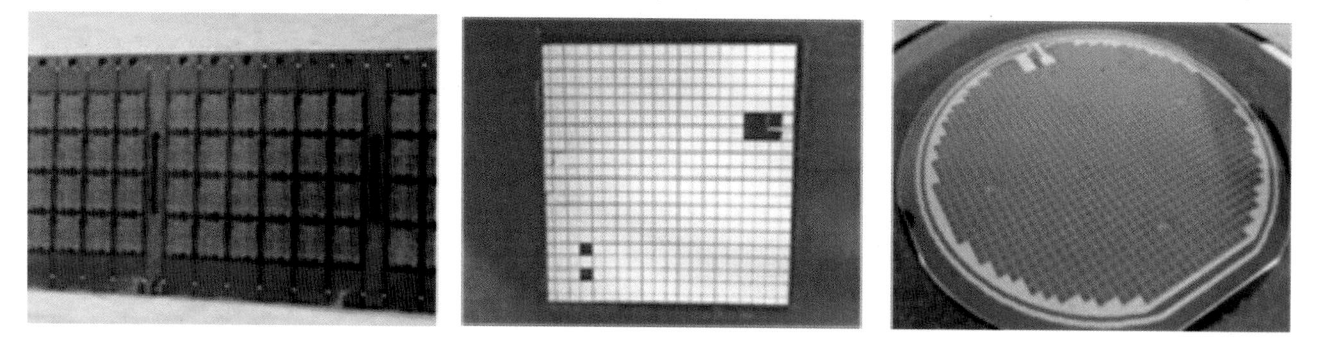

Figure 3. Different Microelectronic substrates

The substrate base material types and thicknesses depend on the end product. They can vary from very hard and brittle materials to very soft and ductile materials or a combination of both. The manufacturing processes of the different devices can be described in a general way as mini electronic boards. The different processes involve photoresist masking, chemical etching, plating, ion etching, and many other unique processes.

The different devices are processed in an array format with a gap between them called "streets" or a dice line. The substrates are mounted and aligned on a dicing saw chuck using vacuum, mounting tape, or other media. A thin diamond dicing blade in different binder matrices and geometries is dicing in the "streets" to separate the devices (Dies). (see fig. 4) The dicing blade and the application side will be discussed in detail.

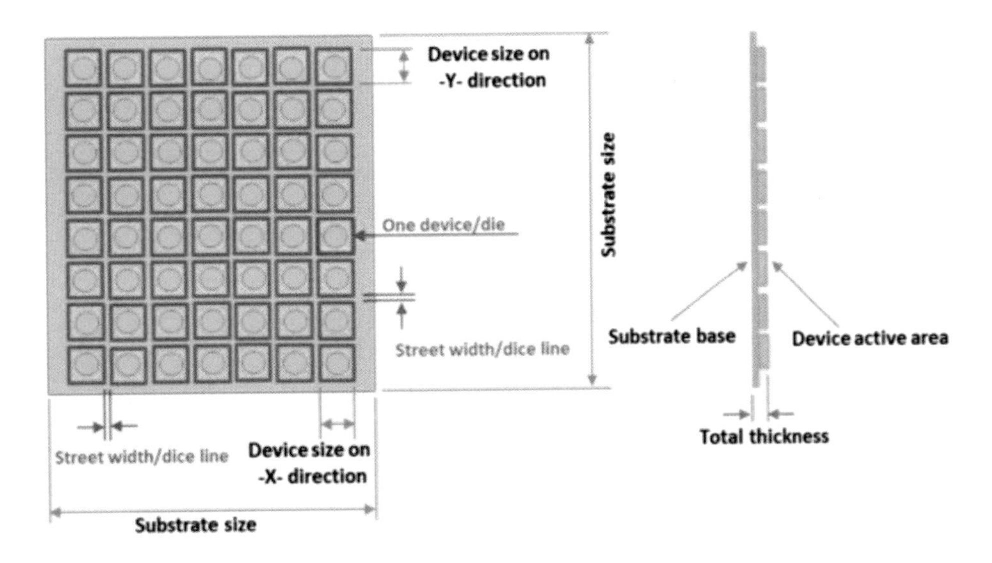

Figure 4. A generic microelectronic substrate geometry.

2.2 - SUBSTRATE MATERIAL ENGINEERING

The need for different substrates requires substrate material engineering development and a much denser substrate geometry design. Many substrates, mainly in the optic industry, also have hard and brittle coatings that are not very "friendly" to the dicing process. The dense geometries of the devices are getting smaller and smaller, which creates mounting and dicing issues requiring special dicing saw features, optimization of dicing parameters, and mainly new state-of-the-art dicing blade matrices. In parallel, the market is continually pushing for better UPH performance while maintaining quality, which in many cases is a real challenge requiring continued process and blade optimization. Below is a generic table just to show a few of the most popular substrate materials used in the microelectronic industry, indicating the material characteristics differences that need to be handled during dicing (see fig 5). More detailed information on substrate materials will be discussed in the application section.

Material	Application	Hardness: Soft	Hardness: Hard	Brittleness: Ductile	Brittleness: Brittle	Common Substrate Thicknesses mm
Sil. / GaAs	Semiconductor	☆			☆	0.10 – 1.00
Hard Al. 96 / 97%	Hybrids, Sensors		☆		☆	0.250 - 3.00
Glass / Silica	Fiber Optics, Image Sensors, Optical Sensors Optical Filters	☆			☆	0.150 - 8.00
Copper w. Ni/Pd or Sn Coating on Epoxy Molding	QFN / MLP for: Automotive, Wireless Analog P. Switching	☆		☆	☆	0.50 - 1.20
PC board	LED, Other base Substrate	☆			☆	0.20 – 2.00
Altec –AlTiCO3 (Different processes)	Magnetic heads		☆		☆	0.10 – 3.00

Figure 5. Popular substrate materials used in the microelectronic industry.

2.3 – DICING MACHINING MECHANISM

General review:

In general, dicing/sawing involves penetrating a solid material by using a thin dicing blade impregnated with diamond particles. The thin blade is machining out a small portion of material from the substrate, creating what is called a "Kerf." To better understand the machining mechanism, let's first discuss other common daily applications of sawing different materials in different industries.

For example, to separate a wood board or a metal rod, we normally use a metal saw with teeth. It can be a straight metal hand operated saw, an electrical band saw, an electrical metal circular saw, or many others. In all cases, hard metal teeth with sharp teeth geometry penetrating the wood or metal are pushing out small wood particles or small metal particles. The ability to easily penetrate into the wood or metal and to produce a nice clean cut with minimum loads and good quality depends on the saw metal material hardness, the teeth geometry, the number of teeth (Inches or millimeter), and the teeth sharpness. See figs. 6, 7, and 8 for a few examples:

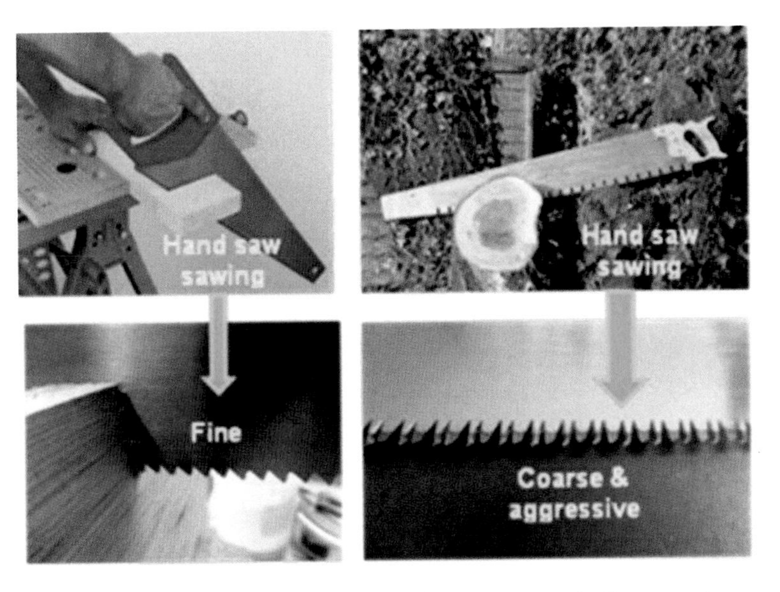

Figure 6. Sawing wood parts using metal saws with fine and coarse teeth

Figure 7. Sawing a round metal bar using a coarse teeth geometry

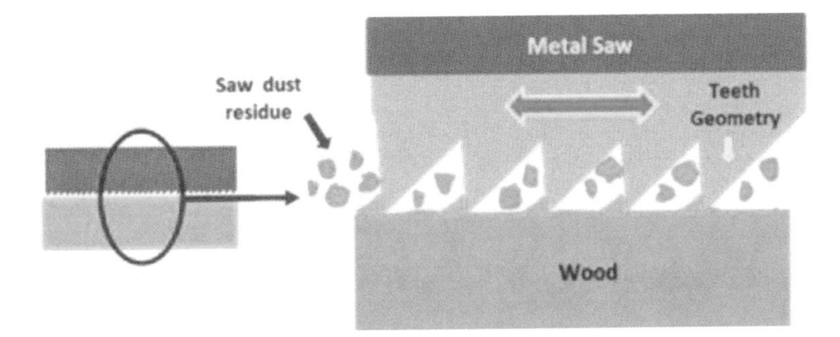

Figure 8. Wood sawing mechanism using a metal saw

Sawing wood and metal parts have their own challenges. However, the microelectronic substrates are, in most cases, much harder, more brittle, and in addition, require much smaller and more accurate geometries. Needless to say, the quality requirements of dicing/sawing are in a different "league." The dicing mechanism is similar; instead of sharp metal teeth, diamonds of different types, different geometries, different %, and different sizes are used. Dicing blades are made using different bond binders to hold the diamonds in place. The diamond type, size, concentration, and the media bond holding the diamonds depends on the type of substrate to be diced. Each diamond is acting as a single cutting tool. This will be discussed in detail later. The diamond ability to penetrate into the substrate material depends on a few parameters and can be called: The Diamond Dicing Machining Mechanism. Let's take a close look at the diamond contact point with the substrate being diced (see fig. 9).

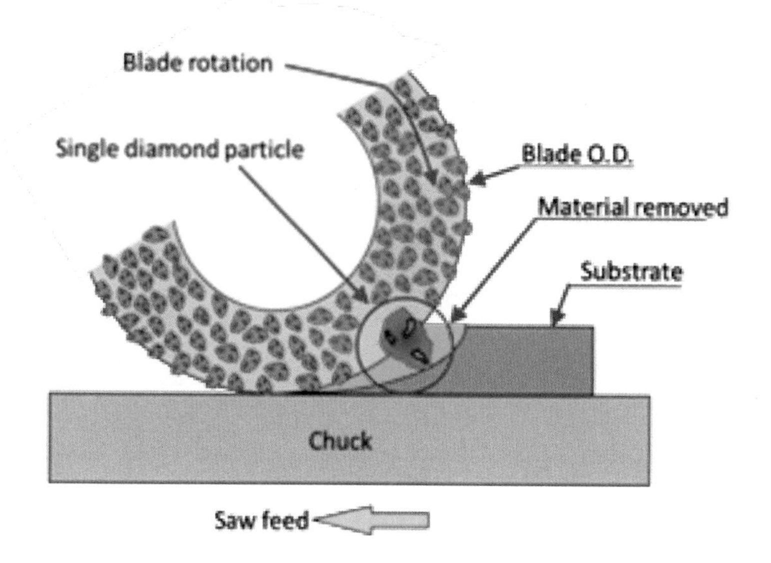

Figure 9. Dicing mechanism of a single diamond penetrating into a substrate

What can be seen in the above sketch is the theoretical material removed by a single diamond on each blade revolution (marked in light green). The amount of material removed is a function of the diamond size, diamond exposure, and feed rate (table speed). Faster table speed means a larger amount of material that each diamond is machining out. This will be discussed in more detail in the application and dressing section. The machinability of a single diamond depends on a good diamond exposure, which will allow it to freely penetrate the substrate with minimum loading, while a poor diamond exposure will result in high loads and poor cut quality (see fig. 10).

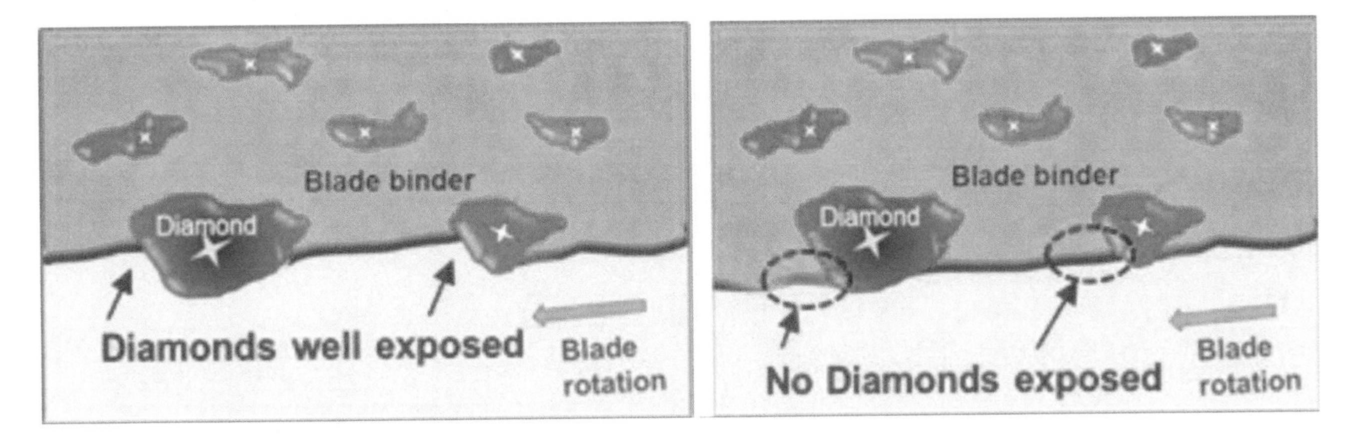

Figure 10. Well-exposed diamonds compare to poorly exposed diamonds

The dicing process in the microelectronic industry using thin diamond-type blades can be called mini "grinding." A thicker grinding wheel with aluminum oxide, silicon carbide, and other abrasive powders bonded together in different binders is used in grinding. Each particle in the grinding wheel penetrates the metal substrates in a similar manner to the diamonds in a thin diamond dicing blade. In both cases, many small particles are machining material out of the surface or out of a thin kerf/grove.

CHAPTER 3
DIAMONDS POWDERS

3.1 - DIAMONDS AND CBN USED IN MANUFACTURING DICING BLADES

General:

When indicating "DIAMONDS" for whatever reason, the first image in mind would be an expensive diamond Jewel.

This is not the case in the manufacturing of industrial diamond tools, and in our case, thin diamond dicing blades. The types of diamonds used for all types of blades are synthetic/polycrystalline diamond (PCD), or some are calling them man-made diamonds. Synthetic diamonds and natural diamonds are made from the same raw material, carbon. Natural diamonds are created by a geological process over millions of years. Synthetic diamonds are created using a HPHT (High Pressure, High Temperature) process. This process was developed in the mid-1950s by GE in the US. Natural diamonds vary in properties as they come from different areas of the globe and vary in properties and purities. Synthetic diamonds are way more consistent on properties like hardness, brittleness, thermal conductivities, electron mobility, and others. Another important advantage of synthetic diamonds is the ability to have perfect control over the manufacturing process parameters. During the years, many different diamonds were developed for new applications requiring different hardness, brittleness, shapes, and special diamond coatings. Some of the different diamonds will be discussed in the blade manufacturing and the application section. The differences and advantages between natural diamonds and synthetic diamonds can be summarized as follows:

- Much better-quality consistency of synthetic diamonds. What you buy today will be the same quality in years to come.
- Way more diamond selection options of synthetic diamonds.
- Similar hardness of the hard-synthetic diamonds compared to natural diamonds. However, natural diamonds are slightly harder.

There is a lot of good information on the internet for everyone who needs to explore more about synthetic diamonds. Diamond powders are manufactured in large and irregular crystals. They are crushed in different tumbling jars using metal and other media balls. This process is time-consuming, and at the end of the process, the diamonds are in a large variety of sizes. The next step is separating the crashed particles into different powder size ranges. One old method of separating the powders into the different size ranges is the elutriation method, which involves running the crushed diamonds from a water slurry main tank with low surface tension water through many glass jars. The first jar is large, and the next jars are smaller and smaller.

The slurry is entering the first jar from the bottom side, continuing to the next jar from the top side, and entering the next jar again from the bottom side. This is being repeated through all the jars, from the first large jar to the last smaller jar. The idea is that on each jar, the larger diamonds are gradually sinking down by gravity, and the smaller diamonds are continuing to the next smaller jar. The end of the process is having a different size of diamonds in every jar, with the larger diamonds in the larger jars and the smaller diamonds in the next smaller jars. Fig. 11 is a generic sketch just to understand the Elutriation process.

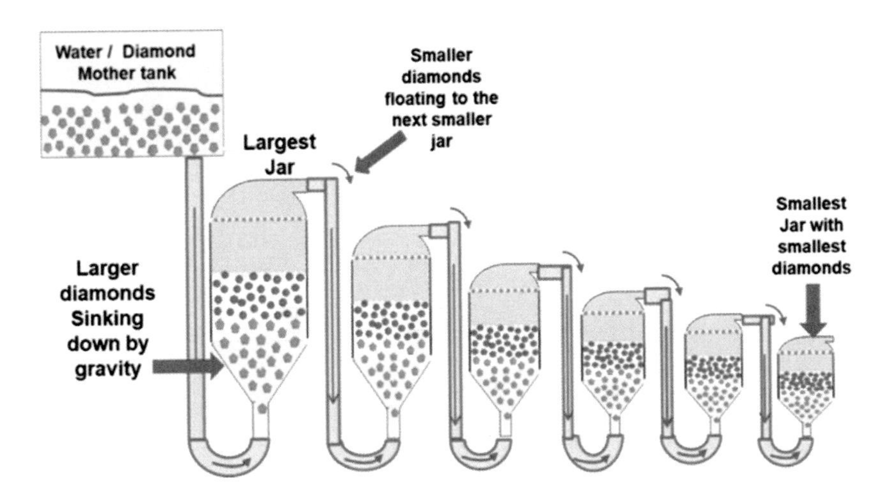

Figure 11. – Diamond micronizing by Elutriation process

In professional language, this process is called micronizing. This old elutriation process sounds simple but requires a lot of expertise and fine tuning. There are many Elutriation articles in the internet to anyone looking for more information. Today, there are different proprietary systems for the micronizing process and, in addition, very sophisticated QC instrumentation. In today's market, there is a strict requirement to get an accurate graph showing the diamond distribution range for each powder batch shipment. There are a few applications requiring a narrow range of the diamond distribution. This is mainly in the smaller micron sizes. In the diamond powder industry, there are standard ranges, for example, a range of 4-8μ has a small range of smaller diamonds of 4-microns and a small range of larger than 8-micron diamonds. The main difficulty is the ability to narrow those out of the 4-8μ range. See Fig. 12. The other important factor is the ability to maintain the narrow diamond distribution in the exact percentage per grit size for years to come, as this has a direct influence on the diamond blade manufacturers' ability to support their customers. The below generic distribution range shows the out-of-range diamonds on the lower side and on the oversize range in a generic 4-8μ grit size. (see fig. 12)

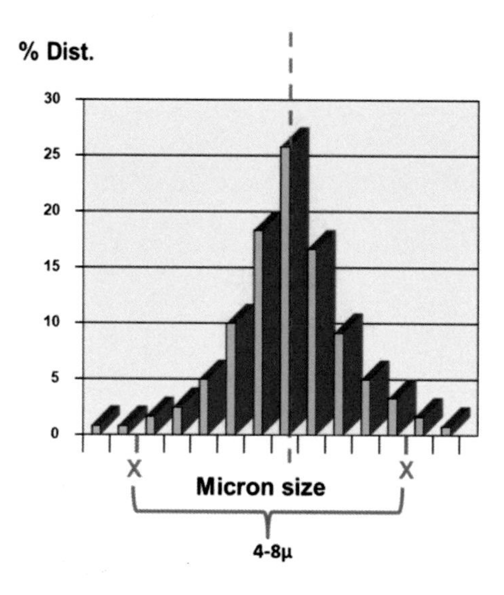

Figure 12. – Normal diamond mean size

Another important parameter to know is that any diamond range can have a different mean size. Let's take the same example of 4-8μ and see the below graph showing a 4-8μ with a mean size on the upper side of the micron range. (see fig. 13).

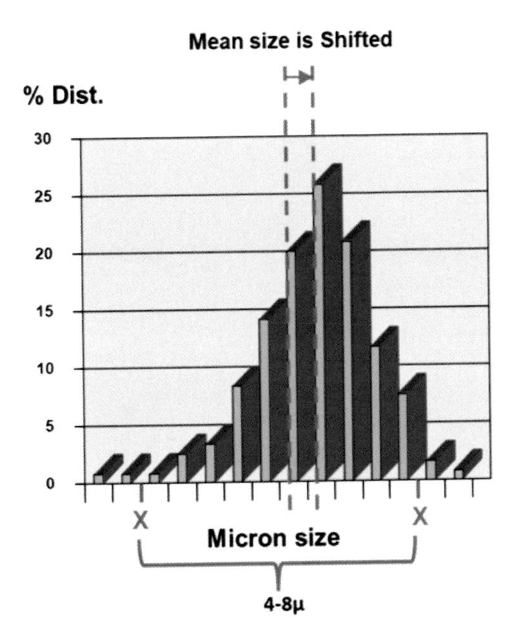

Figure 13. Same diamond grit size with a shifted mean size

It is still called 4-8μ, but it is a very different range from the upper graph in Fig. 12. This example can be a total reject for a specific application requiring a minimum chipping size. There are two main diamond measuring systems. The micron system and the US mesh size system

3.2 - MICRON DIAMOND POWDERS

Micron diamond powders are used for the smaller sizes. They are usually marked in a double-digit dimension, for example, 4-8μ, which means that the majority of particles are deemed to be located between these two sizes. Some manufacturers use single-digit dimensions; in this case, they would refer to a 6μ micron size, which represents the theoretical median size of the product. The size of a single diamond particle is the largest dimension of the particle. (see fig. 14).

Figure 14 . - Measuring Micron grit size

3.3 - MESH SIZE DIAMOND POWDERS

The US mesh system is counting the no. of square opening in one linear inch of screen. If there are 5 squares it means the Mesh size is 5. If there are 100 open squares in one inch = 100 Mesh. (see fig. 15).

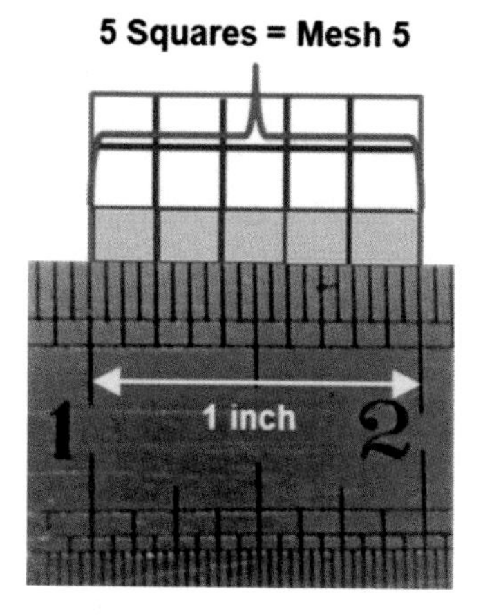

Figure 15. - Measuring in the mesh size methode

The actual sieving process is done on a series of sieves under a vibration unit. This method is used for many industrial powders requiring powder separation. However, with diamonds, the quality control after the mesh process is very critical and important. The higher the mesh number, the smaller the diamond size and the finer the range. The mesh size is not a 100 percent accurate dimension size like the micron system, but on larger diamond grit, this is less important. Below is a comparison of mesh and micron size fig. 16. There are many slightly different comparison charts. However, this is a good one as a reference.

COMPARISON OF MESH & MICRON SIZE

Mesh	Micron	Mesh	Micron
100	140	270	53
110	130	300	46
120	117	325	43
125	113	* 400	35
130	109	500	28
140	107	600	23
150	104	800	18
160	96	1000	13
170	89	1340	10
175	86	2000	6.5
180	84	5000	2.6
200	74	8000	1.6
230	65	10000	1.3
240	63	12700	1.0
250	61		

* Above 400 mesh will be designated in the micron range

Figure 16. - Diamond grit comparison chart

3.4 - CBN, CUBIC BORON NITRIDE - TRADE NAME - BORAZON ®

Another abrasive used in the grinding and dicing industries is CBN (cubic boron nitride). This material is manufactured by the same vendors that make synthetic diamonds. The raw material of diamonds is carbon, which is a problematic material when grinding or dicing ferrous materials. The diamond's sharpness is damaged at high velocity when in contact with the carbon content in ferrous-type materials. The CBN was developed to overcome those problems. CBN is second in hardness to diamond, with twice the hardness and four times the abrasion resistance of conventional abrasives. (see Fig. 17).

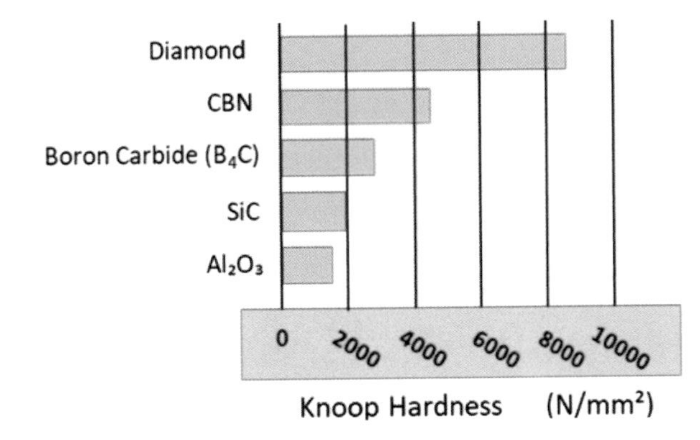

Figure 17. – Hard material comparison chart

The CBN has similar properties to diamond but superior chemical and thermal stability. It also has the same crystalline structure as diamond, but instead of carbon atoms, it is made of alternately bonded atoms of boron and nitrogen. CBN is made in mesh sizes and micron sizes, the same as synthetic diamonds. CBN is also coated with a similar coating to the diamond coating and is available in different crystalline characteristics. Most of the substrates used in the microelectronic industry are made of non-ferrous materials, so the use of CBN in dicing is limited. However, CBN is widely used in the grinding industry. Below are a few SEM of CBN powders (see fig. 18).

Figure 18. – SEM of different CBN powders

CHAPTER 4

GENERAL DICING SAW DESIGN AND A BIT OF THE DICING SAWS HISTORY:

4.1 - SAW CONSTRUCTION REVIEW

Below is a generic sketch presenting a dicing saw with main saw futures. (see fig. 17).

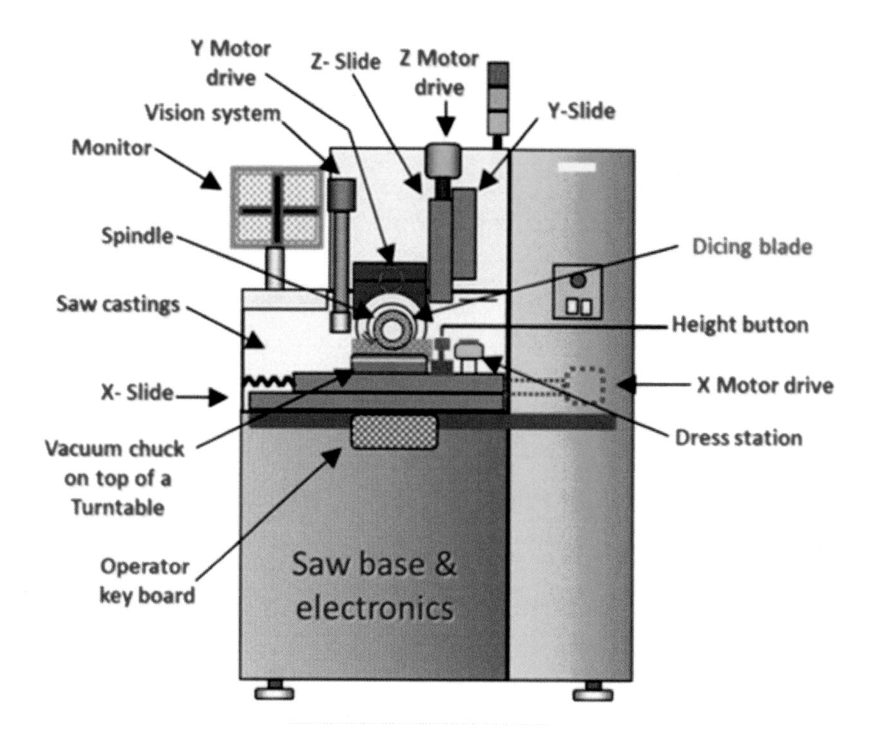

Figure 17. - Generic dicing saw

A dicing saw, in general, is similar in construction to a surface grinding machine. It has four axes—X, Y, Z and T (Theta), a cooling system, an air-bearing spindle, and a vacuum chuck to hold the workpiece. In addition, a dicing saw has many other accurate features:

- Vision system using an optic camera and monitor to view the workpiece
- Optic alignment vision software.
- Accurate vacuum turntable.
- Blade height calibration system.
- Blade dress station.
- Blade wear auto compensation.
- Sub-Micron accuracy using linear encoders on Y and sometimes on Z axis.
- Fully Dicing Automation & Measurements compensation control.
- Dynamic Balancing option.

There are many other features on specific saw systems, depending on the applications and customer requirements. A major feature option on some saws is a fully automated saw that includes a substrate handling, mounting, cleaning, and drying system. This system involves automatically picking a tape-mounted substrate from a cassette, mounting it to a vacuum chuck, aligning the substrate to the correct dicing orientation, inspecting the dicing quality after a rinsing and cleaning process, and demounting the mounted and diced substrate to another cassette. All features and operations of the saw are managed and controlled by a computer and can be saved per application. The accuracy and quality of the dicing are strongly related to the blade mounting accuracy, the blade performance, and the dicing parameters. This will be discussed in detail.

4.2 - DICING SAWS HISTORY

The first saws in the marketplace were not so sophisticated compared to the present dicing systems. All saws were bench-type saws with manual alignment systems and standard microscopes for vision. However, they were accurate small saws with air-bearing spindles and accurate movements on the Y axis, and they met the dicing needs of thirty to forty years ago. Below are a few photos of old bench-type saws: (see fig. 18).

Tempres 602

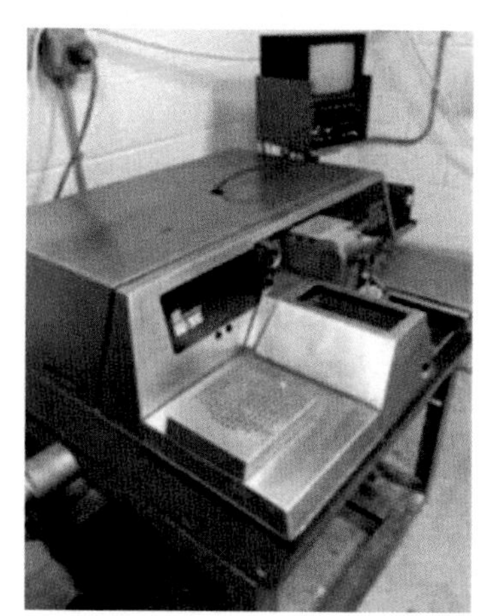

DISCO DAD-2H

Micro Automation 1006

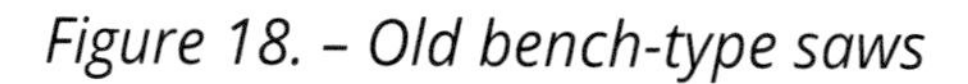

Figure 18. – Old bench-type saws

The first "standalone" dicing saw was the Kulicke & Soffa 775 dicing saw, introduced to the market in 1980. This saw was a major change in the market. It was a standalone dicing system with two spindle options, 2" and 4". It looked more like a mini surface grinding machine. The saw had air-bearing slides on –X and –Y axes and was a very robust dicing system compared to the existing bench-type saws. (see fig. 19). A major historic breakthrough for this saw was the replacement of many small 2" bench-type saws with the K&S 775 – 4" spindle in a few magnetic head (Disk Read/Write Head) houses.

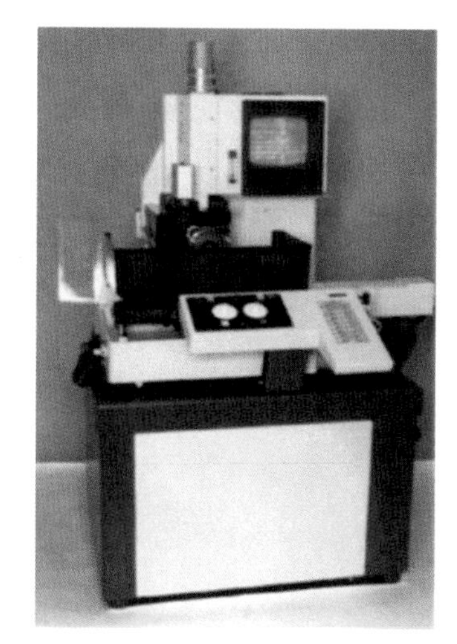

Figure. 19 – First stand alone dicing saw

During the 80s and 90s, new dicing saws were developed by the major dicing saw manufacturers. All the saws were self-standing designs to meet the semiconductor silicon wafer market and the growing microelectronic market, which is also called the hard material market. A major new requirement came when automation was needed. Tape mounting is a must in order to move the mounted substrates from one manufacturing station to the next. For this, a new vacuum chuck design was needed in order to minimize TS and mainly BS chipping. The old chuck design was a ring-type vacuum chuck. The ring chuck design was causing stresses during dicing, which affects mainly backside chipping. The new chuck design is a ceramic or metal porous chuck that drastically minimizes stresses during dicing and improves BS chipping. (see fig. 20).

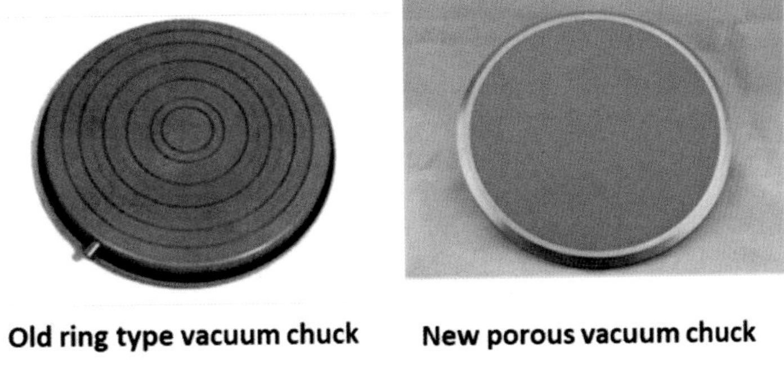

Figure 20. – Different vacuum chuck designs

Below is a sketch showing the stress areas causing backside damage, i.e., chipping at the vacuum groove areas of the old type of chuck. (see fig. 21).

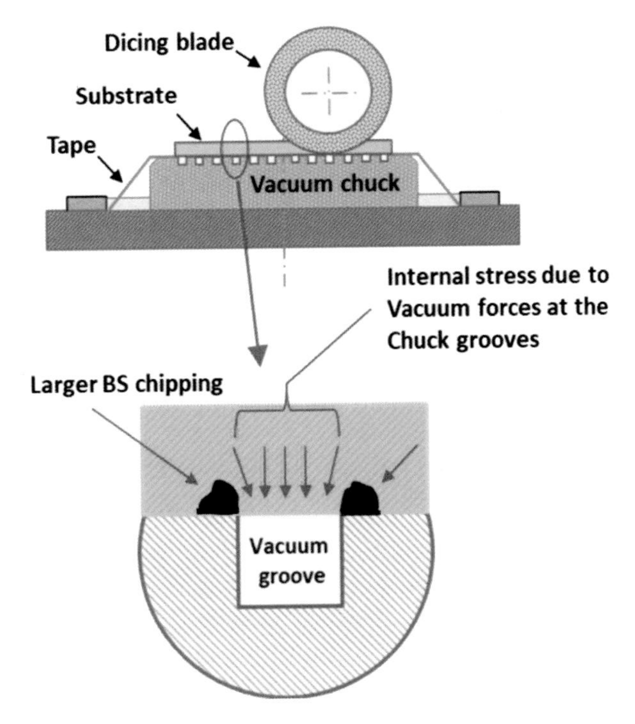

Figure 21. – Backside quality issues using the ring-type chuck

Historically, the blade edge height calibration, or what is called Chuck Zero, is done by rotating the blade at about 30 Krpm and moving down the Z axis till there is a mechanical contact between the blade edge and the chuck surface. At the same time, there is an electrical circuit contact that stops the –Z axis and causes the axis and spindle to move up. The mechanical contact of the blade is causing a small indention in the chuck, which over time is damaging the chuck area. To improve this process and maintain a clean, accurate chuck, a side height button was developed that is calibrated to the same height as the vacuum chuck. The height calibration is done on the height button to perform the same Chuck Zero calibration process without any damage to the chuck. Most dicing saw manufacturers have this system. (see fig. 22).

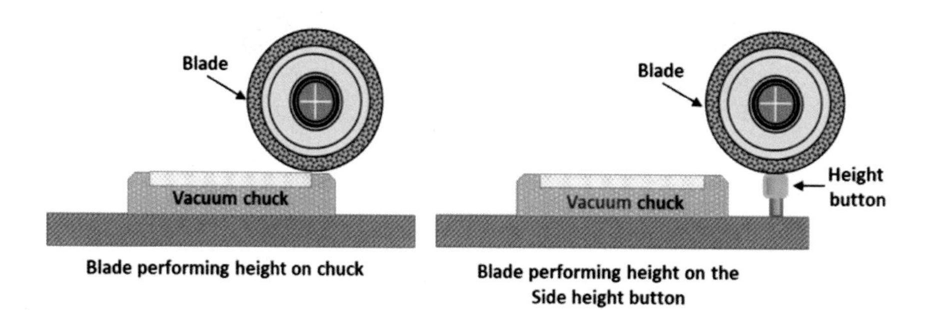

Figure 22. – Different locations for height calibration

Another improvement on the blade height calibration process was developed and is called; non-contact height (NCH). The NCH is a small optic/ Laser beam that can detect the blade edge. This system is calibrated with the chuck via the software. The –Z – axis / spindle is rotating at about 30Krpm and is moving down till the spindle / blade edge gets in contact with the NCH. At the contact the –Z – axis is moving up and the saw software recognizes this point as the chuck zero point. Most new saws today do have this option. Dressing on the saw was always and is a must for every new blade mounted on the spindle and during the dicing process. These processes required in the past to stop the dicing, dismount the wafer and mount a dressing media on the chuck. Today most dicing saw manufacturers developed a dress station located aside of the chuck to perform the dressing process without taking off the production substrate. This is a production time saver and leans towards automation. See fig. 23 showing the old method and the new dress station method. Dressing will be discussed in detail.

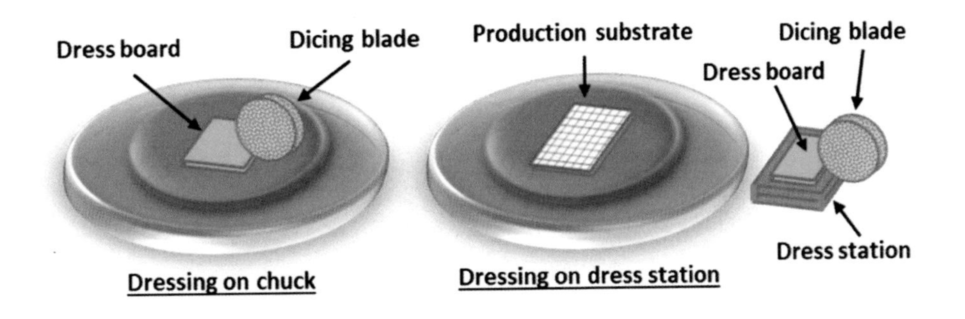

Figure 23. – Different dress mounting locations.

The above developments were a function of the market needs like improved accuracy, higher quality requirements, and better UPH requirements. The above improvements resulted in new and much better dicing systems. Below are a few of those improved dicing systems made by the major dicing saw manufacturers (see fig. 24).

Figure 24. – Newer dicing saws generations

4.3 - FULLY AUTOMATED DICING SAWS

Another major development over the years was the development of fully automated dicing systems. The need for those systems came from the need to automate mass production of products and the need for better UPH. Those systems have loading and unloading systems to load the tape mounted wafers to the chuck, and from the chuck, backload them to a cassette after dicing. The new systems include high-speed spindles up to 60 Krpm, automated aligning software, water rinsing & drying the wafers, special software to perform a pre-cut/initial dressing for new mounted blades, automated height calibration to compensate for blade wear, BBD broken blade detector, automated kerf check, automatic Y offset, touch screen monitor, UV curing station, and a few more. Below are photos representing some major dicing systems on the market (see fig. 25).

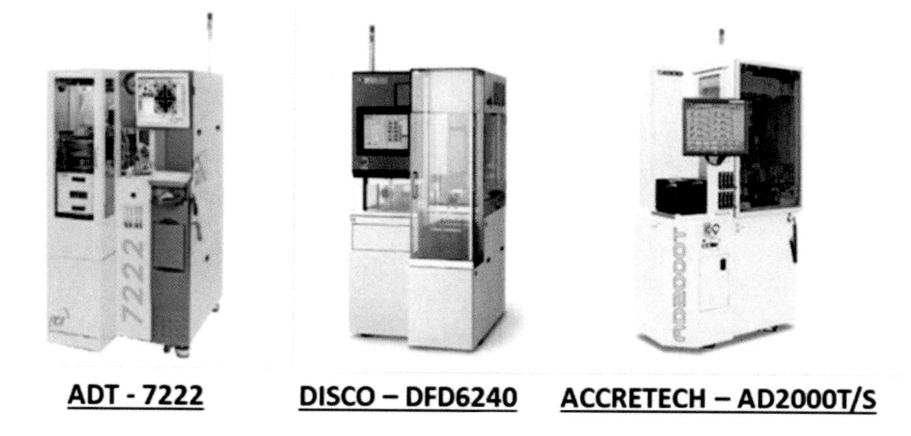

Figure No. – 25 – Fully Automated dicing systems

As the market was pushing for more innovations to improve UPH, dual spindle saws were developed using fully automated dicing systems. Two spindle options were developed: two parallel spindles design and two facing spindles design (see fig. 26).

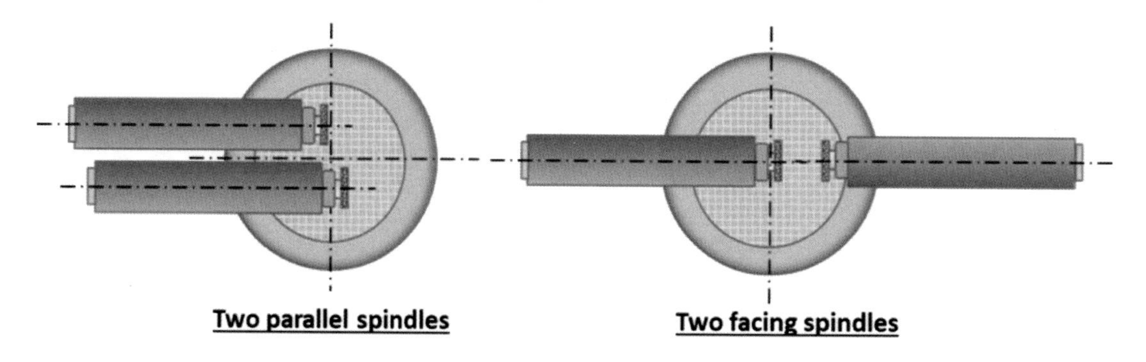

Figure 26. – Different dual spindle designs

The two fully automated systems using the two spindle options are widely used in the larger manufacturing houses, mainly in the silicon/semiconductor, LED, and other manufacturing facilities (see fig. 27).

Figure 27. – Fully automated dual spindle systems

The above short history review of dicing saws covers the basics of past saw designs and a few major feature developments over the years. Needless to say, many more specific developments were developed regarding special software per customer needs, special cooling systems, special accuracy requirements, special vibration control requirements, and many more.

CHAPTER 5

DICING BLADES MANUFACTURING TECHNOLOGIES

(NICKEL BOND, ANNULAR BLADE ELECTROFORMING, HUB BLADES, RESIN BOND, VITRIFIED BOND, METAL SINTERED, WC=WOLFRAM CARBIDE, TUNGSTEN CARBIDE SAW BLADES)

It is obvious that to dice accurately microelectronic substrates, a dicing saw is needed. However, without the right diamond dicing blade, even the simplest and most sophisticated saw will not be enough. During the years, many diamond dicing blades were developed to meet the many different microelectronic type substrates. A bit was covered in the dicing mechanism section, but it was only generic information. Making blades is partly technology and partly art. It is a combination of specific technological know-how, common sense, and experience in the dicing world. It should be stated that some of the processes took years to develop and optimize. Due to confidential related information, it is quite difficult to deeply dive into the different manufacturing processes. To cover and understand the dicing world, it is important to cover the different diamond dicing manufacturing techniques. I will do my best to highlight the different processes without getting into too much confidential information. Following are the major technologies involved in making diamond dicing blades:

- Nickel Electroforming
- Phenolic Resin Thermosetting process
- Metal Sintering
- Vitrified bond
- WC + Wolfram carbide, tungsten carbide saw blades.

5.1 - NICKEL ELECTROFORMING PRINCIPLES

This process in general is a nickel-plating process. Nickel plating is a commercially viable and versatile surface finishing process and can be used for the following three applications:

- Decorative
- Functional / protection
- Electroforming used for manufacturing diamond dicing products and other complicated geometries not related to diamond dicing products.

A typical nickel electroforming plating tank consists of a nickel-plating solution [Nickel Sulfamate or other], a nickel anode, a cathode made of a conductive material, and a DC power supply. When the power supply is turned on, positive ions in the plating solution are attracted to the negatively based cathode. The nickel ions that reach the cathode gain electrons and are deposited/plated onto the surface of the cathode, forming the electroformed nickel plating (see the sketch below fig. 28).

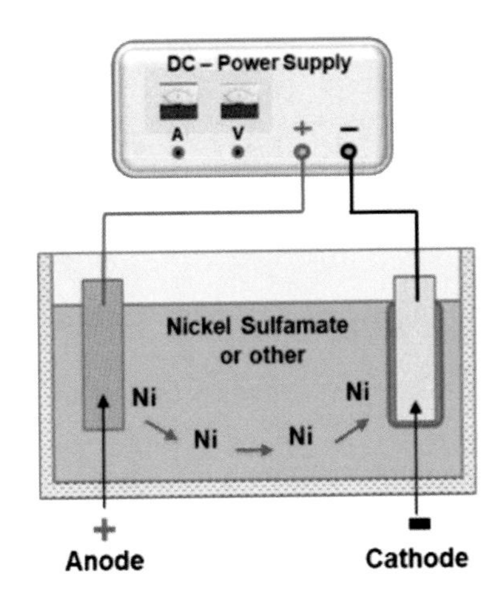

Figure 28. – Nickel sulfamate plating principle

In general, there are two methods of depositing nickel & diamond particles:

- Single layer of diamonds method, used for many industrial products like dental tools, diamond files, hole drills, special routers for stone and marble and many others (see fig. 29).
- Multilayer of diamonds method, used for making diamond dicing blades and similar.

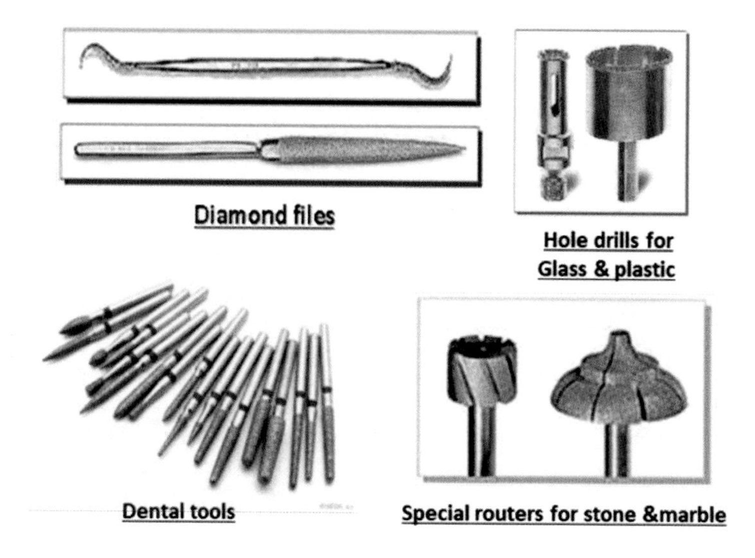

Figure 29. – Different diamond nickel-plated products

Below is a short walk through the process of making single diamond layer products. This is just to compare and explain the complex process of making thin diamond dicing blades without exposing confidential information. The main difference between the two processes is the fact that single diamond layer products are plated to a solid metal and adhered to a metal base till the end of its life. The other main difference is the size of the diamond. Single diamond layer products are usually made with very large diamonds of 40–200 microns; dicing blades use much smaller diamonds of 2–4 microns up to 70 microns. The end life of those tools is when the single diamond layer wears out. In comparison, the multilayer diamond dicing blade is a thin, electroformed product not connected to any solid metal base. The problematic issues with the multilayer, thin blades will be discussed. Any single diamond layer product starts with many steps of metal preparation: Etching, cleaning, and initial thin base nickel plating are performed prior to the single diamond layer plating. After the initial steps, the metal base product (the cathode) is put in a plating solution loaded with diamond powder. When the power supply is on, the diamonds are bonded on the initial thin nickel layer. After the initial plating with the diamonds, the metal parts (the cathode) are put in a clean nickel solution bath to cover the diamonds in order to better hold the diamonds in place. The plating is covering the diamonds only to about 50% of their theoretical diameter. This leaves the diamonds well exposed but well bonded to the base product. Below are a few sketches of a simple exercise to better understand this process (see fig. 30).

The final product for the exercise

Figure 30. Straight diamond plated metal pin

Let's skip the initial surface preparations of etching and cleaning. The next step is a thin layer of nickel plating (see fig. 31).

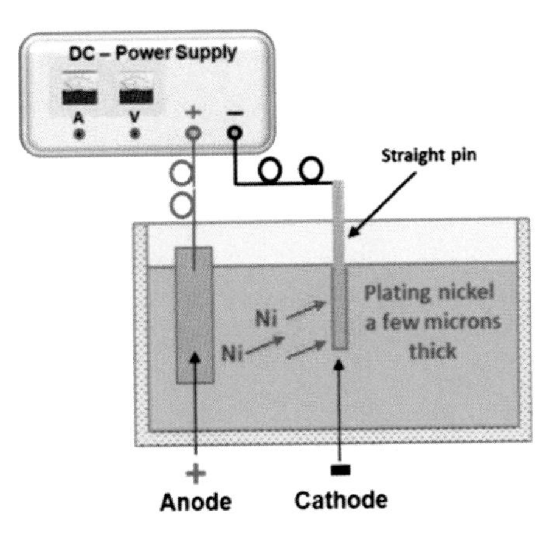

Figure 31. – Initial nickel plating

Initial diamond plating on the pin:

This process is just to bond the diamonds to the thin nickel layer.

It has been done in a nickel-plating tank with heavy loaded "diamond mud."

The diamonds are bonded to the steel pin by only a few microns of nickel plating. (see fig. 32)

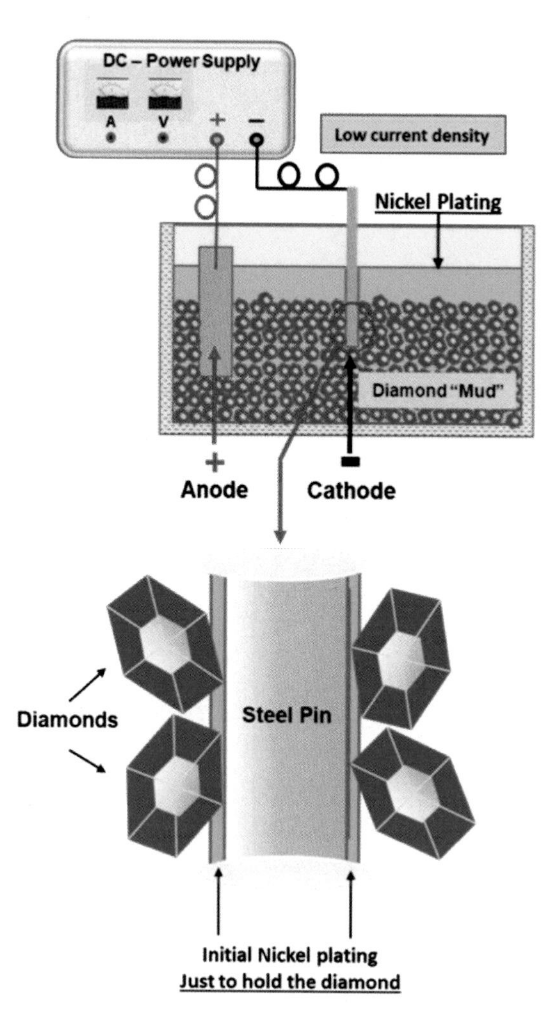

Figure 32. – Initial diamonds plated/bonded on the thin nickel layer

Final nickel plating:

This process is done in a clean nickel-plating tank to better cover the diamonds, which are a bit over their theoretical diameter (see fig. 33).

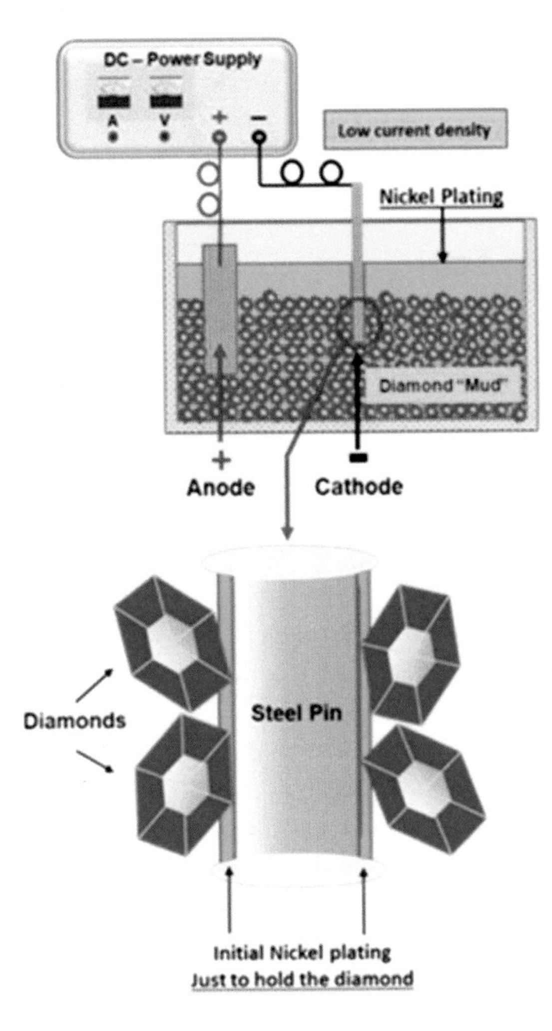

Figure 33. – Final nickel plating to firmly hold the diamonds in place

The single diamond layer process consists of many steps but is considered a relatively simple technology with relatively easy process parameters to control. The multilayer process is way more sensitive and requires state-of-the-art plating setup design and many process parameters to control. Blow is a general process comparison of important parameters between the multilayer process and the single layer process:

PROCESS PARAMETERS OF IMPORTANCE

Process Parameter	Multilayer process	Single Layer process
Thickness accuracy	Very important	Not so critical
Maintaining the overall geometry accuracy	Very important	Less critical. The metal base is controlling the final geometry
Stresses in the plating	Very important in order to maintain flatness	Way less critical
Hardness of plating	Very important with a few ranges of hardness	Less critical
Diamond concentration	Very important with a few diamonds' concentration ranges	Not so critical. It is a function of the Initial diamond plating in the diamond "MUD" plating
Time of plating	Depending on blade thickness, can be quite long. Accurate plating time is important	Not as critical
Chemical specks of the plating tank	Very important	Less important
Physical cleanliness of the plating tank	Very important	Less important

After understanding a generic industrial diamond tool plating process, let's concentrate a bit on the critical parameters of making thin diamond dicing blades using the nickel electroforming process.

In any plating setup, there is what professional plating shops are calling the edge effect. The area of the cathode closer to the anode gets more current density/higher power lines, and the plating gets thicker at this area. This phenomenon in any plating process is unacceptable and requires a special design of the plating tank geometry and the tooling holding the product to be plated (the cathode) in order to improve the uniformity of the power lines (see fig. 34).

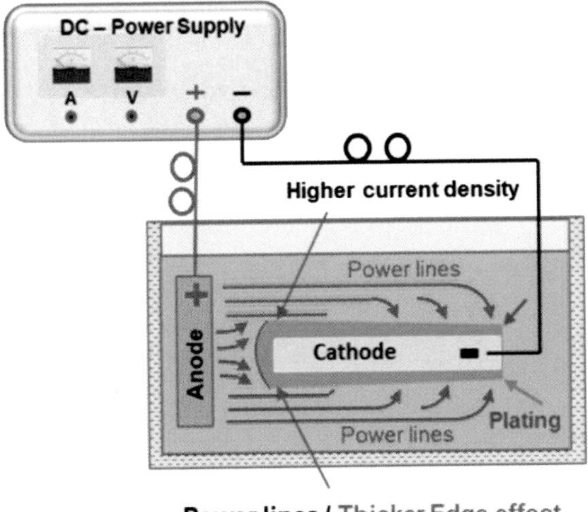

Figure 34. – Power lines affecting the plating thickness uniformity

In the manufacturing of dicing blades, the above is very critical and took years to develop and optimize. Below are critical parameters to control:

Blade thickness is a major issue, especially on thin nickel blades. Maintaining all the critical process parameters will result in a final accurate thickness. Thicker blades are normally plated oversize and then lapped to a very accurate thickness (see fig. 60 in the resin blade section). In order to plate accurately thin blades, precise and Constant Current Rectifiers are used, and the thickness is controlled by measuring amp/hours. finishing.com is a good site for more information on the process of precise nickel electroplating.

The chemistry of any nickel-plating process needs to be continually controlled. The plating process of making diamond dicing blades is way more critical. There are close to ten different chemical parameters that need to be closely monitored and corrected to meet the specs.

In order to meet a few specific hardness ranges per specific dicing applications, a specific additive is added to the plating solution and monitored. More important is to develop an accurate quality control process to measure the right parameters online and offline. Normally, there are three hardness ranges on nickel dicing blades: hard, medium hard, and soft. The hardness can be measured offline using special tooling on a Vickers hardness unit (see fig. 35).

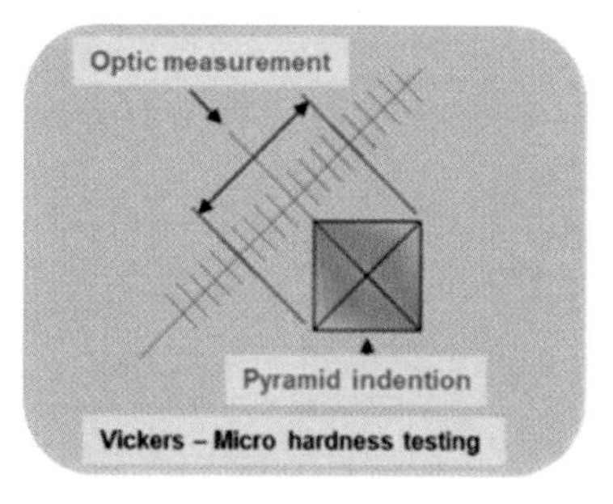

Figure 35. – Hardness measurement on a Vickers tester

5.1.1 - DIAMONDS USED IN THE NICKEL ELECTROFORMING PROCESS.

The types of diamonds used on nickel electroformed blades are hard and blocky. The nickel electroformed bond is the hardest of all bonds in diamond dicing blades. For this reason, it is important to use hard and blocky diamonds in order to maintain the sharp edges for a long time (see fig.38). Using non-sharp and friable diamonds will result in overloading, short life, and poor cut quality.

Source - Hyperion Materials & Technologies

Figure 38. – Hard and Blocky diamond geometry

A major potential contamination in a nickel diamond plating setup is diamond size contamination. The main potential problem is large diamonds getting into a small diamond blade. This is a total plating tank reject. Small diamonds getting into a large diamond blade is not good, but not a total reject. This requires very clean tooling and special attention in the manufacturing process. Fig. 39 shows a large diamond in a small diamond blade.

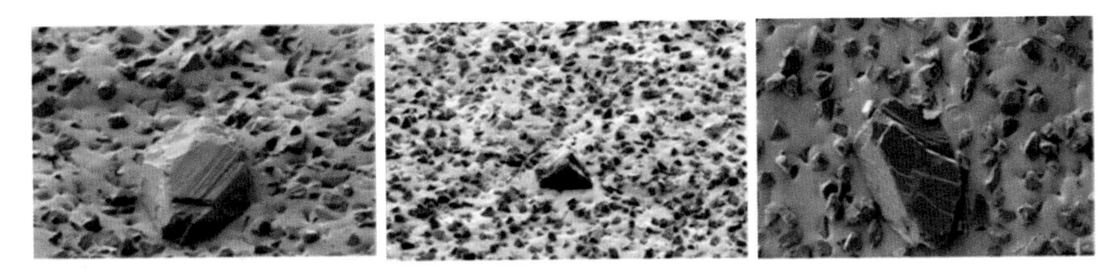

SEM of Diamond contaminations

Source: Electron Microscopy Sciences

Figure 39. – Large diamond contamination in a fine diamond product

There are three major electroformed diamond nickel blade geometries:

- Annular blades with different diameters used in a large variety of applications.
- Hub type blades used mainly for semiconductor silicon wafers.
- Steel core type blades

5.2 - ANNULAR NICKEL BLADES

There are many different geometries in this category. Following are the main geometries and the relevant diamond grit:

Diameter	Thicknesses (mm)	Majority thicknesses (mm)	Diamond grits (um)	Remarks
2" (50mm)	0.015 - 0.250	0.020-0.100	2-4, 3-6, 10, 17, 25, 30, 40 ,50,70	
3" (75mm)	0.075-0.250	0.075-0.200	10, 17, 25, 30, 40, 50,70	
4"-5" (100-125mm)	0.050-0.250	0.050-0.250	3-6, 10, 12, 17, 20, 30, 40, 50,70	Thin blades for the Magnetic head application

The above are generic numbers and may change slightly between dicing blade vendors. Annular blades are plated on a unique base substrate that acts as a manufacturing tool. The final thicknesses of thin 2" blades, 0.050 mm (.002 inch), are achieved by the plating parameter that can maintain an accuracy of ± 0.002 mm (.0001 inch). The final diameters are also controlled by the plating parameters and just need minor correction machining. Thicker blade geometries on all diameters are machined by different machining techniques to achieve the final thickness and the inner and outer diameters. A unique annular blade is 4"–4.5" (100–114 mm) O.D. x 0.060–0.100 mm thick and requires very unique manufacturing process parameters. This will be discussed in the Magnetic Head application session. Another important geometry parameter is the serrations on the blade edge. The serrations are machined after the plating process and can be machined to many different numbers of serrations and many different serration depths and widths, all depending on the application. This will be discussed in the application session. Fig. 40 shows a sample photo of a serrated blade.

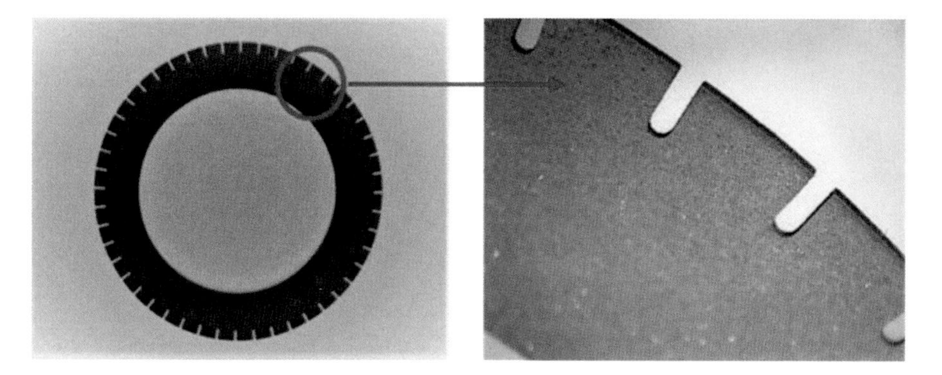

Figure. 40 – Metal serrated diamond blade

The outside and inside diameters of thicker blades are machined using the EDM process. This process is also used for machining any serration geometry. This process is discussed in more detail in the metal sintering process (see also figs. 72 and 73). The final O.D. is grinded on cylindrical grinding machines and will be discussed in the dressing section. The inner diameter needs only a small amount to be grinded and is grinded using a unique process.

5.3 - HUB BLADES

The Hub blade is the main blade type used for dicing silicon wafers in the semiconductor industry. The Hub blade is a thin and very stiff nickel diamond blade. The hub blade has a small exposure, enough to penetrate through the silicon wafer and dice into the mounting tape holding the silicon wafer. The nickel diamond plating process is similar to the annular diamond plating process. There is a major difference between making annular blades and Hub type blades. Annular blades are self-standing blades that are mounted on an accurate flange set, and Hub blades do have an annular diamond/nickel area that is adhered strongly to a very accurate aluminum Hub. In addition, a small rim of the diamond nickel area is exposed, and this is the active dicing area. See the below sketches in fig. 41.

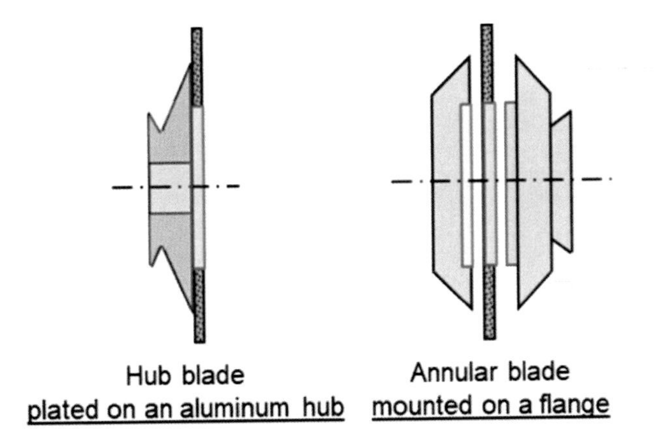

Figure 41. – Hub and Annular blade geometry

The application side of the Hub blades will be discussed; however, it is important to understand in general how Hub blades are made and what are the main important parameters to control. The initial manufacturing is machining a very accurate aluminum alloy hub. (see fig. 42)

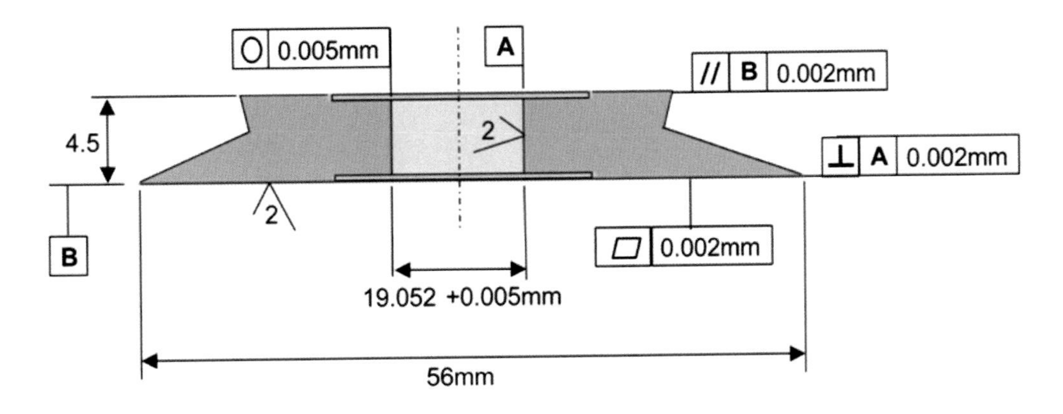

Figure 42 . – Aluminum hub blank geometry

The machining of hub blade blanks is performed on accurate CNC machines in large quantities of many thousands. Some important dimensions do require extra machining in order to meet the tight dimensions and the required surface finish. It is important to understand that a perfect diamond plating process without a perfect machined aluminum hub will result in a rejected product. For example, if the surface on the aluminum hub to be diamond nickel plated is out of tolerance in flatness and out of the 0.002 mm 90° to the inner diameter, it results in wider kerfs (see fig. 43).

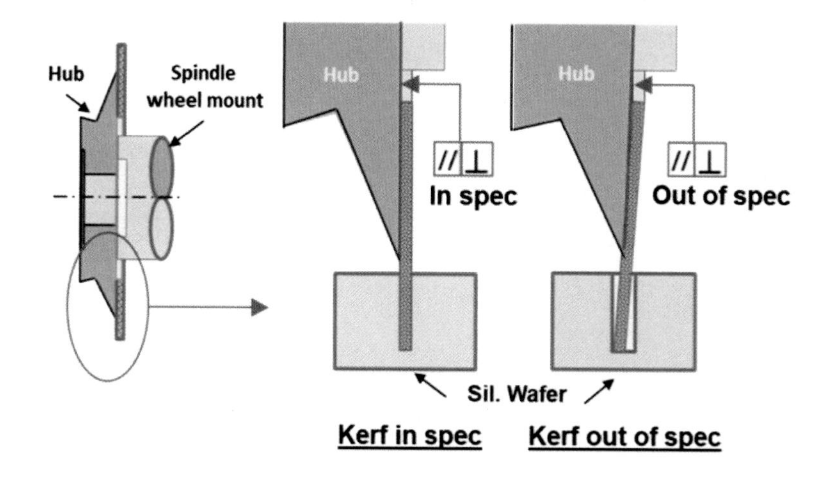

Figure 43. – The important of the aluminum hub accuracy

The tough requirement of machining the aluminum hub blank requires a tight QC before releasing to the diamond plating process. Using an out-of-spec hub blank will result in a big loss of many production steps to be performed on the hub blanks. The manufacturing process is performed in batches of many hub blanks. Below is a generic process flow:

PROCESS FLOW

- Chemically cleaning the hub blanks after the machining process.
- Mounting the hub blanks in a special jig.
- Chemically etching and water rinsing the future plating area.
- Chemical preparation of the future plating area (How to plate nickel on aluminum can be found in finishing.com).
- Water rinsing after the chemical preparation.
- Diamond/nickel plating.
- Water rinsing after the plating.
- Disassembling the plated hub blanks. At this point, the hub has no blade exposure.

The plated diamond/nickel edge has an irregular shape and needs to be edge grounded on a cylindrical grinding machine using silicon carbide grinding wheels (see fig. 44).

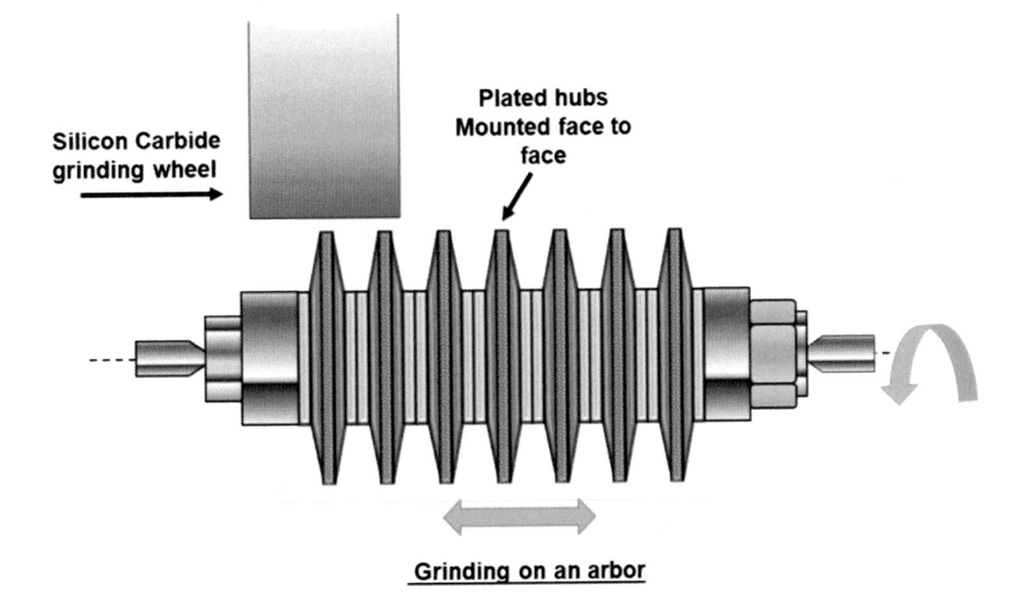

Figure 44. – Outside diameter grinding on cylindrical grinding machine

The result of this process is a nice flat edge, meeting the final outsider diameter of 2.187 inch, which is the standard hub O.D. (see fig. 45).

Figure 45 – Flat blade edge after Outside grinding

Mounting the plated hubs in new jigs to machine off a portion of the aluminum hub and creating the blade exposure. This process is very delicate because of the very thin diamond / nickel layer. A too aggressive process will deflect the diamond/nickel layer. Performing the blade exposure needs to meet a relevant blade thickness per a part number. Machining the exposure is done partly by mechanical machining and finally by a chemical etching process, so there is no final mechanical contact with the thin plated blade (see fig. 46).

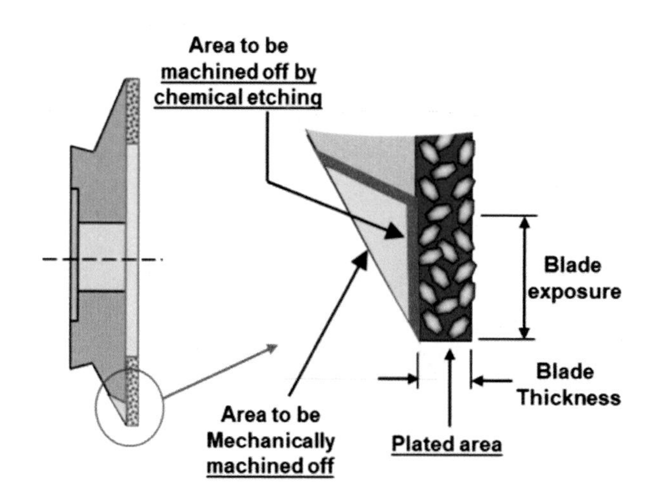

Figure 46. – Initial mechanical machining and final chemical etching to perform the exposure.

After exposing the thin diamond/nickel area from the aluminum hub, is another process of exposing the diamonds, mainly on the previous aluminum side. Having the diamonds exposed evenly on both sides of the blades is extremely important and a must in order to perform well when dicing silicon wafers (see fig. 47). The application side will be discussed in the application discussion.

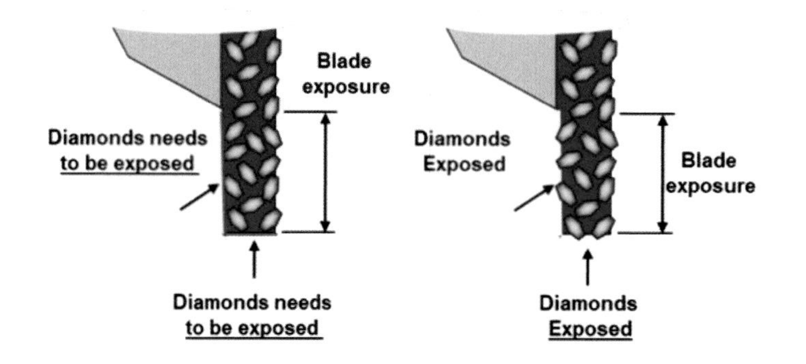

Figure 47. – Diamond exposing by Electropolishing / nickel etching Hub blade QC:

The above generic process flow shows the difficulties involved in the delicate manufacturing process. In the process, there are many "milestones" to meet, and some of them are part of the QC. However, every hub needs to be 100 percent inspected in order to make sure it will perform well in the marketplace. This is right for all blade types but is more critical with hub blades, mainly because of the mass production semiconductor houses dicing silicon wafers with narrow streets and tough quality specs. There are a few methods of making sure every hub blade will perform well; it is a time-consuming process, and it varies between the different hub blade manufacturers.

5.4 - STEEL CORE NICKEL BLADES

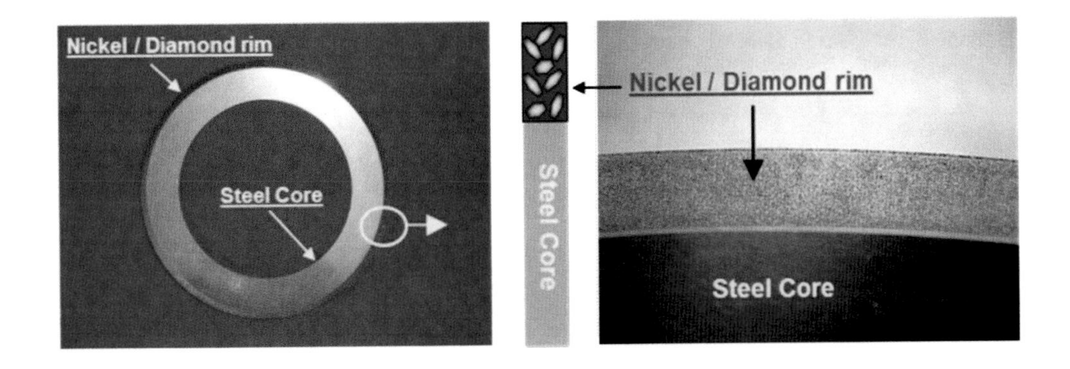

Figure 48. – Steel core nickel blade, front and cross section

From all the nickel/diamond dicing blades, the steel core blade is the least popular. The application side will be discussed in the application section. This blade geometry starts from a metal steel core with a nickel/diamond rim plated to the steel core edge (see fig. 48). The steel core blade can be made in 2″ up to 5″ diameter and in almost any thickness on the upper side up to .040″ (1.00 mm). The limitation is on the lower side; the minimum thickness is normally .005″ (0.127 mm). The lower size limitation is mainly due to manufacturing limitations. The nickel/diamond plating process is similar to the annular blades process, but the mechanical side of the jigs tooling and plating orientation is very different and requires major changes. The preparation of plating the nickel/diamond rim on the steel core edge requires special mechanical geometry machining and a few chemical preparations in order to get a good and stiff bond between the steel core to the nickel/diamond rim. Another difficulty in the manufacturing process is minimizing stresses in order to maintain the flatness of the final product. To summarize, making steel core blades is a complicated and time-consuming process. The blades can be made in 15–60 mic. grits depending on the blade thickness. The majority of blades in the marketplace are 30 and 50mic. grits. The nickel/diamond rim length can be plated to different sizes depending on the application and requires different tooling. The main advantages of the steel core blades are their stiffness and very low wear.

CHAPTER 6

PHENOLIC RESIN THERMOSETTING PROCESS

Resin blades are on the opposite side of nickel blades in terms of hardness. In general, it is a molding process using phenolic resin powder, or, in simpler terms, plastic powder molding made from a resin fine powder (see fig. 49).

Figure 49. – Resin blades made of phenolic resin powder

In 1907, Leo Hendrik Baekeland, a Belgian-born American chemist, applied for a patent on a phenol-formaldehyde thermoset that eventually became known by the trade name Bakelite, also known as phenolic resins. Phenol-formaldehyde polymers were the first completely synthetic polymers to be commercialized. During the years, many types of phenolic resins were developed for many different products used in our daily life. Over 60 percent of the different resins are used in the wood adhesive industry for manufacturing plywood for gluing the layers and on pressed lumber products. More information on the scientific side of phenolic resins can be found at the end of this publication under the reference list. A variety of phenolic resins are available, and some of them are used in manufacturing diamond grinding wheels and diamond resin blades. Phenolic resins are characterized by low melt viscosities, moderate molding temperatures, and moderate heat resistance in the cured stage. The specific type of phenolic resin used for thin dicing blades needs to have maximum resistance to stress. This is required in order to perform with good stiffness on thin blades of ".003" (0.075 mm) to .0200" (0.50

mm). Thicker blades are less sensitive and can be less stiff, but they are also in the good stiffness range. Normally, resin vendors will indicate that the resin is designed for cutoff wheels. This is based on the vast experience of making thick and wide diamond resin grinding wheels. There is a variety of resins to choose from with different properties for stiffer harder blades and for softer and more wearing blades. This will be discussed in the application section.

6.1 - DIAMONDS USED IN THE RESIN MANUFACTURING PROCESS

Friable diamonds are normally used; they are characterized as irregular, semi-blocky shapes with a rough surface and many re-entrant angles. There are two major diamond types used, uncoated and coated with nickel alloy in a few percentages by weight: 30% 56% & 60%. The nickel coating has a rough surface which is performing with much better bond to the resin matrix (see fig. 50). Fig. 51 shows the different wear between the different diamonds.

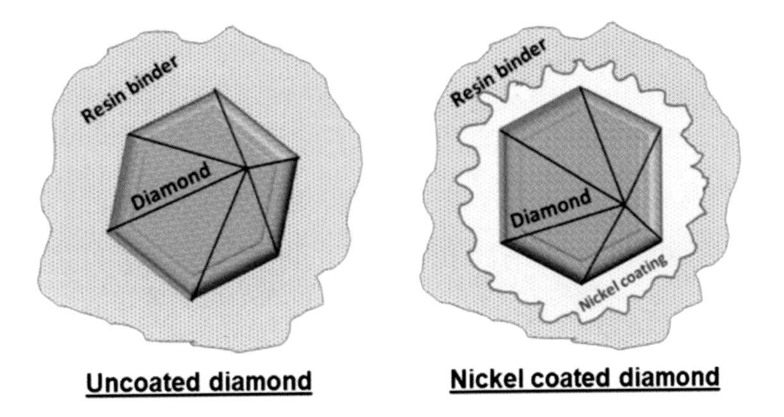

Figure 50 – Different diamond powders used with resin bonds.

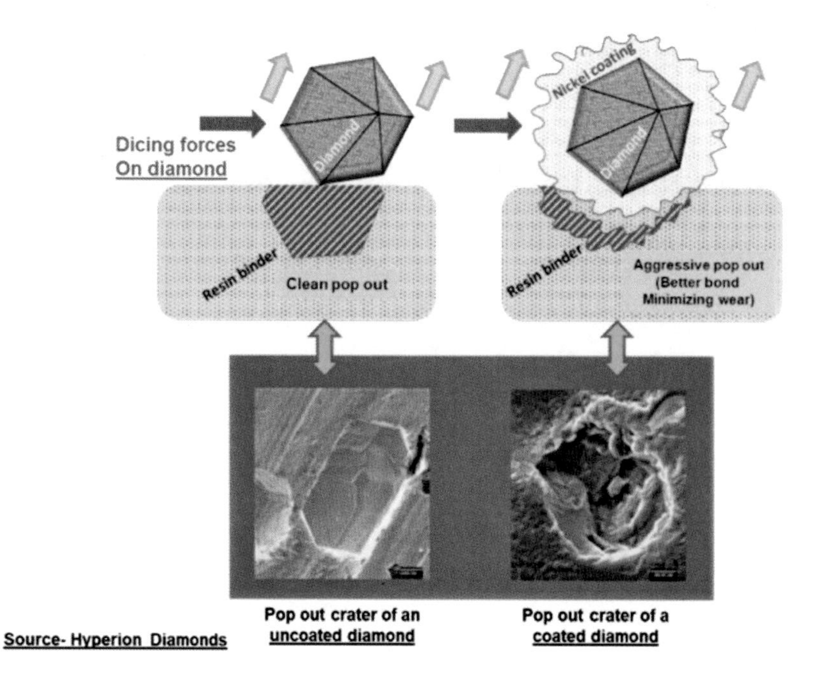

Figure 51. – Different wear characteristics between coated and uncoated diamonds

Fig. 52 shows the Hyperion SEM photos of the two major diamonds used in the diamond resin blade process. The left SEM is the traditional RVG friable diamond. The right SEM is the same RVG friable diamond but with a nickel coating. This changes the name to RVG-W. The RVG–W is available in different nickel % by weight design per application.

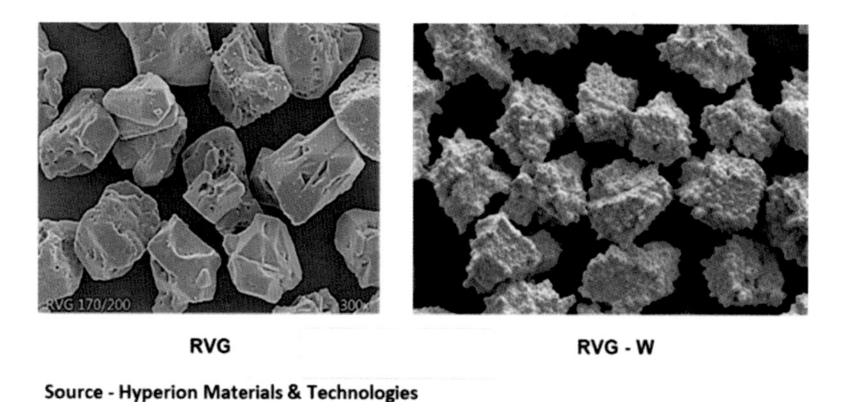

Figure 52. – Friable uncoated diamonds and Friable nickel coated diamonds

6.2 FILLERS IN THE RESIN BOND

Conductive fillers:

On almost all dicing saws, height calibration, or what is called Chuck Zero, is required in order to keep the right cut depth during the dicing process. As discussed, this is performed by an electrical contact between the spindle and the chuck via the rotating blade edge. In order to perform the Chuck Zero process, the blade needs to be electrically conductive. The phenolic resin binder is not conductive, and conductive filler needs to be added in order to make the blade conductive. The fillers added to the blade are normally carbon powders of different types and sizes. They do affect the blade wear, mainly causing higher wear, so the right amount needs to be optimized in order to minimize the wear and at the same time meet the right conductivity. Conductivity is measured on the blade surface as resistance; the resistance should be < 10 K'Ω. Normally, the resistance is much lower.

Other fillers to improve blade wear and cut quality:

The main reason to add fillers in resin blades is the market's push to improve blade life by minimizing blade wear and, on the other side, to improve cut quality. Minimizing wear or increasing stiffness is, in general, hardening the matrix. Making softer matrixes to improve cut quality on brittle materials will have the opposite effect and affect wear. Common fillers in the grinding world are Sic. for hardening and stiffening and aluminum oxide for softening the matrix binders. In thin resin dicing blades, there are many other options for different metal powders and other additives. The idea is to make a harder bond or to better lubricate the blade to minimize friction. Another option is adding Teflon to the matrix, again to minimize frictions for better cut quality. There are a few applications where a mixture of large diamonds mixed with small diamonds in order to minimize blade wear is a good option. The above different options to improve blade performance is really an art and requires a good understanding of the material diced, good dicing knowledge, and sometimes out-of-the-box creative ideas. The market is pushing all blade vendors to continually improve the above factors. In order to meet those tough goals, a very close relation between the blade vendors and the many different application houses is a must.

6.3 - PREPARING THE RESIN POWDER MIX

General:

All powders involved in preparing the resin mix need to have a written specification and a strict incoming quality inspection. This is critical for the consistent quality of the end product. The resin powder is the main powder in the mix and is very sensitive to humidity;

therefore, it must be stored in a controlled storage environment. In some matrices, the resin powder needs to be treated before mixing it with all the other powders.

Weighing and mixing the powders:

In a clean environment, all the different powders involved in a specific resin matrix are weighed on a digital scale and added to a mixing jar. The diamond type and quantity per the designated diamond percentage is also weighed and added to the powder mix. There are a few mixing options in different mixing equipment. The idea is to get a uniform powder mix, dry and ready for the next molding process. As for the diamond concentration, there is an international standard for the concentration calculation. This standard is used worldwide in grinding wheel manufacturing and is also used in diamond dicing manufacturing (see the formula below fig. 53).

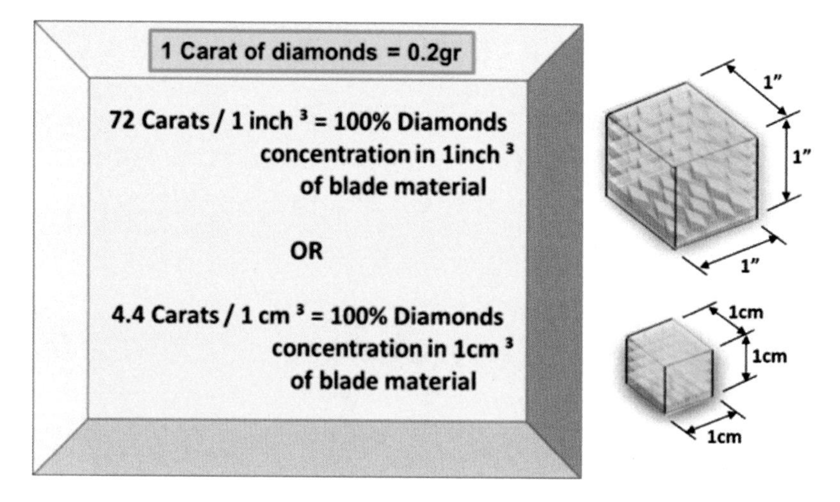

Figure 53. – Diamond concentration formula

Common concentrations used in the market are calculated by volumes and weight of the blade matrix (see fig. 54).

Diamond Concentration	Diamond Volume in the blade matrix	Diamond Weight in Carats / cm ^{3}
25%	6.25%	1.1
50%	12.50%	2.2
75%	18.75%	3.3
100%	25%	4.4
125%	31.25%	5.5
* 150%	37.50%	6.6

Figure 54. – Diamond concentration compared to Diamond volume and diamond weights

* 150% is a theoretical % and cannot be used mainly on large diamonds as the diamonds will not have enough matrix bonds to hold the diamonds in place.

6.4 - PREPARING THE POWDER MIX FOR MOLDING

In the manufacturing process of grinding wheels, the use of large molding jigs is common practice. Most diamond grinding wheels have an aluminum core, and only a few millimeters on the outer diameter are a molding of phenolic resin with diamonds. In order to understand how complex the process of molding thin diamond resin blades is, it would help to understand the molding process of manufacturing standard resin diamond grinding wheels. Following is a generic sketch of a molding setup for making resin diamond grinding wheels (see fig. 55).

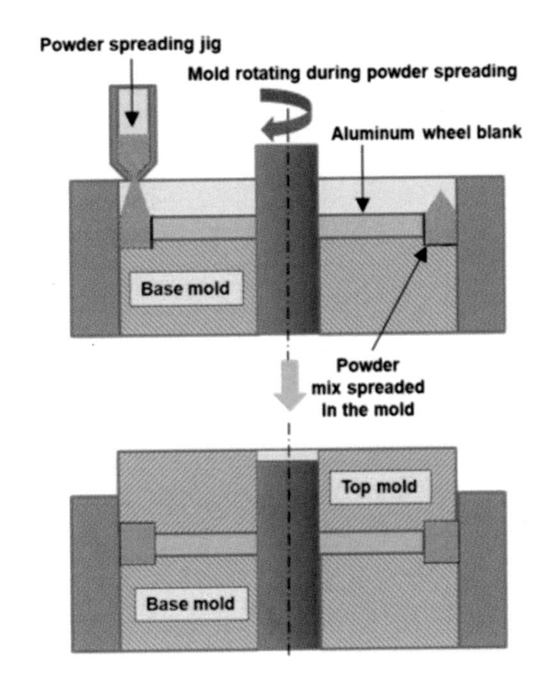

Figure 55. – Diamond Resin grinding wheel molding set-up

Some thick annular resin blades are still made using similar molding designs; however, it is very difficult to release thin resin blades from the mold after the molding process without breaking them. For this reason, other unique molding and powder spreading designs are used for manufacturing annular thin resin blades.

6.5 - MOLDING PROCESS:

After the powder mix is loaded into the mold, the mold is placed in a hydraulic press for 20–30-ton pressure depending on the blade geometry. The mold is pressed between two heated plates to about 150–200°C, which is a common temperature range for phenolic resin. This Sycle takes a few minutes for the initial curing (see fig. 56).

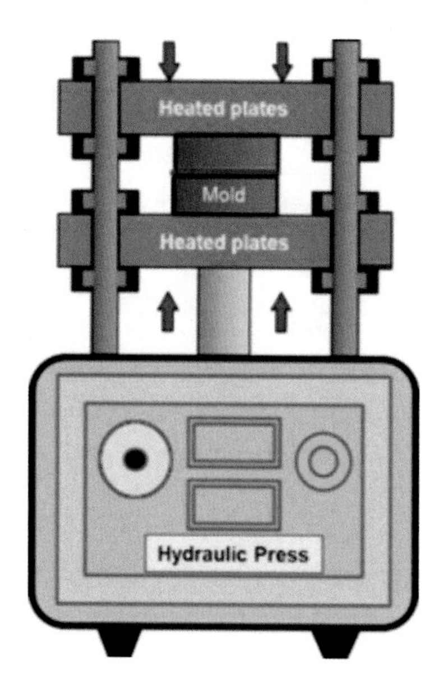

Figure 56. – Hydraulic hot press used in the resin molding process

After the heated pressing process is completed, the mold is cooled in a controlled cooling cycle jig to minimize stresses/warpage. The mold is then disassembled, and the blade is released from the mold. Most phenolic resin processes are using two cycles of curing, so another curing cycle is performed in a curing oven for a few hours. For those who like to explore more information on phenolic resin molding process parameters, there is a lot of good information on the internet.

6.6 - RESIN BLADE GEOMETRIES

Many blade diameters are made to fit different saws and different applications. Making different diameters requires different mold and powder spreading tooling. In addition, different punching tooling is required for each diameter. Punching is one option for getting the geometry O.D. and I.D., but there are other options. Fig. 57 shows a generic punching tool.

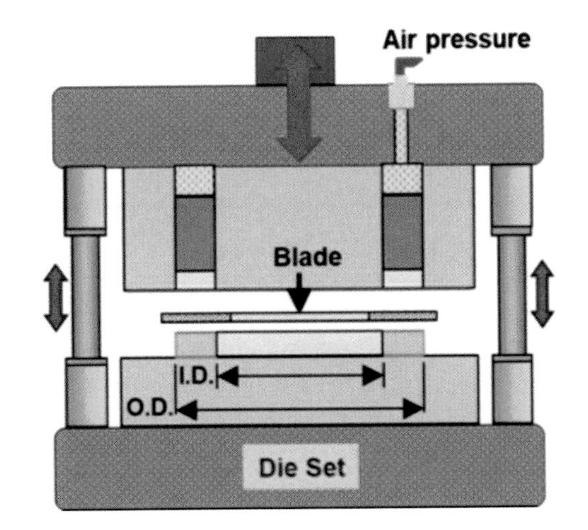

Figure 57. – Blade Geometry punching set-up

Some of the geometries with thicker blades are done using the closed mold design. On all geometry manufacturing options, additional O.D. and I.D. grinding is needed to meet the final dimensions. Fig. 58 shows the O.D. grinding process, and fig. 59 shows the I.D. grinding process.

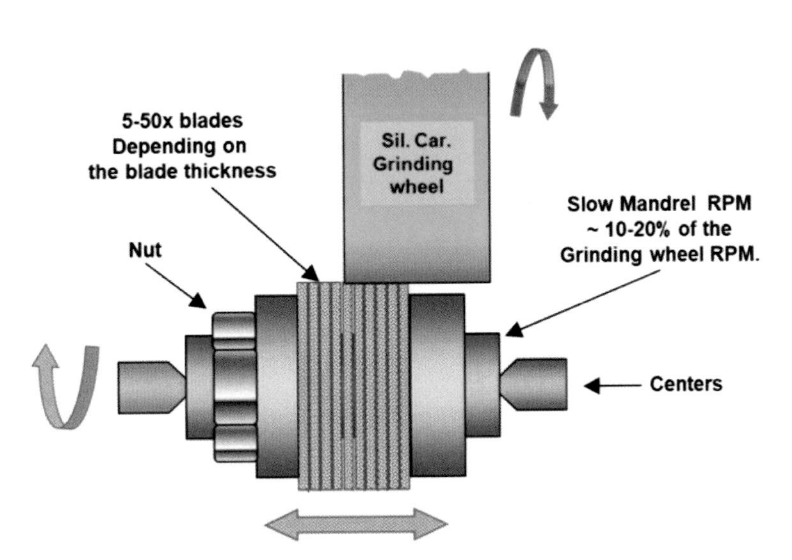

Figure 58. – O.D. grinding to the final O.D. geometry

Grinding the O.D. is also important in order to get a nice flat edge, which is a must to perform well in the dicing process. Grinding a resin blade O.D. is a relatively fast process using a Sil. Carbide grinding wheel. However, grinding the I.D. requires grinding a small amount, but the process is more difficult to perform mainly because the grinding wheel has a much smaller diameter compared to the large O.D. grinding wheel (see fig. 59).

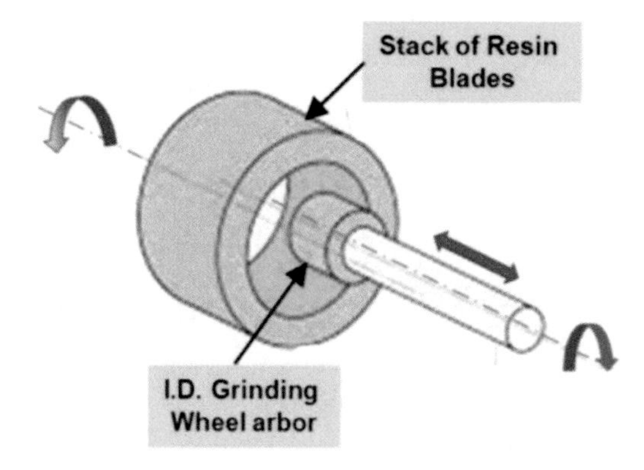

Figure 59. – I.D. grinding to meet the dicing blade spec

The grinding process will be discussed more in the dressing section.

6.7 - BLADE THICKNESS

The standard thickness tolerance of resin blades is + – .0002" up to + – .0005" (0.005 – 0.0127mm) depending on the blade thickness. Those tolerances can be achieved with a good molding process and good tooling. If a more precise thickness tolerance is needed of + – .0001" (0.0025mm) a lapping process using a dual lapping machine is needed or a rotary surface grinder. (See fig. 60).

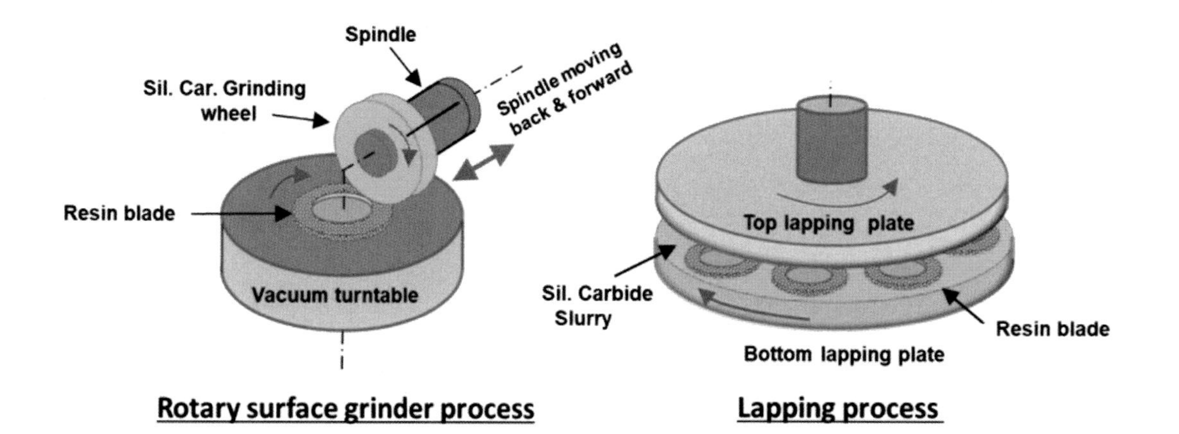

Figure 60. – Different methods of machining the blade thickness

Both processes are very delicate, requiring professional skill attention.

CHAPTER 7

VITRIFIED DICING BLADES.

Vitrified bond is very popular in the grinding industry for standard abrasives like aluminum oxide, silicon carbide, and other grit materials for grinding in the metal industry. It is also used with diamond grits and CBN (Cubic Boron Nitride) grits but with special modified bonds. The diamond concentration on grinding wheels and diamond dicing blades is similar to the resin bond products (see fig. 61).

Diamond Concentration	Diamond Volume in the blade matrix	Diamond Weight in Carats / cm ³
25%	6.25%	1.1
50%	12.50%	2.2
75%	18.75%	3.3
100%	25%	4.4
125%	31.25%	5.5
* 150%	37.50%	6.6

Figure 61. – CBN diamond concentration similar to the diamond calculations

Below is a sample of vitrified grinding wheel products. (see fig. 62).

Source - Xinxiang New Zuan Diamond Tools

Figure 62. – Samples of CBN grinding wheels

Vitrified bond is less popular with thin diamond dicing blades on standard applications but is used on thick, tough, and hard materials where a strong and resilient bond is needed to minimize wear and maintain good cut quality. The Vitrified bond dicing blades are in between the Resin and the Metal Sintered bond. Vitrified diamond dicing blades feature high strength, high abrasion resistance, and higher rigidity compared to resin blades, and at the same time, acts with self-sharpening and relatively easy dressing. The bonding strength is better than resin diamond blades and the diamond machinability is better than metal sintered blades. The bond does not load easily in hard materials; it performs with small thermal expansion, which helps to maintain accurate kerf dimensions. In a generic view, it looks that the Vitrified matrix is similar to the resin matrix. However, there are major differences in the technology and the manufacturing process.

7.1 - MANUFACTURING PROCESS REVIEW

Some manufacturers are calling the Vitrified bond ceramic wheels. Traditional Vitrified bond is a mixture of the following or part of the following materials: loess geology, clay, bentonite, Borax Glass, and Zirconia, all in very fine powder forms. For example, the loess can be found in the residue of heavy glacier friction load on the rocks below the glacier, resulting in an extremely fine powder causing the rivers close to the glacier to have a light milky blue color. Other materials found in the manufacturing and the optimization of Vitrified bond are fluorite, quartz, borax glass, talc, feldspar, and other fillers, including conductive fillers needed for the chuck zero calibration. The type and quantities vary between manufacturing vendors and per the different applications. The process in general is a water wet mixing process to achieve homogeneous mixture and a molding process using closed mold jigs. The final geometry results after the molding are close to the final dimensions but require machining of the thickness and the inside and outside diameters. The machining process of the geometry is similar to what was described in the resin geometry process. The accurate geometries are important, as most microelectronic applications require very accurate blade dimension tolerances. Following is a generic process flow of the Vitrified process: (see fig. 63)

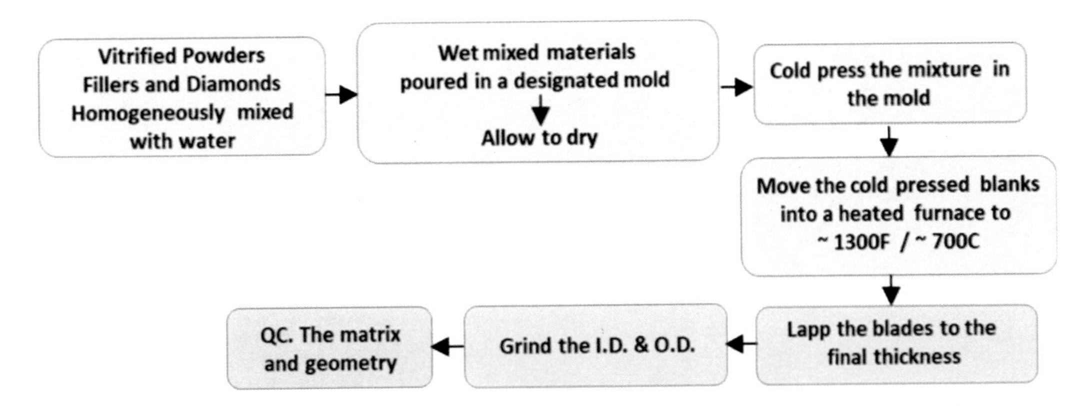

Figure 63. – Generic process flow of manufacturing CBN grinding & dicing blades

Vitrified diamond blades are made in annular geometries starting from 2-inch (50 mm) O.D. up to 6-inch (152 mm) O.D. The other popular product is the steel core Vitrified. Those blades are for thicker substrates requiring larger exposures with stiffness requirements (see fig. 64).

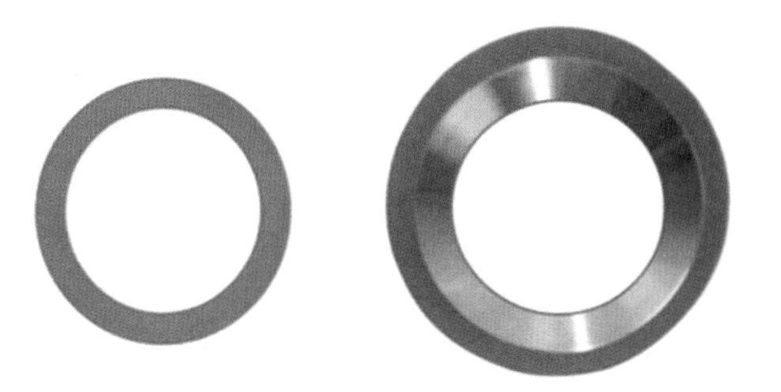

Figure 64. – Annular and steel core CBN dicing blades

The main applications requiring a Vitrified bond are High loading materials like sapphire and hard materials from the ceramic families. For more very good and informative information on the Vitrified process, go to the Xinxiang New Zuan Diamond Tools site that can be found in the reference section.

CHAPTER 8

METAL SINTERED BLADES.

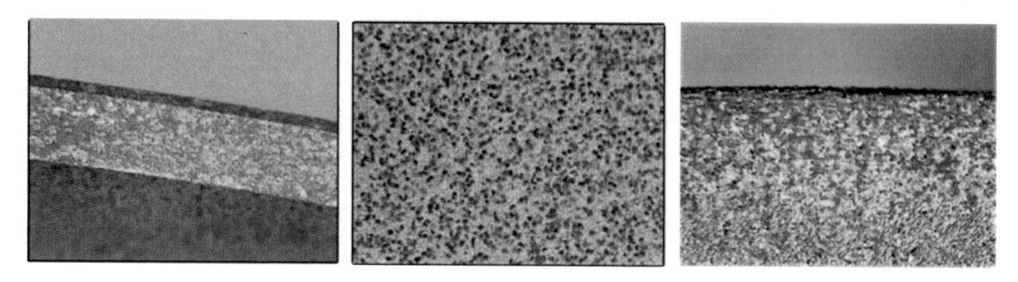

Figure 65. – Metal sintered blade's edge and surface

Metal Sintered blades are composed of metal powders, diamonds, and other powder fillers. The blade bond strength is a result of the metal's sintering process. The sintering process fuses the metal powders together without truly generating a homogeneous alloy however it becomes a solid matrix holding the diamonds in place. The mechanical properties of the blade are varying by the metal powders used and the sintering temperature and all other pressure during the sintering process. Bronze is the most commonly used alloy. Bronze is an alloy of copper and tin with the most common alloy of 80/20 copper-tin. However, the percentage varies depending on blade properties required for specific applications. Bronze is the major metal powder used however many other metal powders are used like, iron, copper, tin, zinc, cobalt, nickel, and some pre alloyed metal powders. Bronze bonds generate softer bonds performing with freer cutting while other metal powders generate harder bonds performing with less wearing matrixes. Harder metal powders can be added to the bronze matrix in order to perform a tougher bond strength. Other nonmetal fillers like different ceramic powders are added to create a more friable bond. The right bond design matrix relates to the application requirements. Different synthetic diamonds are used in metal sintered blades; the majorities are hard, blocky, and cubo-octahedral uncoated diamonds, but some with special very thin hard coatings for better bond between the diamonds and the bond binder. See fig. 66 and more information in the reference section.

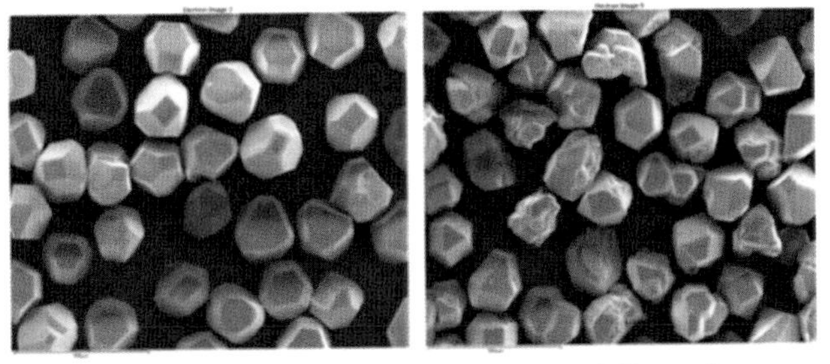

Hyperion hard cubo-octahedral uncoated diamonds

Source - Hyperion Materials & Technologies

Figure 66. – SEM of diamonds used in metal sintered blades

The diamond concentration used in sintered diamond dicing blades is calculated the same as with the resin blades. (See fig. no. 53 & 54).

8.1 - PREPARING THE SINTERED POWDER MIX

Metal powders have a good flow and are relatively easy to handle compared to resin powders that are sensitive to humidity. The metal powders, diamonds and fillers are accurately weighed and mixed in a designated mixer. There are many mixers in the market for a large variety of different applications. Many manufacturers of grinding and dicing blades are designing and making their own mixers using special hard media inside the mixers to help the mixing process. See a generic sketch fig. no. 67.

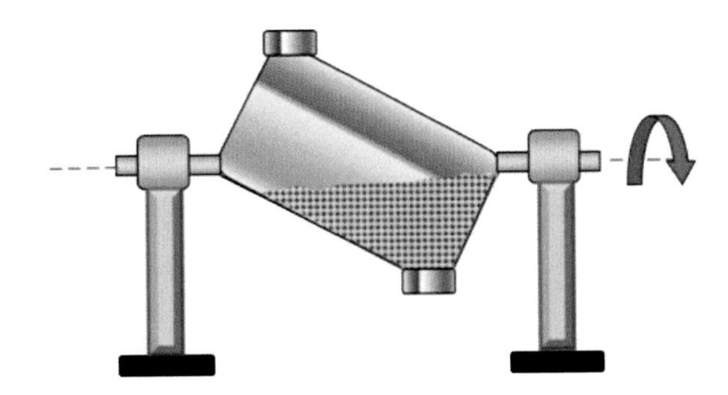

Figure 67. – A generic powder mixing jig

The main reasons for the self-designing and making special mixing jigs are due to the need for optimum optimization per manufacturer. Among many powder mixing units is one very popular, the TURBULA mixer that is widely used for many applications. See fig. 68 and more info in the reference section.

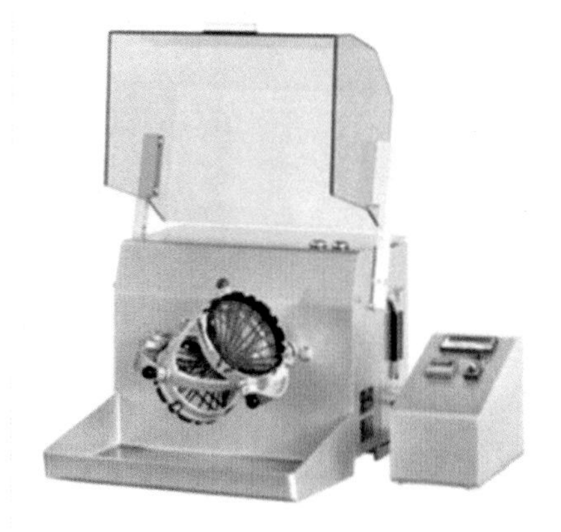

Source - TURBULA® T2F from Willy A. Bachofen AG, Switzerland

Figure 68. – TURBULA powder mixing unit, widely used in the dicing blades manufacturing

8.2 - MOLDING HANDLING

Molding metal sintered blades is done in closed molds made mainly in dense graphite molds for good heat transfer conductivity and in some processes using iron molds mainly used for some processes using pre cold pressing prior to the sintering process. The molds are powder loaded using special jigs to spread the powder mix homogeneously.

Loading the molds evenly is important in order to perform a homogeneous matrix on the blank which is critical to perform well on quality and even wear rate. The blade geometry is sintered to oversize dimensions and machined to the final geometry after the sintering process.

8.3 - THE SINTERING PROCESS

The basic principles of induction heating were understood and applied to manufacturing testing during the 1920s. The practical manufacturing process started in the 1940s mainly for hardening metal parts, annealing metal parts, brazing metal parts, manufacturing machines, and other industrial metal geometries and diamond grinding wheels. Many metal powders and combinations of metal powders are used depending on the end product requirements. The end products can be iron, stainless steel on the hard side and bronze type on the softer side. The induction process occurs by inducing eddy currents in graphite molds. The induction coil geometry is designed for the part or molding geometry. The induction system is operated by a high frequency power supply and a coolant refrigerated system for cooling both the coil and the power supply. The system is mounted on an accurate hydraulic press which is important to maintain accurate and dense geometries (see fig. 69).

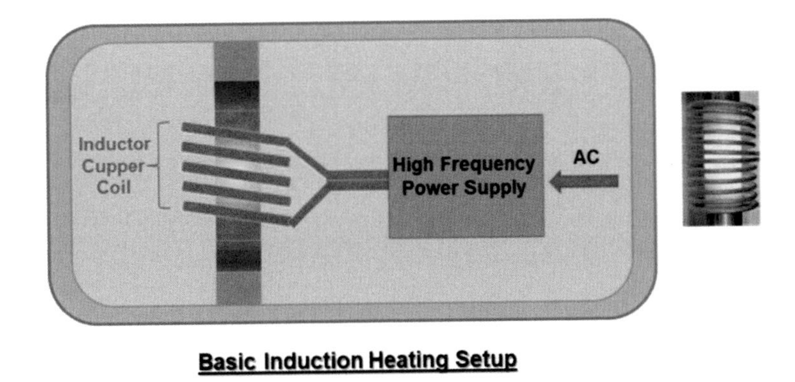

Figure 69 – A generic Induction heating set-up used in some blade manufacturing.

The coil heating method has the disadvantage of performing with higher temperatures on the outer side of the part or mold, however, this is minimized and compensated by the heating elevation graph which can be optimized. The other newer and more controlled sintering process uses a different heating/sintering technology of Direct Resistance Heating (DH), which is a type of resistance heating that passes electric current directly through an electrically conductive heated object to directly heat the object by Joule heat utilizing the internal resistance of the object. This method can be used in a vacuum chamber or an inert environment using different gas media. The main advantage of this technology is a much more even heating on the blank work piece from the outside to the inside. In addition, the vacuumed or gas-inerting environment reduces oxidation of the sintered blank (see fig. 70).

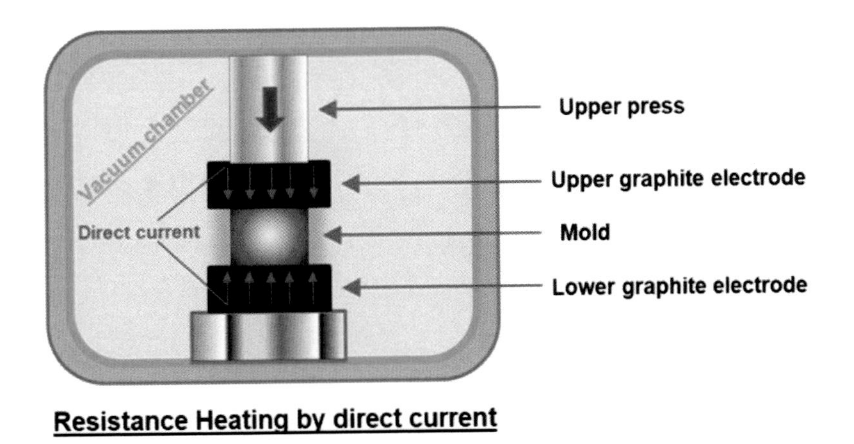

Figure 70. – Resistance heating set-up

Some sintered metal parts and diamond tools are made by using a cold pressing process in metal molds and then sintering in a direct resistance heating oven or in a "free" sintering process without any pressure. For more accurate parts like the thin diamond dicing blades, the sintering needs to be done under pressure in order to achieve a denser and stiffer product. This process is called "hot press sintering."

8.4 - MACHINING THE SINTERED BLANKS TO THE FINAL GEOMETRY

The blank after the sintering process has rough edges on the O.D. & I.D. and is oversize on the thickness. In order to accurately mount the blanks on any machining device in a stack mode, the thickness needs to be machined / lapped to be parallel, flat, and accurate on the thickness dimension. (see Fig. 71)

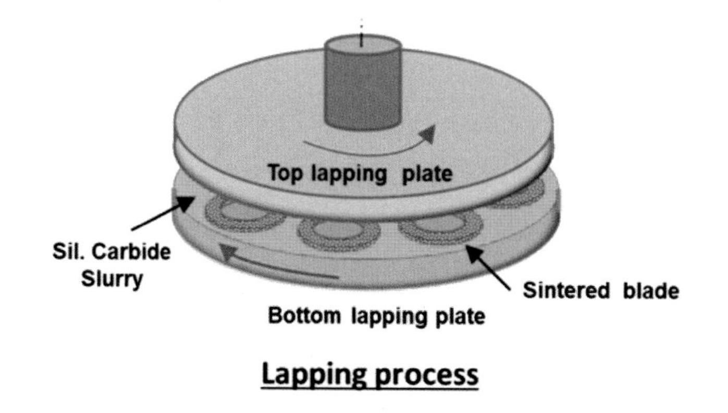

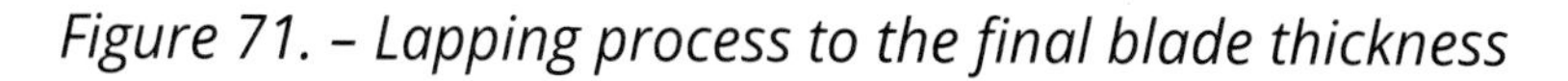

Figure 71. – Lapping process to the final blade thickness

Inside and outside diameter are machined in batches after an initial lapping process that is needed to accurately mount the blade together. The diameters are machined by EDM (Electrical Discharge Machining – Spark Erosion) (see fig. 72).

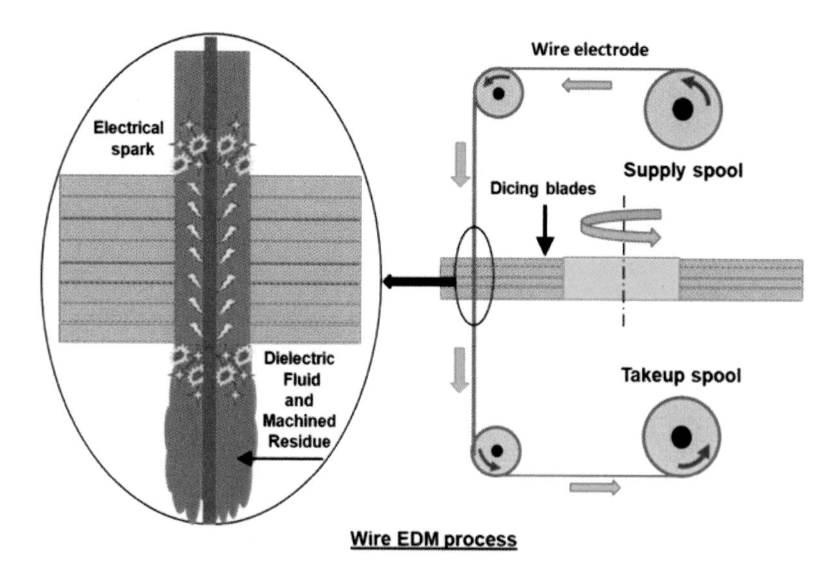

Figure 72. – EDM machining close to the final blade I.D. and close to the final blade O.D.

The EDM process is also used to perform other geometries like the serrated edge. Any geometry or no. of slots can be made. (see fig. 73)

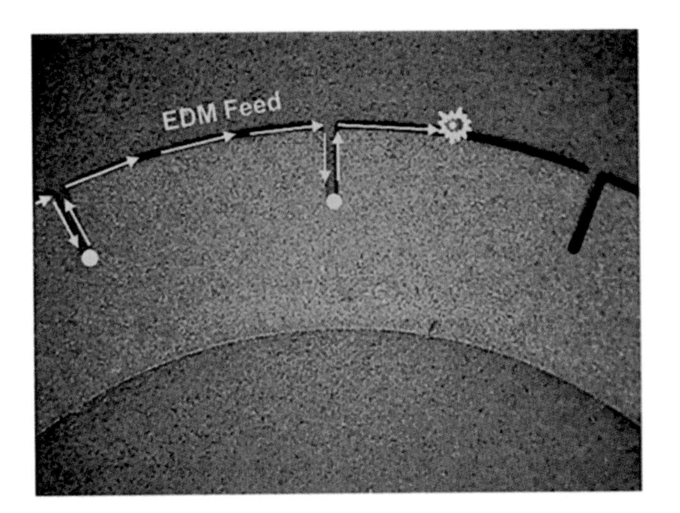

Figure 73. – EDM process flow to machine the edge geometry

The EDM process is leaving a slightly rough burned edge, so it is machined a bit oversize and then O.D. & I.D. grinded. (see fig. 74)

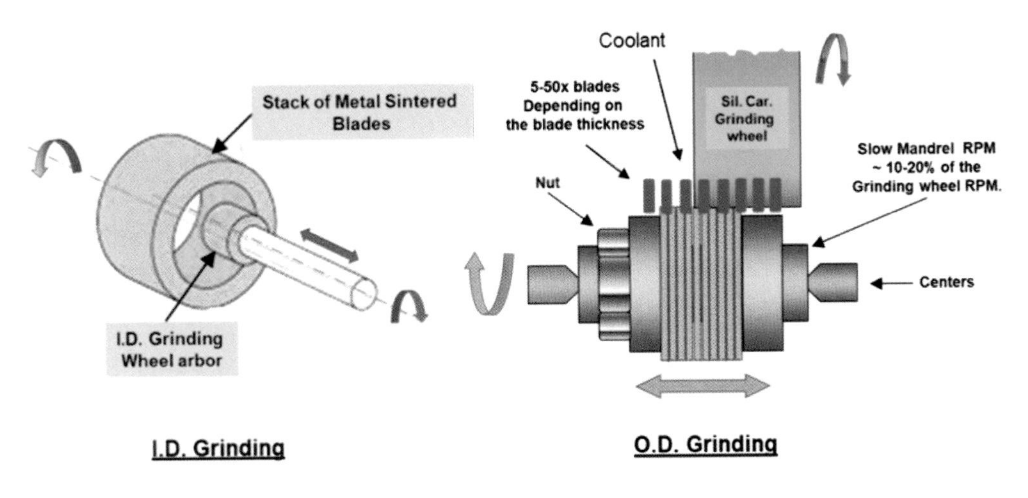

Figure 74. – Final O.D. and I.D. grinding

CHAPTER - 9

WC=WOLFRAM CARBIDE, TUNGSTEN CARBIDE SAW BLADES

A unique dicing blade design used in a small market niche that will be discussed in the application section. This is a very different blade compared to the diamond type blade designs. The saw blade is more like a cutting tool in the metal industry using hard metal and sharp teeth instead of diamonds. A few different types of thin metal dicing saw blades are available: HSS (high speed steel), WC (tungsten carbide) some with different thin hard coating, but the one type that is used in the microelectronic industry is the WC saw blade. The raw material manufacturing process is a carbide sintering process using carbide powders and cobalt as a binder. An improved sintering process used for manufacturing the WC saw blade blanks is the HIP process (hot isostatic pressing). This process is performed with improved dense material, which is also used for different small routers used for drilling PCB and others. The internet is loaded with informative information on the HIP process. An increase in cobalt content results in increased toughness TRS (transverse rupture strength), while hardness and wear resistance are reduced. These opposite parameters of the two desirable parameters, hardness and toughness, can be optimized by reducing the carbide grain size to an ultra-fine carbide powder. The result is increased hardness. It also creates high toughness. In general, superfine grain carbide grades offer increased hardness while maintaining toughness.

9.1 - BLADE GEOMETRY

Beside the raw material, the teeth geometry, the number of teeth, and the surface finish of the teeth are the next important parameters. The geometry and the number of teeth is a function of the material to be diced and the blade diameter. Following is a generic sketch of a teeth geometry (see fig. 75).

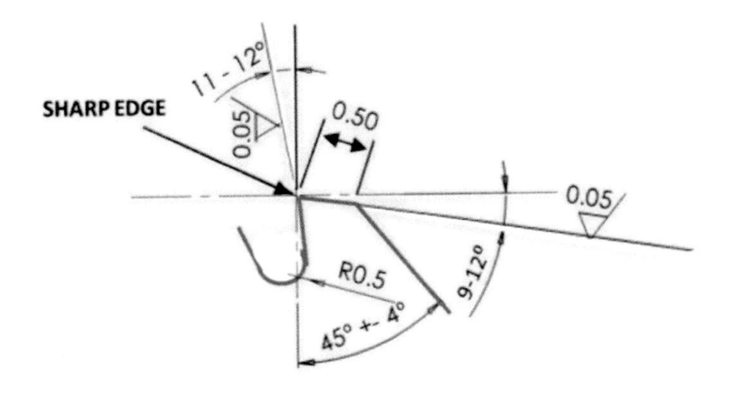

Figure 75. – Generic Teeth geometry of a Tungsten Carbide saw blades

The common blade diameters are ranging from 50mm (2") up to 152mm (6"). The majority are 114.3mm (4.5"). The thicknesses available starting at 0.127mm .005" going up to 0.5mm (.020"). The thinner blades are very tricky to handle as they are brittle and tend to break easy. The blade life is a function of the WC material, the surface finish, and the ability to maintain the sharp edge as long as possible. The teeth angles need to be optimized per application. One important angle is a very small side release on both sides of the blade a bit over the full cut depth. (see fig. 76)

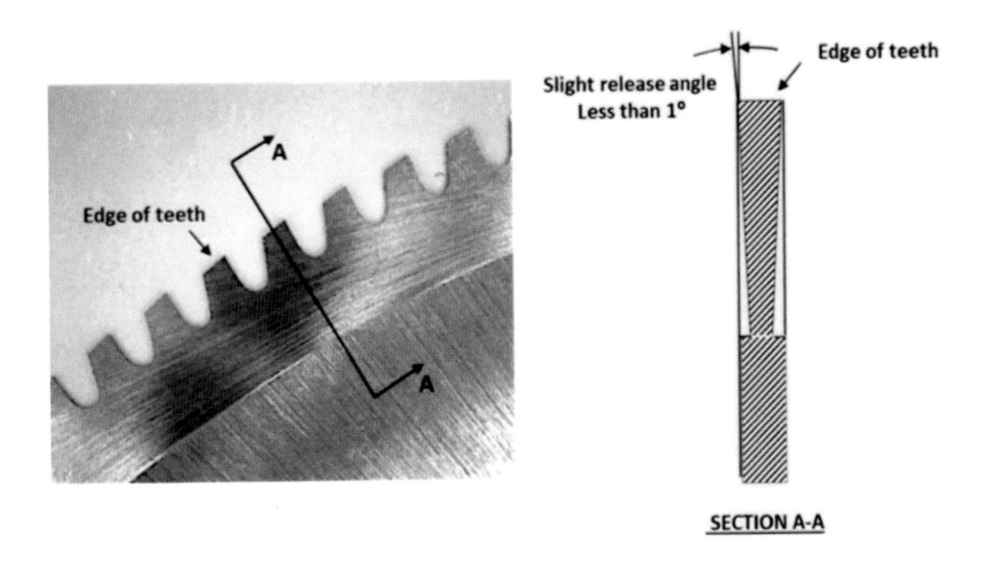

Figure 76. – Teeth edge and teeth cross section

The surface finish of the blade is of extreme importance and is a function of the raw material and the grinding process. The grinding process is performed on special grinding machines that are calibrated and indexed to perform the teeth geometry. There are many different grinding processes, and the final grinding process uses very fine diamonds in a resin bond. Below are photos showing different teeth geometries and different surface finishes on the teeth (see fig. 77).

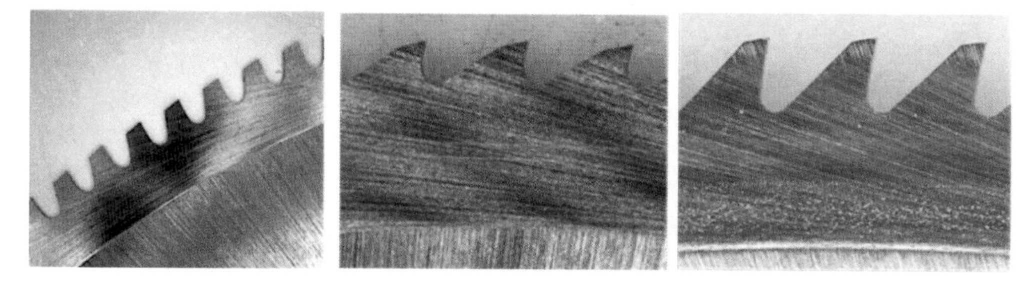

Figure 77. – Different teeth geometries

WC blades need to be mounted on the mounting flange only in one direction: the direction of the blade rotation into the substrate (see fig. 78).

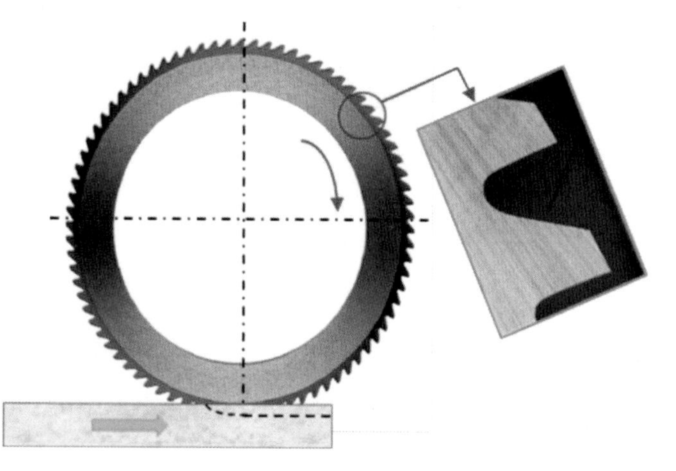

Figure 78. – Teeth sharp edge rotating into the substrate

CHAPTER 10

BLADES QUALITY CONTROL(QC)

Today's requirements in the different product manufacturing lines are demanding high yields close to 100 percent. Those demands require the diamond dicing blade manufacturers to perform at their best for cut quality and blade life. In addition, performance consistency from batch to batch is a must. Most geometry QC are the same for all blade types. Following are the main geometry parameters to be tested and the most common options used in the industry:

10.1 - BLADE THICKNESS

Most blade thickness tolerance is ±0.005mm (±.0002inch). In some cases, the tolerance can be even tighter and, in some thicker blades, a bit more open. The main simple measuring method is using an accurate micrometer; however, only one type of micrometer should be used, a non-spindle rotating type. From experience, most customers I met over the years were not aware of how the thickness measurement should be done, so the following information elaborates on this process in more detail (see fig. 79).

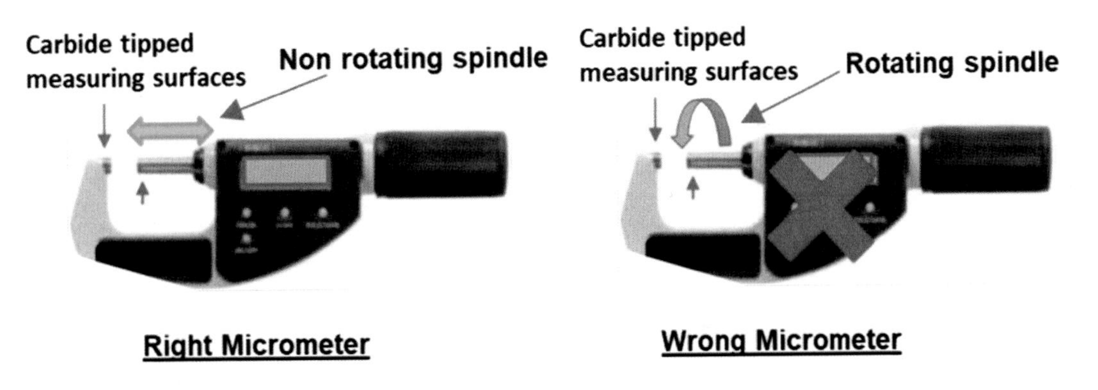

Figure 79. – Different micrometers design

Using the wrong micrometer with the rotating anvil/spindle is actually causing the anvil to be in contact with the blade while the anvil surface has a rotating/rubbing movement on the diamonds of the blade surface. This creates a machining action, and over time, the carbide anvil surfaces are actually losing flatness and parallelism (see fig. 80).

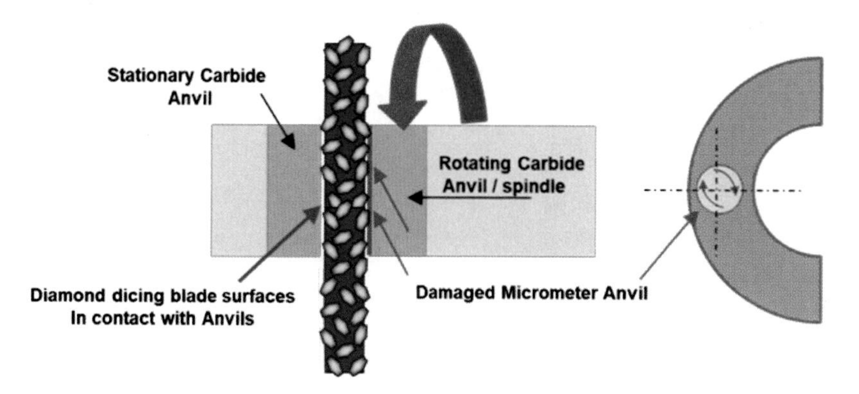

Figure 80. – Using the wrong micrometer with the rotating anvil / spindle

Fig. 81. shows the right sliding micrometer anvil eliminating the rotation contact of the anvil with the diamond surface on the blades.

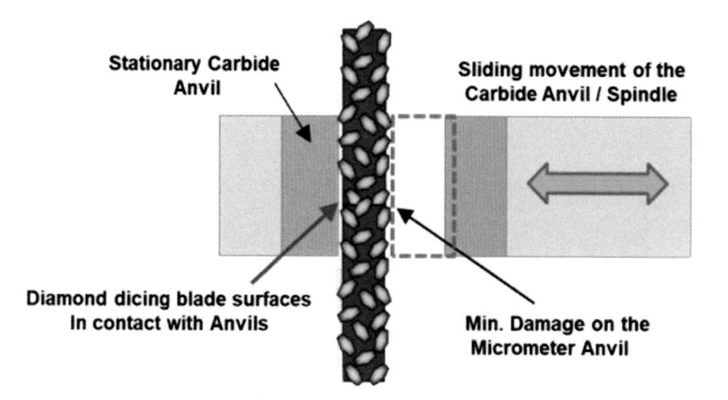

Figure 81. – Using the right micrometer with the sliding anvil / spindle movement

All major Q.C. Instrumentation manufacturers are making the non-rotating spindle micrometer design. Following is a typical micrometer specification:

Resolution: 0.001 mm or .00005"/0.001 mm

Flatness of anvil: 0.3 µm /.000012"

Parallelism of both anvil surfaces: 2 µm /.00008"

Anvil tips: Carbide

The above specification needs to be maintained. Following is a suggested process of how to measure/inspect the micrometer:

10.2 - MICROMETER SURFACES FLATNESS AND PARALLELISM INSPECTION

Period of the test – Depending on the amount of micrometer usage, once or twice a week.

Tools to be used: Synthetic Ruby ball 5 or 6mm diameter / Grade 25 mounted to a metal rod for easy handling.

Cleaning the anvil carbide surfaces with alcohol.

Close the micrometer anvils on a piece of paper.

While the anvils are clamping the paper, push the paper out.

Repeat the above process till you get a stable reading on the micrometer.

"Zero" the micrometer reading.

Clean the Rubi ball.

Close the micrometer till a contact with the anvils on the Rubi ball and zero the micrometer to 0.000mm.

Using the Ruby ball, make 10-12 measurements on the micrometer surfaces, 5-6 on the outer diameter of the micrometer anvil surfaces and 5-6 on the inner side of the anvil surfaces. (see fig. 82).

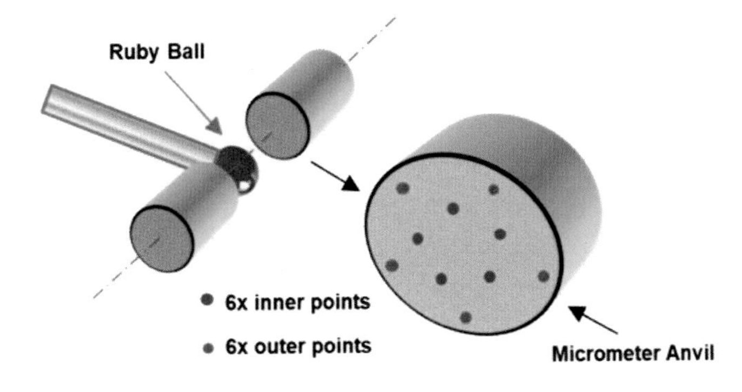

Figure 82. – Measuring the accuracy of the anvil using a Synthetic Ruby ball

Maximum variations between the reading = **0.002mm**. If out of spec, send out to a qualified vendor. It will probably require lapping of the micrometer surfaces.

10.3 - MICROMETER DIMENSIONAL INSPECTION

Period of test - Depending on the amount of micrometer usage - once a week up to once a month.

Tools for QC test - Johansson Gauge blocks in thicknesses of - 0.100, 0.300 and 0.700 mm.

After checking the micrometer surfaces flatness and parallelism clean the Micrometer surfaces.

Close the micrometer surfaces and push - Origin - to get a 0.000 mm reading.

Clean the Johansson blocks with cleaning solvent.

At the center of the Johansson blocks measure the thickness. Max variations from the nominal size should not exceed + 0.001 mm. If the measurement on one of the 3 blocks exceeds the + 0.001 mm, send the micrometer to a qualified vendor for calibration (see fig. 83).

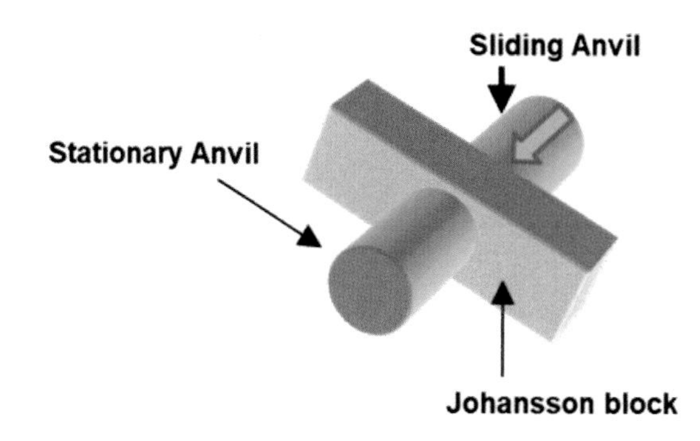

Figure 83. - Measuring the micrometer accuracy using a Johansson block

There are other common methods for blade thickness measurements. Using a dial indicator is one of them. There are real accurate indicators available however the measurement blade handling jig is the important part of this measurement set-up. It should include a rotating jig to move the blade below the indicator in an easy and accurate mode (see fig. 84).

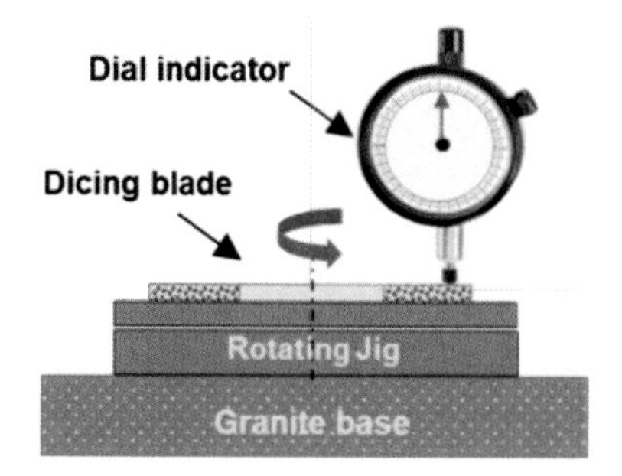

Figure 84. - Measuring blade thickness using a dial indicator on a rotating jig

10.4 - OTHER THICKNESS MEASUREMENTS OPTION

Another more advanced method used mainly for very thin nickel blades is non-contact measurement using laser displacement sensors. This method is widely used in the semi-conductor and microelectronic industries for mass production applications. It is an option for measuring blade thickness in large volumes. An illustration sketch showing the idea in fig. 85 by KEYENCE Corporation. See additional information in the reference.

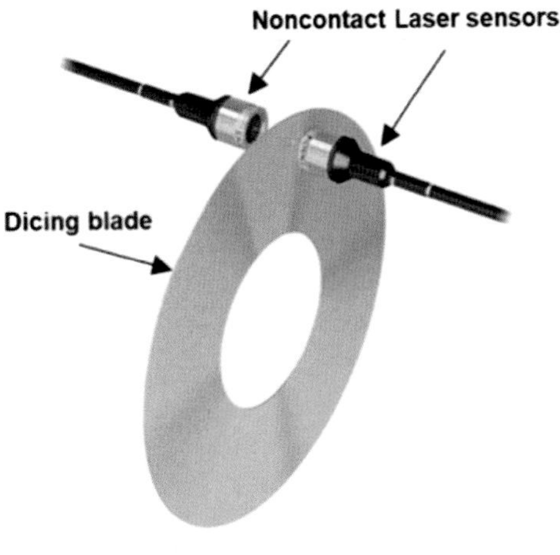

Figure 85. – Blade thickness measurement using a non contact laser displacement sensors

10.5 - INSIDE DIAMETER

The easiest and best method of measuring the inside diameter is by using hard and grounded GO & NO-GO Gages. The best gauges are grinded to undersize diameters, hard chrome plated to oversize diameters and grounded to the final dimensions. The principle of how to decide what dimension and tolerance to use can be found in many internet sites: Tolerance Calculator for Cylindrical I.D. Gages. Following is a generic illustration of a Go/No-Go gauge setup using separate gauges (see fig. 86).

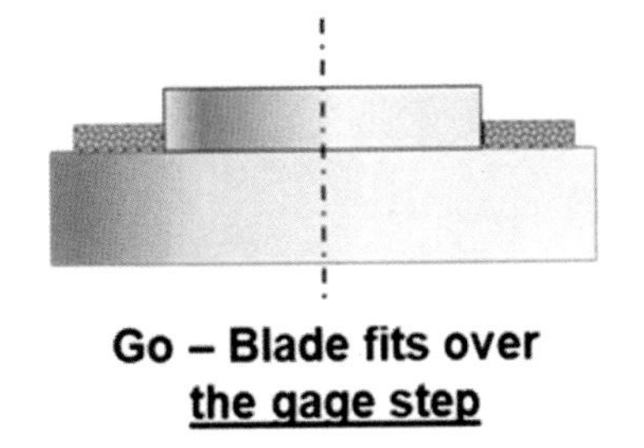

Figure 86 – Measuring blade I.D. using a go, no/go gauge

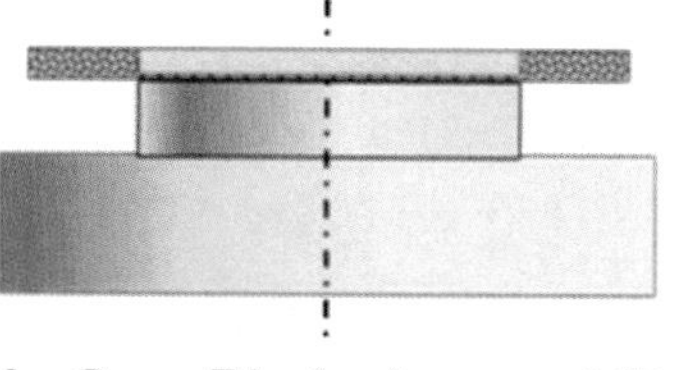

Figure 87. – Go & No go gage on one jig

10.6 - OUTSIDE DIAMETER

In a production mode of manufacturing blades, the easiest method to measure blade O.D. is by using a caliper or micrometer during the O.D. grinding (see fig. 88).

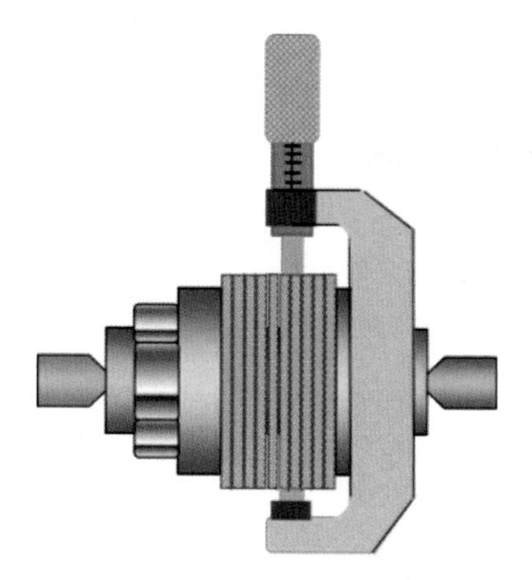

Figure 88. – Measuring blade O.D. on a grinding arbore during O.D. grinding

At a customer house the safest way is to measure the blade O.D. while mounted in a flange (see fig. 89). For very thin blades it would be safer to measure the O.D under a measuring microscope.

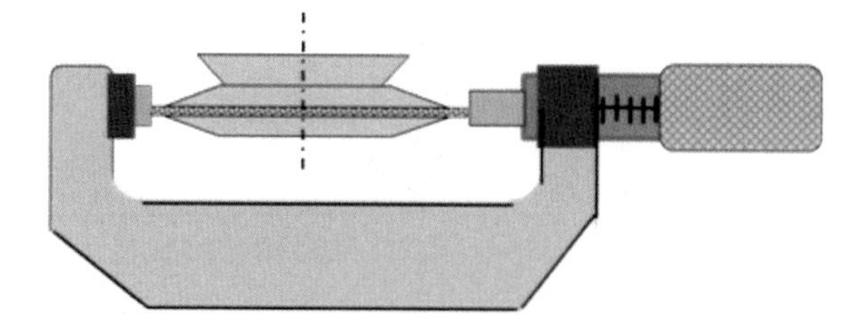

Figure 89. – Using a micrometer to measure O.D. of a single blade mounted in a flange ass.

10.7 - BLADE FLATNESS

Blade flatness is critical in some applications, mainly when using large blade exposures on relatively thin blades. The easiest way to measure flatness is to lay the blade on a clean and high-grade granite QC block and slide feeler gauges in different thickness steps in the gap between the blade and the solid granite block (see fig. 90).

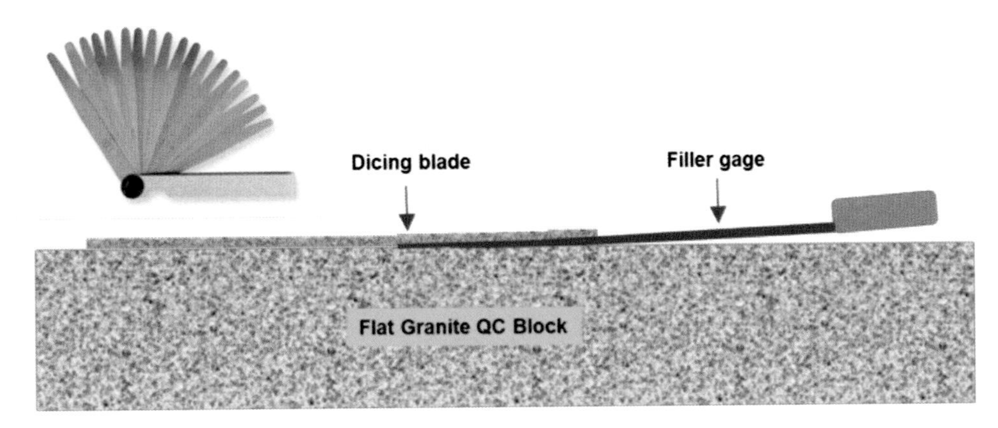

Figure 90. – Using feeler gage on a granite block to measure blade flatness

There are other more accurate ways to measure the blade flatness, but they are more time consuming and do not justify the effort.

10.8 - OTHER BLADE QC PROCEDURES RELATED TO EACH BLADE MATRIX (BLADE TYPE)

10.8.1 - NICKEL BOND:

Unlike other matrices, diamond concentration on nickel electroformed blades is tricky to measure. The diamond concentration is controlled during the plating process by the process parameters. The different process parameters were already discussed in the nickel section. Measuring the final diamond concentration in the final product can easily be observed under an optic microscope or an SEM (Scanning Electron Microscope). However, to get a more accurate reading, it requires other methods. One option is to use a carbon determinator. Diamonds are made of carbon, so they form as carbon atoms, and the carbon determinator will analyze the carbon content of a nickel dicing blade. This will reflect the diamond content. The process is a destruction process, meaning that it will reject the sample being tested. Another method to determine the diamond concentration on a nickel blade is to use the Archimedes principle. A blade is weighted on an accurate scale and then weighted while immersed in water. The accuracy of the scale needs to be 0.001

gr. It is a must in order to get an accurate measurement. The Archimedes method and the calculated formula is described on many good internet sites. A few sites are indicated in the reference section.

10.8.2 - HUB BLADES:

Checking the aluminum machined hub blade is a major initial step. The geometry of the aluminum hub has a direct impact on the performance of the hub during the plating process and the dicing process (see fig. 91).

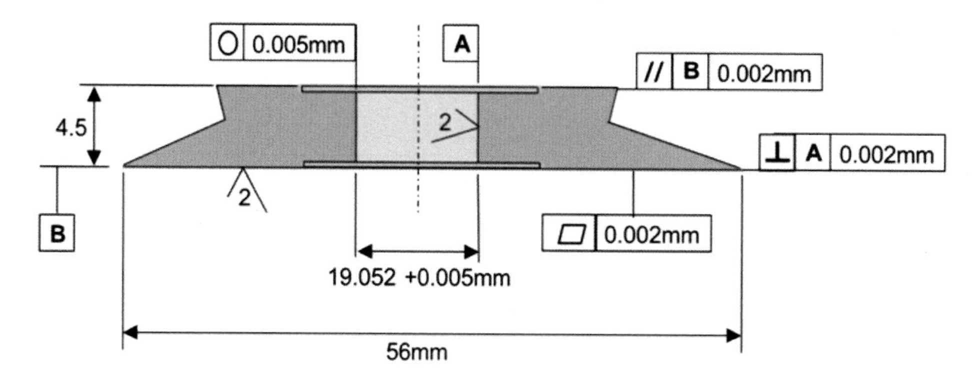

Figure 91. – Aluminum hub cross section dimensions

Top-of-the-line QC instruments are needed and calibrated periodically. Diamond concentration and bond hardness is controlled in the plating process. This was already discussed in the nickel section. With today's different application requirements, there are a few diamond concentrations and hardness requirements, so those parameters need to be monitored and controlled continuously. Beside the geometry inspection, there is another important functional geometry inspection of the blade side runout. This test is performed after the blade is exposed from the aluminum hub. This QC inspection is performed with another high-power microscope, checking the blade edge runout while rotating on an accurate spindle (see fig. 92).

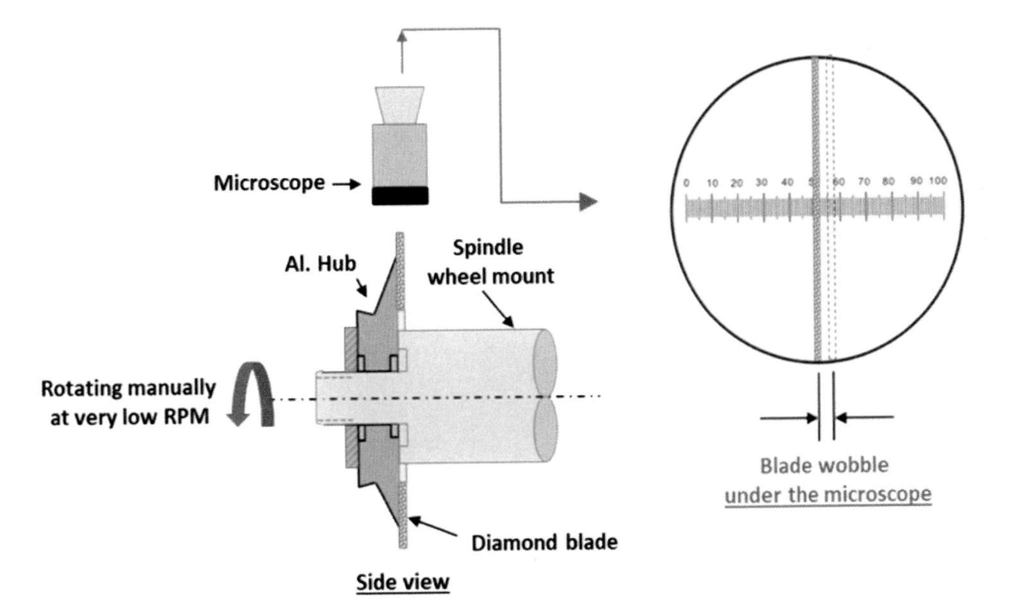

Figure 92. – Measuring blade wobble under a microscope

In many Hub manufacturing processes, a special dressing and cut quality process is performed as part of the QC process. This is needed in order to make sure the hub blades will perform in the cut quality specs when arriving at the customers manufacturing sites.

10.8.3 - RESIN AND VITRIFIED BOND:

The main QC requirements are geometry and flatness. In addition, the resin blades and the Vitrified blades being made of a non-conductive material with additional carbon or other conductive materials need to be checked for conductivity. The conductivity is measured as max resistivity using an AVO meter and the result should be less than 10 K Ώ. Resin blades usually perform with good cut quality; however, wear can be critical, and in some cases, a wear test is performed per batch of manufacturing. Diamond concentration is easy to control as it is part of the powder mix preparation, and the amount of diamonds is weighed to the exact amount needed.

10.8.4 - METAL SINTERED BOND:

Same as the resin and vitrified blades, the geometry and flatness are the main parameters to QC. The same is true for diamond concentration; the powder mix is accurately controlled, and the diamonds are weighted to the exact amount. A metal sintered blade is a powder metallurgy process that needs to be controlled using tensile, bending, and other mechanical instruments. SEM (Scanning electron microscopy) is also used to investigate material and surface properties.

10.8.5 - WC (TUNGSTEN CARBIDE) BLADES:

GEOMETRY

Inside diameter is measured the same way by using GO – & NO-GO gauges.

Blade thickness is measured using the same micrometer however it is a bit more difficult as the teeth have a slight side relief, and they are brittle and sensitive to breakage especially on the very sharp edge. (see Fig. 93).

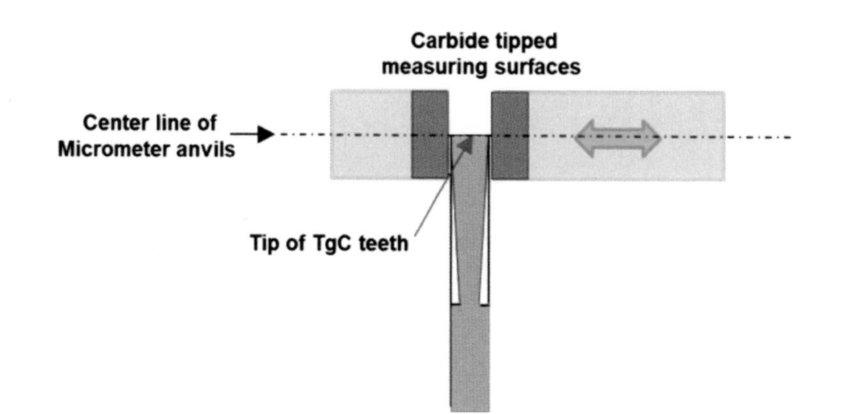

Figure 93. – Measuring Tungsten Carbide teeth edge thickness

Measuring the teeth profile angles can be performed on a XYZ measuring system or on a measuring microscope. The blade needs to be centered and aligned prior to measuring the teeth geometry. (see fig. 94).

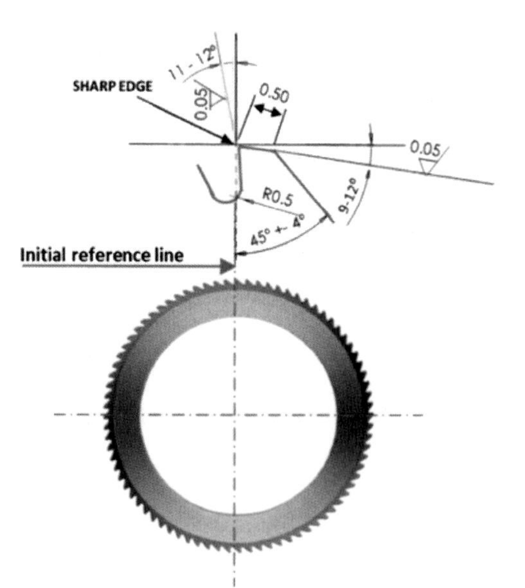

Figure 94. – Measuring the teeth geometry under a profile projector or a Microscope

Surface finish which is very important is measured using a roughness tester. (see fig. 95).

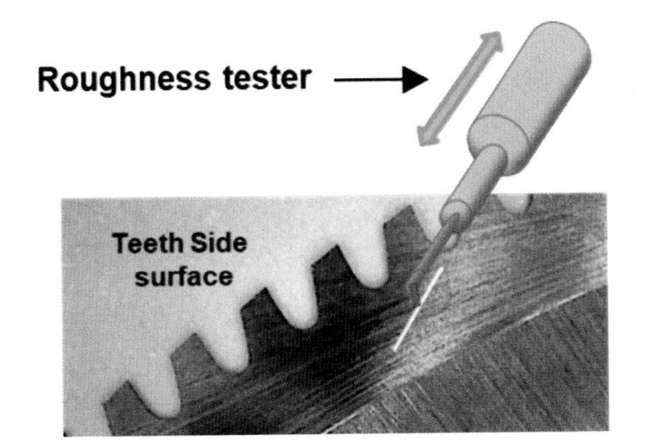

Figure 95. – Measuring surface finish / roughness

Blade flatness is a common issue on WC blades due internal stresses. One unique stress known with the WC blades is "oil canning" stress. It can be observed by holding the blade using both hands and trying to bend it a bit. If the "oil canning" stress exists, it can be easily observed by a slight "click effect." This effect is a blade rejection (see fig. 96).

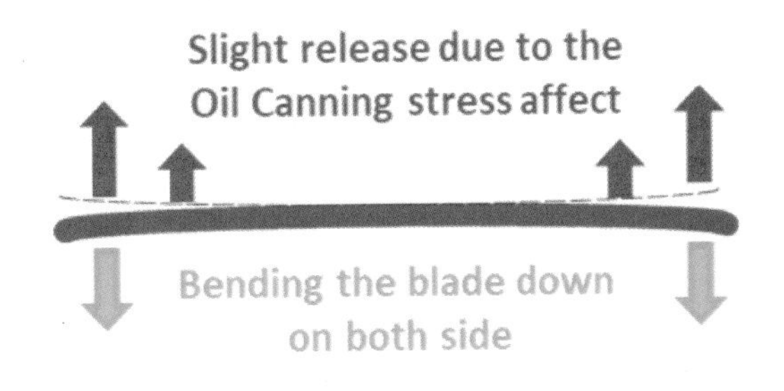

Figure 96. – Oil canning stress inspection

In the industry a common problematic case of the "oil canning" effect can be found in the metal roofing industry. There is a lot of good information on this subject on the internet, one of them is mentioned in the reference section. Normal flatness like on other blades is measured using the filler gauge on a flat granite stone.

Visual inspection of the teeth edge sharpness and the side surface finish can be inspected on a high-power microscope or on an SEM (Scanning Electron Microscope) (see fig. 97). showing the sharpness of the teeth edge and the surface finish using an SEM.

Figure 97. – SEM showing the teeth sharpness and the side surface finish

The above QC procedures are close to what the industry is using, they can be used as good guidelines for incoming inspection for all the dicing blades users.

CHAPTER 11

DICING MACHINE IMPORTANT MECHANICAL SPECS INVOLVED IN CUT QUALITY.

11.1 SPINDLE ACCURACY

The best flange and the best blade in all aspects of geometry and matrix will perform poorly if mounted on an out of spec spindle. Before discussing the flange design parameters, it is important to view a few mechanical aspects on the dicing saw. The first one is the spindle wheel mount at the mounting position of the flange. See fig. 98 an example of one of Colibri 's air bearing spindles. See more info at the reference section.

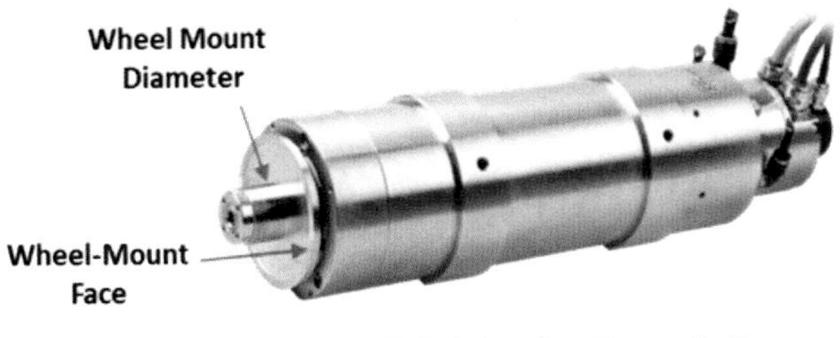

Figure 98. – Front spindle Wheel Mount

Any spindle on a dicing saw regardless if it is a 2" spindle with a ¾" shaft dia. or a 4" spindle with a 1¼" shaft dia. need to meet real tight dimension specs. (See fig. 99). The below spec is usually used on the assembled dicing saws; however, the spindle vendors are specifying even a tighter tolerance of 0.001 mm.

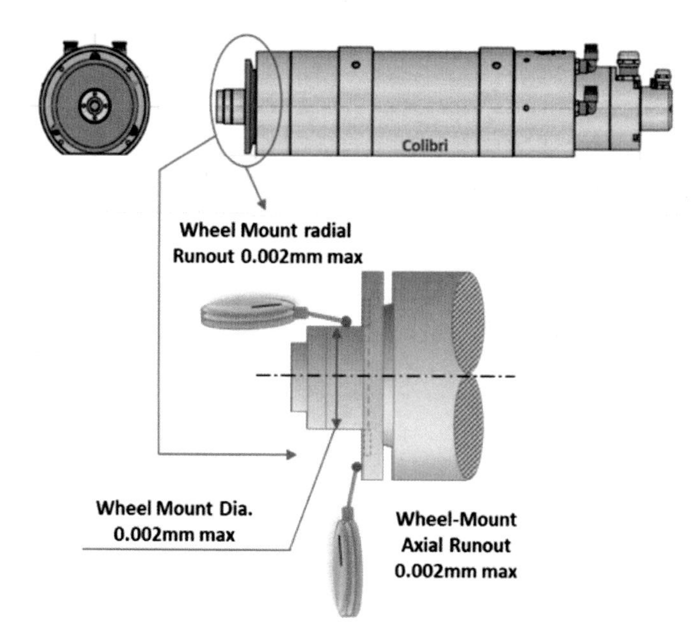

Figure 99. – Wheel Mount accuracy measurements

Any new dicing saw should be checked for the above dimensions. They should also be checked periodically. Another important parameter that needs to be checked periodically is the alignment of the spindle centerline to the –X – movement which should be exactly 90 degrees (see fig. 100).

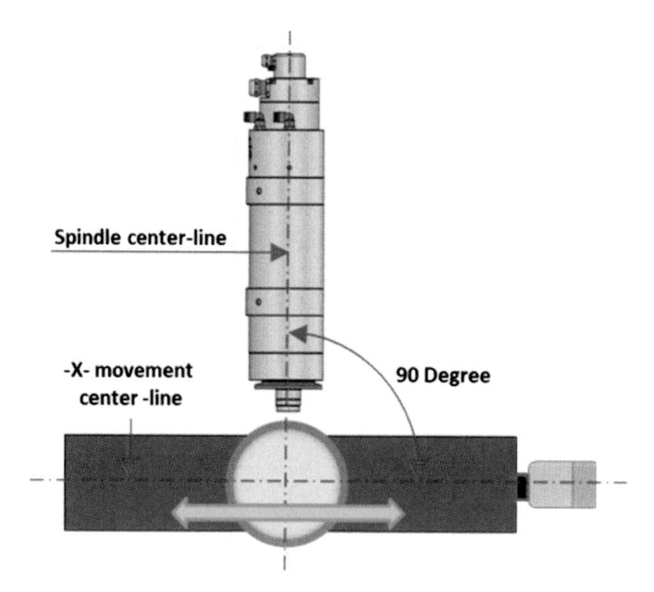

Figure 100. – Alignment inspection of spindle to – X – movement

Following is a process procedure how this is done in a generic mode which should be similar on most dicing saws in the marketplace. A dial test indicator needs to run across the wheel-mount face by moving the – X – slide from side to side (see fig. 101).

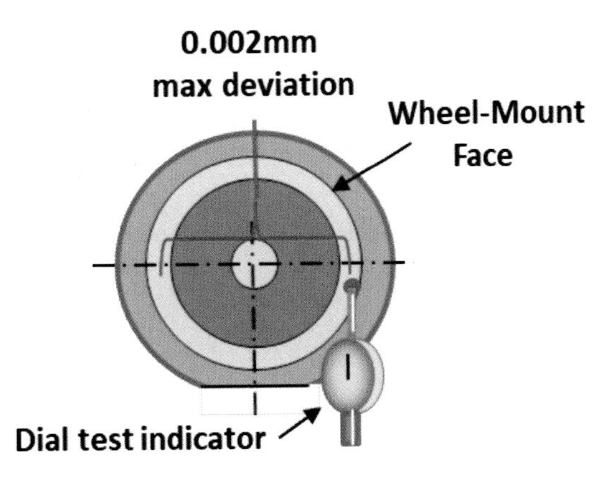

Figure 101. – Moving a dial indicator across the wheel mount by using the – X – movement

An easy way is to use a straight and perfect parallel bar mounted to the wheel-mount and run the Dial test indicator along the parallel bar. (see Fig. 102)

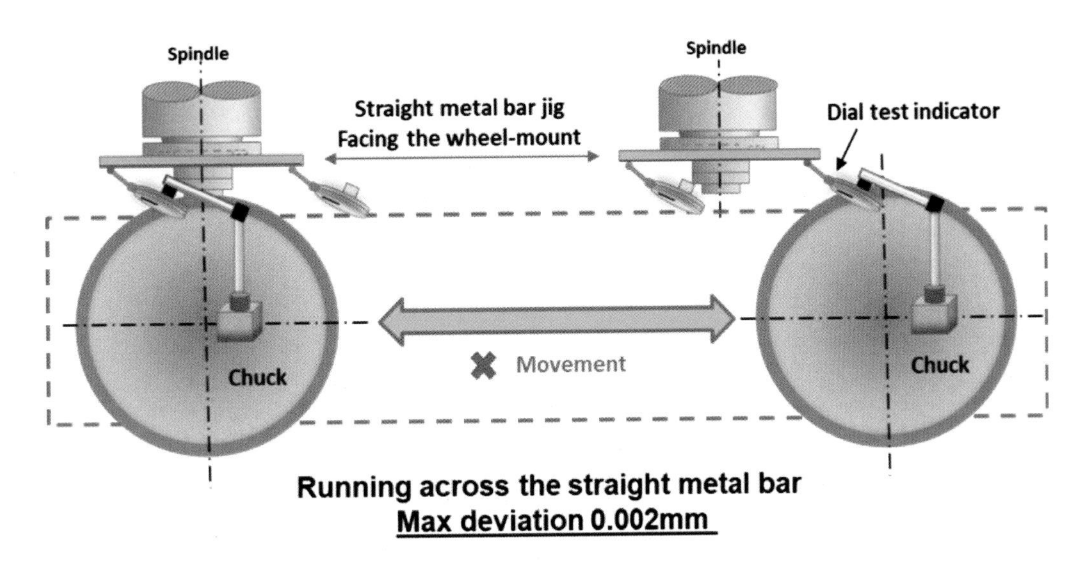

Figure 102. – Top view of dial indicator running across the wheel mount

This process needs to be checked on a new arriving saw and periodically by the maintenance group. If the spindle/wheel-mount is out of specification, it will result in wider cuts and more chipping (see fig. 103).

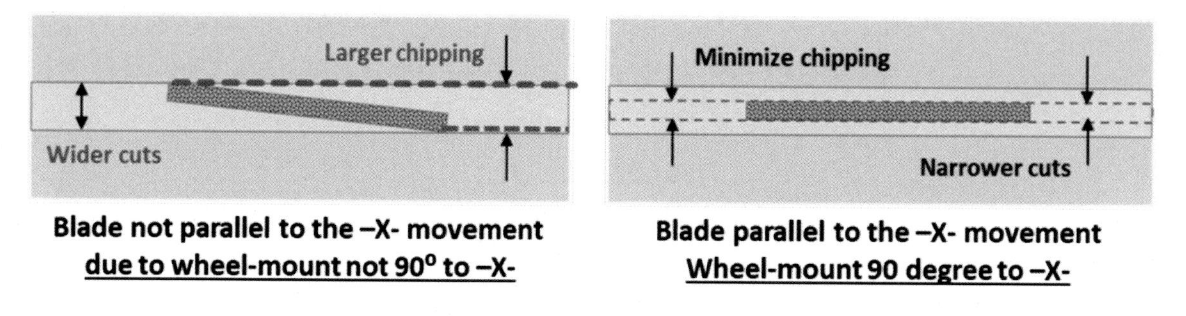

Figure 103. – Kerf quality issues if the spindle is not 90 degree to the – X – movement

The blade edge can also be damaged; losing profile if the wheel-mount is out of the 90° to the –X – slide.

11.2 - SPINDLE PERFORMANCE:

There are a few good air bearing spindle vendors for 2" & 4" diameter dicing blades. A good spindle beside the mechanical accuracy needs to perform with min. vibrations and to maintain torque at all speeds during dicing. There are other important specs to monitor, see in the reference list. Another important parameter related to the spindle mounting is to mount the spindle as close as possible to the front side. This will minimize vibrations mainly at high spindle rpm (see fig. 104).

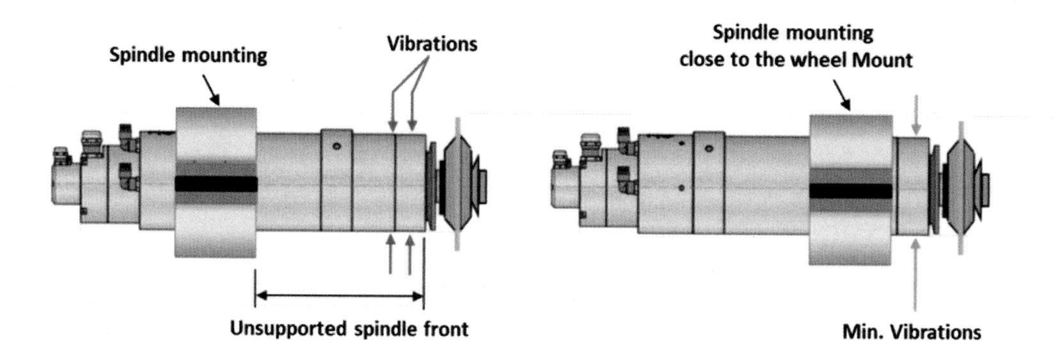

Figure 104. – Spindle mounting options

11.3 - CHUCK

Today most chucks are vacuum chucks made of porous ceramic. The chucks are lapped to a very tight flatness Spec. regarding flatness and parallelism. However, mounting the chuck on the dicing saw needs to be monitored and inspected on a new saw and periodically in a production environment. The specs are individual per the dicing saw manufacturer;

however, they are similar. Following is a generic inspection method checking the chuck parallelism on the axes – X – and – Y – (see figs. 105 and 106).

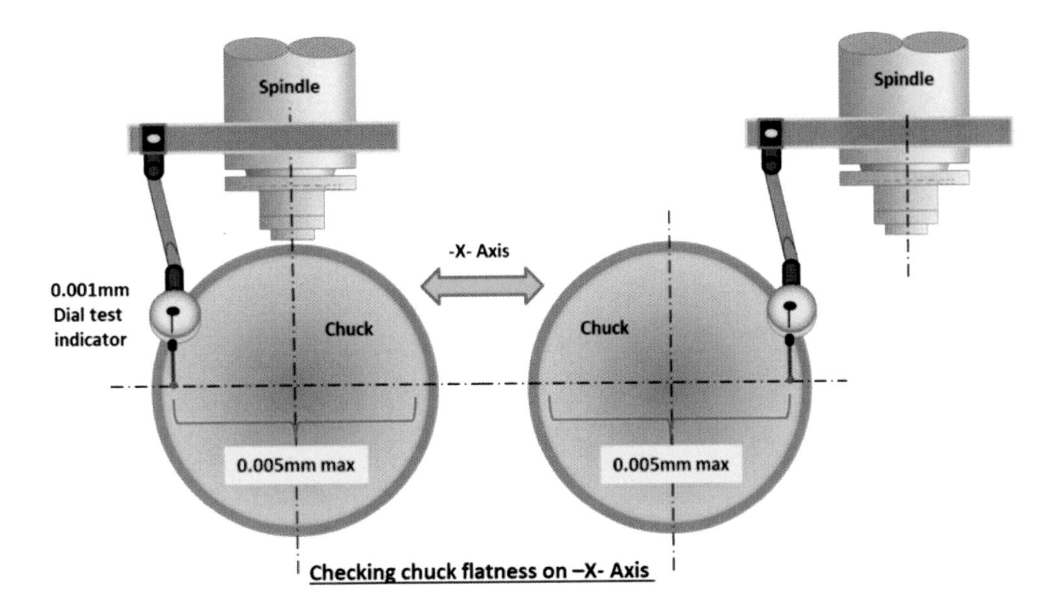

Figure 105. – Checking the chuck flatness using a dial indicator while moving the – X – axis

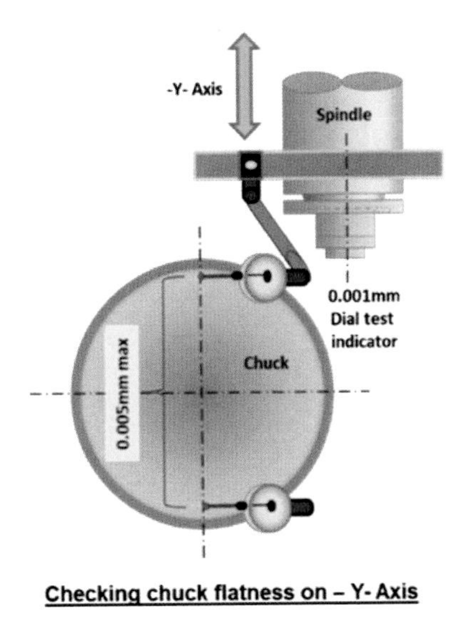

Figure 106. – Checking the chuck flatness using a dial indicator while moving the – Y – axis

CHAPTER 12

COOLANT ON THE DICING SAW

Any dicing process requires a good coolant system in order to cool down the blade and the diced substrate. Another purpose is to wash away the dust particles created during the dicing. Following are important coolant parameters to be discussed:

- Type of coolant
- Cooling nozzle geometry
- Cooling nozzle location
- Cooling pressure
- Chilling the coolant
- Coolant treatment
- Special cooling nozzles
- Air knife phenomena in the cooling block

12.1 - COOLANT TYPE

In most dicing processes, water is the coolant and D.I. water is the majority. The reason for using D.I. water is to eliminate contamination of the substrates being diced. There are some applications requiring only water softening treatment which is lower cost effective. When dicing semiconductor silicon or GaAs wafers the D.I. spec resistivity is in the range of 10–18MΩ /cm. The high resistivity is causing on some substrates ESD (electrostatic discharge) that can damage the integrated dies. This will be discussed later.

12.2 - COOLING NOZZLES GEOMETRY AND LOCATION

Cooling nozzle design is an art. There are different geometries and different locations of mounting the cooling nozzle on the wheel mount and on the front blade guard. It all depends on the saw type, the application, and the water utility pressure at the customer house. The very basic standard cooling nozzle is a round S.S. pipe in diameters ranging from 1–3 mm as standard, sometimes with larger diameters depending on the applications.

The very basic nozzle location is aiming the cooling nozzle to the blade contact with the substrate (see fig. 107).

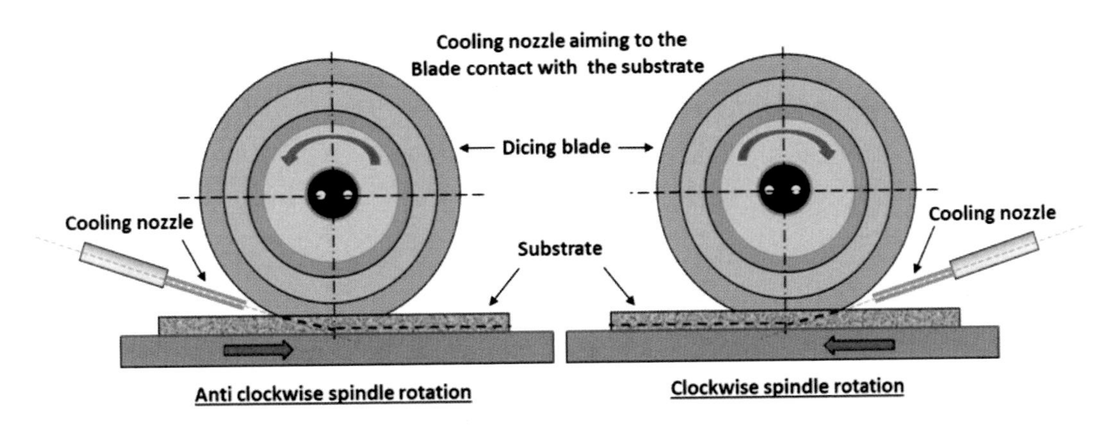

Figure 107. – Standard cooling nozzle location at the blade contact with the substrate

The decision on what nozzle diameter to use depends on the blade used and the utility water pressure at the customer house. A small diameter nozzle of 1 mm produces very high pressure that can deflect or vibrate a thin blade. This is more critical when using a thin nickel or hub blade. However, in a facility using many saws with low utility water pressure the 1mm nozzle may be an option. Another option using a much larger diameter cooling nozzle is on delicate applications requiring stiff blades with min. vibrations. A good example is a 4–6 mm diameter nozzle that has special round knee geometry to eliminate water turbulence and to perform with a lot of water at low pressure resulting in a nice clean water stream (see fig. 108).

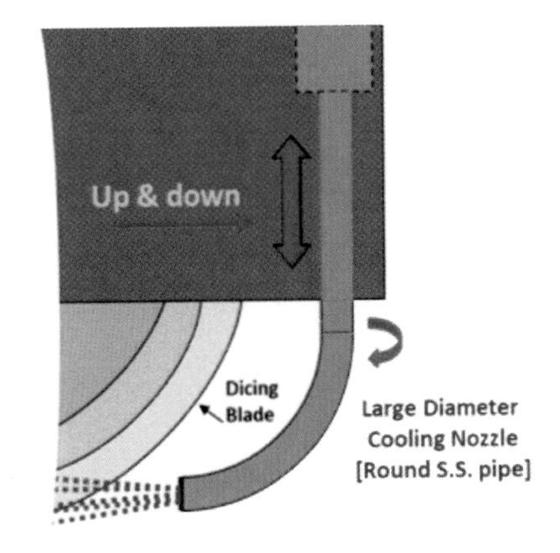

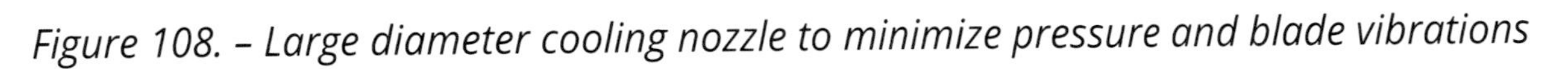

Figure 108. – Large diameter cooling nozzle to minimize pressure and blade vibrations

A mass production dicing environment requires optimization of all parameters in order to improve quality performance. During the years, a unique cooling nozzle was developed that requires no side nozzle alignments. A relatively larger diameter nozzle is used with a wide slot on top of the nozzle. This provides a gap for the blade edge to be inside the cooling nozzle with a continuous water coolant contact (see fig. 109).

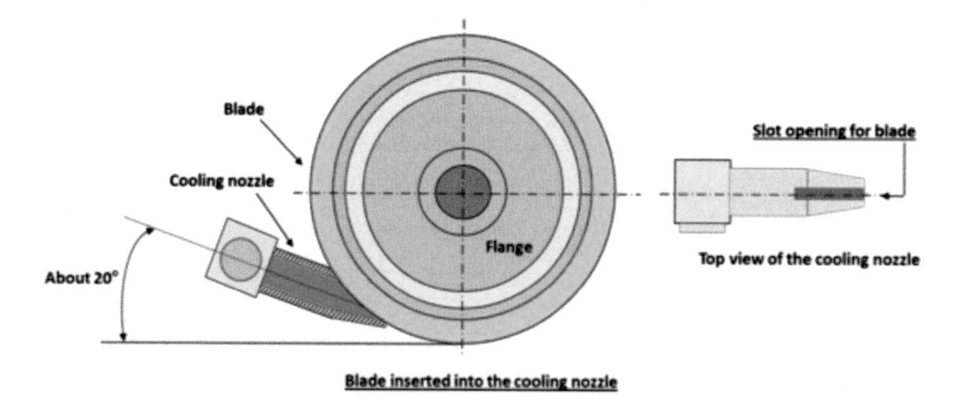

Figure 109. – Blade inside the cooling nozzle

The slot nozzle can be made in different geometries all depending on what is best for a specific application. The cooling water used in most 1–3 mm diameter cooling nozzles is in the range of 0.5 to 3.5 L/min. and again it depends on all the above-mentioned parameters. On many applications side cooling jets or what is called side spray nozzle are used. In some cases, they are a Go or No go to meet the application spec. The water pressure on the side spray nozzle is lower compared to the main nozzle mainly to minimize blade vibrations (see fig. 110).

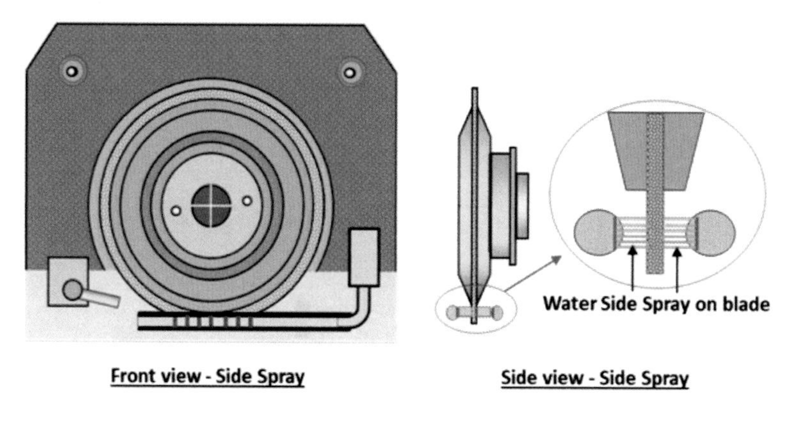

Figure 110. – Blade cooled with side water sprays

Another important phenomenon in any dicing system is what is called "the air knife." The air knife is a stiff air pressure around the blade edge caused by the high spindle rpm of 30k and over. The air knife can deflect the water stream from the blade edge (see fig. 111).

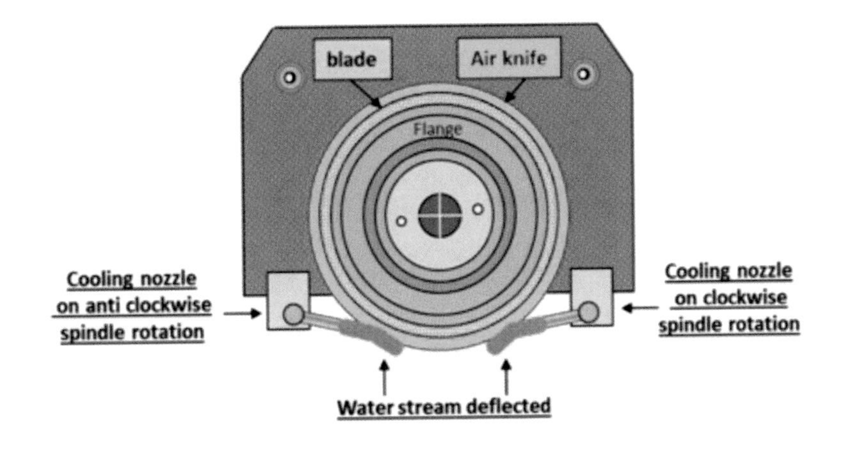

Figure 111. – "Air Knife" eliminating the coolant to hit the blade edge

In many cases, this is a major issue to be solved mainly on thin blades but not only. For this reason, many cooling nozzle geometries and location designs were developed by all saw vendors. Some of the designs use other industrial nozzle designs and some are changing the standard nozzle location to move away from the traditional 7 or 5 o'clock to 9 o'clock or to add additional nozzles to the 12 o'clock location. The idea of the different nozzle location is to help overcome the air knife issue (see fig. 112).

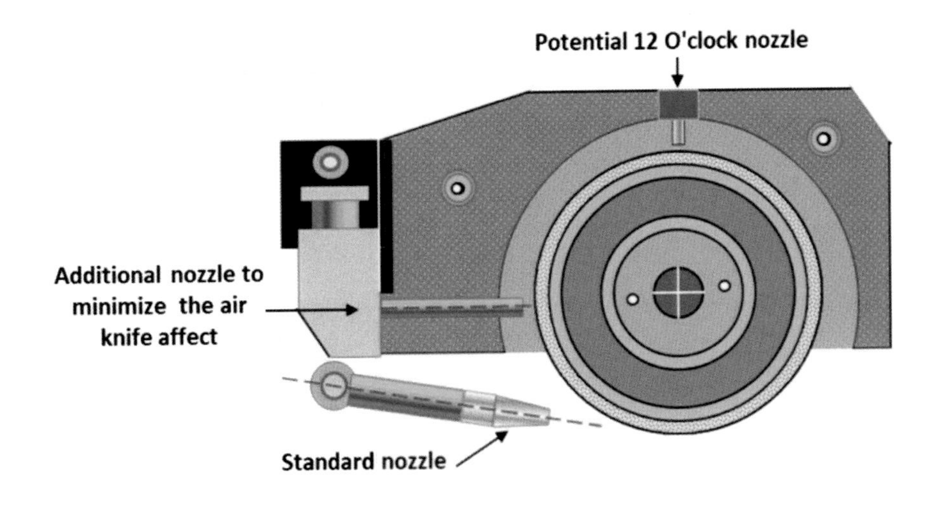

Figure 112. – Different coolant nozzle locations to overcome the "Air Knife"

In the grinding industry the air knife is a well-known phenomenon affecting the surface finish on surface grinders and other grinding machines. This problem was solved many years ago by adding an air knife breaker (see. fig. 113).

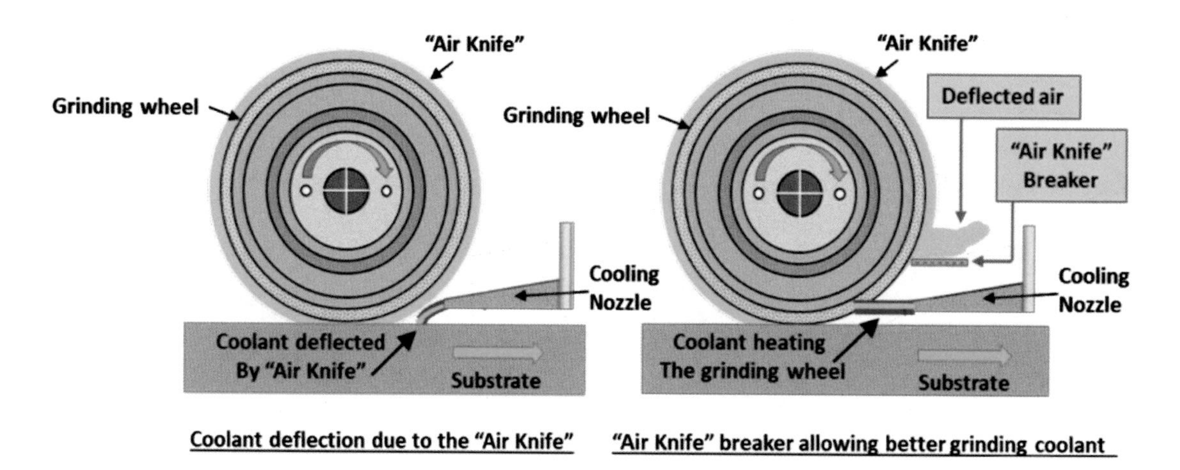

Figure 113. – "Air knife" Breaker to cut the "Air knife" and help the coolant to hit the grinding wheel

In the dicing industry I do not recall such a device on the saw to break the "Air Knife." Below is my "two cents" on how this idea can be implemented (see fig. 114).

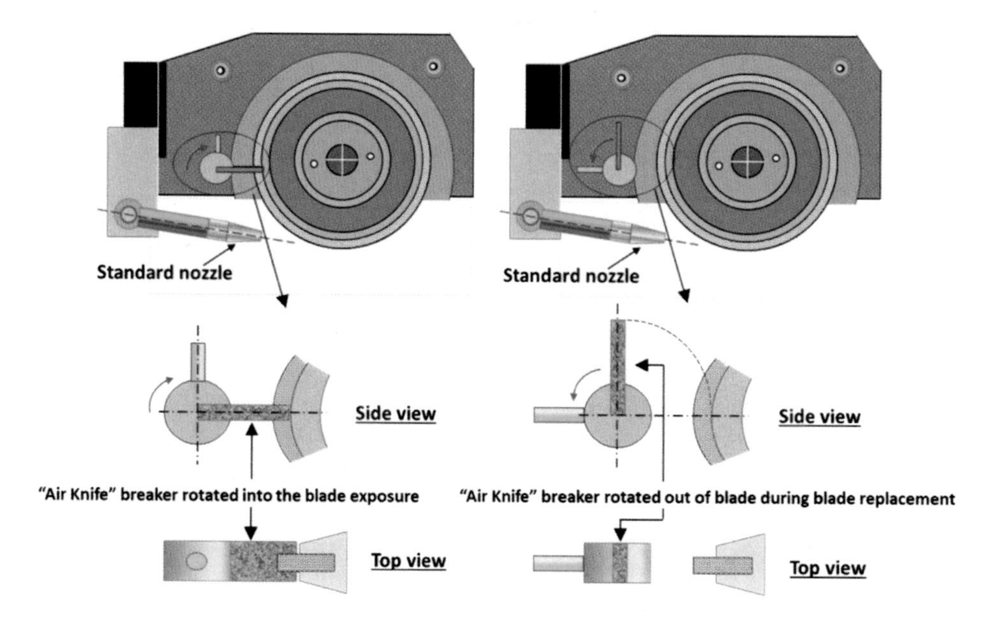

Figure 114. – "Air Knife" breaker idea that can be implemented on existing saws

The above sketch shows the location of the breaker at 9 o'clock. However, the location depends on the spindle rotation, clockwise or counterclockwise and the location of the main cooling nozzle. It is important to have the breaker close and prior to the main cooling nozzle. The idea is to make a "friendly" blade breaker like a 1 mm thick dressing board or similar that can rotate into the blade while the blade is rotating and out of the blade when a blade change needs to be done. It can be a manual rotating or a fine solenoid slow movement. The main idea is to break the "Air Knife" around the blade edge so the coolant can properly hit the blade and the substrate.

12.3 - COOLING WASHERS (CLEANERS)

On some dicing systems a washing pipe designed to clean the diced substrate is an important tool. This washing pipe is used mainly when dicing silicon wafers where the dicing residue is a very fine silicon powder that is contaminating the sensitive devices. This washing jet is used in other applications as well. The idea is to use a long SS pipe with many small hole diameters or slots covering the entire chuck area (see fig. 115).

Figure 115. - Wash pipe cleaning the wafer during dicing

Other standard cooling nozzle designs are also introduced to dicing. (see Fig. 116)

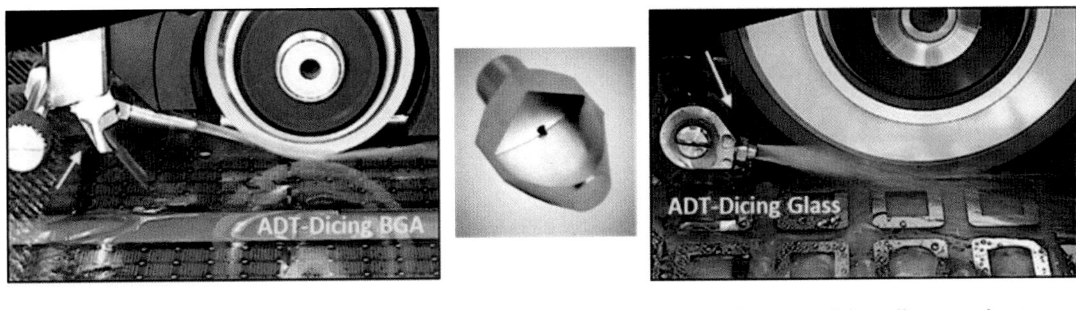

Figure 116. - Commercial cleaning and cooling nozzles on existing saws

12.4 - CHILLING THE COOLANT

Some applications require lower coolant temperatures in order to minimize blade loading and to perform with better cut quality and accurate cut placement. Most dicing manufacturers are offering such systems, the systems are working in a close loop and the coolant is being filtered to take out the powder residue created during the dicing. The systems in most cases can handle 2 dicing saws while controlling the temperatures to ranges of

5° – 25° C with an accuracy of + – 1° C. The water flow is also controlled even if the water supply in the facility is low which is an advantage. Most customers are using D.I. water however an additive can be added to the coolant and the concentration is controlled. Adding additives will be discussed in the next additive discussion. In some large manufacturing houses, using many dicing saws for the same application a large system with a mother coolant tank is used. The main advantage is that all dicing saws get the same coolant quality and pressure.

12.5 – USING ADDITIVE IN THE COOLANT

A major problem in some dicing applications is the ability of the coolant to penetrate the kerf and to wash away the powder residue created during the dicing. The main reason for having coolant problems is the high surface tension of the water used during the dicing. Water has a high surface tension because the water molecules on the surface are pulled together by strong hydrogen bonds. Let's discuss the differences of high surface tension and low surface tension: Water has a higher surface tension (72.8 millinewtons (mN) per meter at 20°C) than most other liquids. Water has the tendency to shrink into minimum surface area possible because of the relatively high attraction of water molecules to each other. In simple dicing wording, standard non-treated water has not an easy/nice stream flow. Following are a few interesting examples of water with high surface tension (see fig. 117).

Figure 117. – High surface tension behavior samples in the nature

Reducing the water surface tension can be done by adding soap and commercial wetting agents. It has a dramatic effect on the water surface tension. Following is an example of what can be done with low water surface tension lowered by soap (see fig. 118).

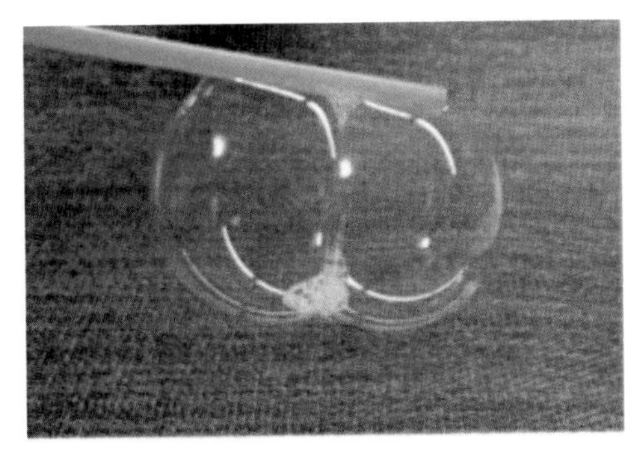

Figure 118. – Air bobble created using soapy low surface tension water

In the dicing word there are a few wetting agents available to reduce the surface tension of the coolant water during the dicing process. They are soapy type wetting agents with the following benefits:

- Helping the coolant to penetrate the kerf, reducing friction, lowering temperature elevation during the dicing and better cleaning swarf/particles during the dicing process.
- Minimizing loads and extended blade life.
- With some products minimizing ESD (Electrical Discharge) which is critical on semiconductor silicon wafers.
- Helping to minimize corrosion on the diced devices (see fig. 119).

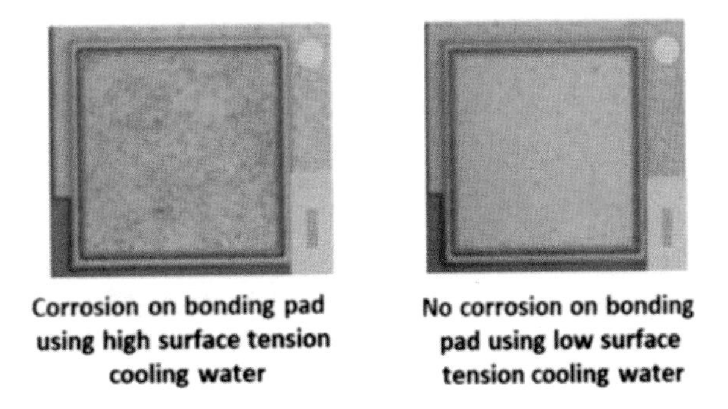

Figure 119. – Bonding pads results using coolant additive and using standard collant

Another good option of adding additives to the coolant is using a non-electric water-driven proportional dosing pump used for agriculture fertilizing. It is a simple setup to use, it can be adjusted to the right % of additive and is relatively inexpensive. The only disadvantage (maybe an advantage) is the fact that the coolant with the additive is not saved after the dicing as it is drained out. As most additives today are "green," there are no issues to dump the mixture and there is no special need for special treatments (see fig. 120).

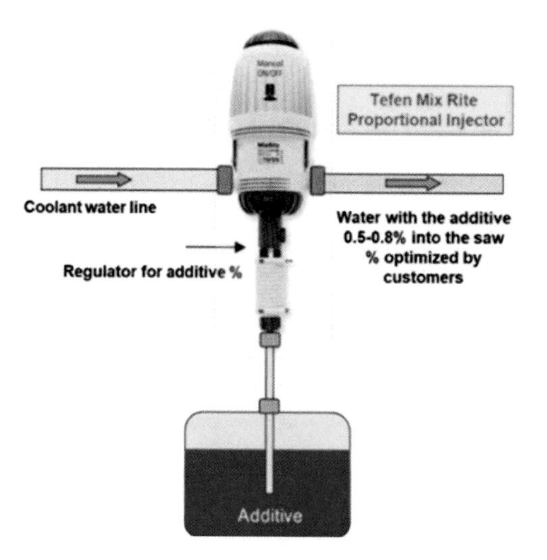

Figure 120. – Using a Dosing agriculture pump for adding coolant additive

12.6 - COOLANT NOZZLE ALIGNMENT

With most of the above discussed nozzle options, one important parameter is aligning the nozzle not only to the contact of the blade with the substrate but also making sure the water stream is centered to the blade when looking from the side. (see fig. 121)

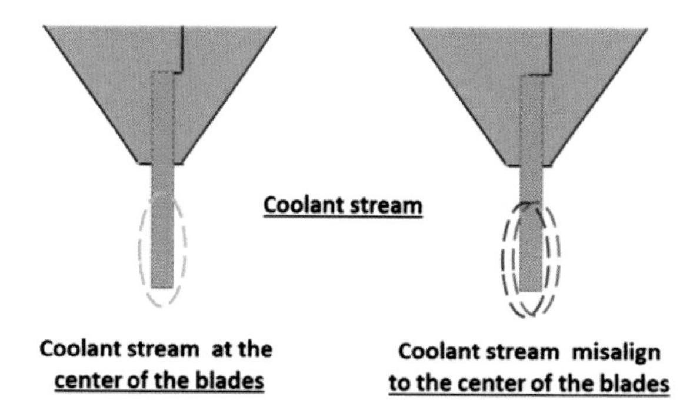

Figure 121. – Coolant alignment to the center of the blade

Another option is using dual cooling nozzles however this requires also aligning both nozzles to be centered to the blade (see fig. 122).

ADT dual cooling nozzle

Figure 122. – Using dual coolant nozzles on some special applications

CHAPTER 13

CO^2 AIR BUBBLER

DI water is the coolant used in dicing silicon & GaAs semiconductor wafers. The purity of the coolant is required to maintain strict quality requirements of the wafer dies. The parameter defining the purity of DI water is Resistivity. The measurement unit is MΩ-cm. DI water resistivity in the semiconductor industry is in the range of 10–18 MΩ-cm normally closer to the upper side. The high resistivity of DI water coupled with pressure cleaning can create static charges which can lead to ESD (Electrostatic Discharge) failure. For this reason, a CO2 Air Bubbler is used. The CO2 air bubbler lowers the resistivity of D.I. water by dissolving carbon dioxide gas into the dicing cutting water. This can bring down the resistivity to about 0.1–1.0 MΩ-cm while maintaining the purity of the water. The normal resistivity used is between 0.5–1. In addition, adding CO2 to the coolant is lowering the PH, causing it to be more acidic which can affect higher wear of the nickel blade binder. The above requires a close monitoring of the CO2 air bubbler set-up and the static electricity. There are many static electricity sensors which are used already by the dicing vendors and customer users (see fig. 123).

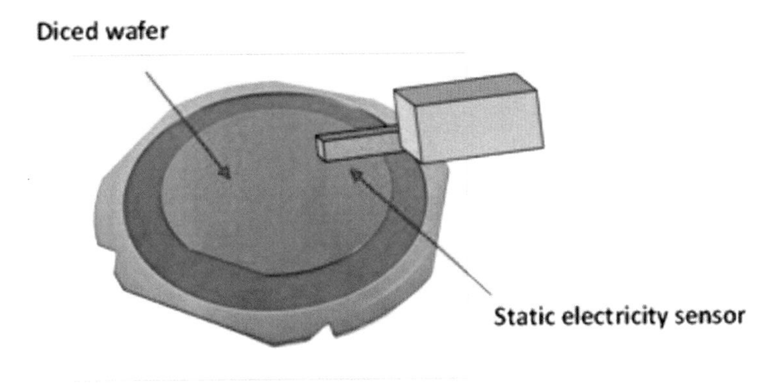

Figure 123. – Static electricity measurement on a semiconductor wafer

CHAPTER 14

FLANGES

Flanges are one of the most important tools in any dicing process. Many parameters are involved in the flange design. Following are the main parameters:

- Flange principle
- Flange design
- Flange material
- Flange accuracy
- Flange balancing
- Flange inspection and maintenance

14.1 - PRINCIPLE AND LAYOUT OF FLANGES ON THE WHEEL-MOUNT

The flange and blade are one mechanical assembly mounted on the spindle. Following are the parts involved: spindle, the wheel-mount, the back flange, dicing blade, front flange, flange nut, and spindle nut. There are differences of the flange and mounting design between the different manufacturers. The main differences are that the spindle wheel mount is part of the back flange mounted on a tapered cone compared to a cylindrical wheel mount, which adds one part to the assembly. Figs. 123 and 124 show a generic sketch view of both designs.

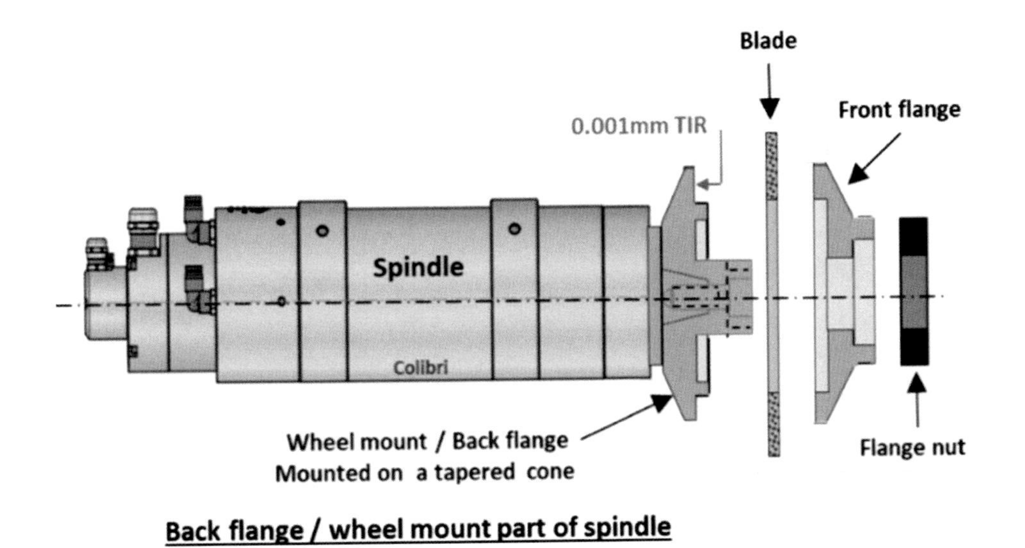

Figure 123. – Wheel mount design with a back flange part of the spindle

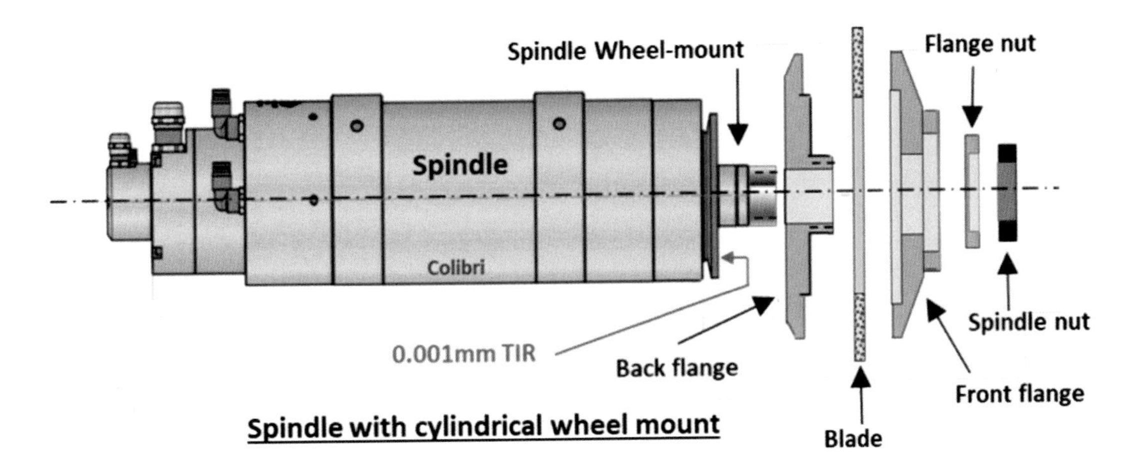

Figure 124. – Standard wheel mount design without the back flange

The back flange being one part of the spindle wheel mount is mostly on 2" and 3" spindles. The cylindrical spindle using separate back and front flanges is more on 4" spindles. There are a few advantages and disadvantages of both systems, mainly on 2" spindles. See the below table:

Flange type	Advantages	Disadvantages
Cylindrical type	• **Easy down-sizing for different exposures.** • **Easier to compensate for blade wear.** • **Easy to dismount from the wheel-mount.**	• **Higher no. of parts.** • **More difficult to maintain dynamic balancing.** • **More difficult to maintain axial & radial runout.**
Cone type	• **Better control on axial & radial runout.** • **Less parts / better control on balancing.**	• **More complicate to dismount the back flange from the cone wheel-mount.** • **More Difficult to downsize for different exposures.** • **More difficult to compensate on blade wear.** • **When dismounting the back flange to change flange O.D. a potential diamond left on the cone can cause a deep scratch that will damage the cone when mounting a new flange size.**

14.2 - FLANGE MATERIALS

The most common materials used in making flanges are stainless steel, aircraft titanium, and aircraft aluminum. All 2″ and 3″ flanges are made of SS and titanium. 4″ flanges are made in most cases of back side SS and front side of Aluminum with anodized coating. In some cases, 4″ flanges are made of SS both on the front and back side. Advantages and disadvantages of the different materials:

- **Stainless steel advantages:** – Non-Corrosive, medium hardness, relatively easy to machine, can be machined to accurate dimensions.
- **Stainless steel disadvantages:** – Heavy, more difficult to maintain dynamic balance.
- **Titanium advantages:** – lighter than Stainless steel, Maintains better dynamic balancing.
- **Titanium disadvantages:** – More difficult to machine, requires special tooling and knowhow to machine, softer than SS, tends easier to be damaged.
- **Aluminum advantages:** – (Used mainly on 4" front flanges). Easy to machine, lighter and easier to maintain dynamic balancing.
- **Aluminum Disadvantages:** – Softer than SS, tends easier to be damaged, requires anodizing to protect against corrosion.

14.3 - FLANGE DESIGN AND IMPORTANT PARAMETERS

Regardless if the flange is a standard one or if a special flange needs to be designed, there are a few important technical geometry points to take in consideration. The flange needs to be accurate to meet radial and axial run-out on specific application requirements and for special new applications. At the same time, the flange needs to be easy to handle. Following are a few important points:

Flange edge geometry: – The edge of the flange is the area where the blade is being clamped. Following are a few generic roles; however, the roles may be different between the different vendors and the different flange diameters (see fig. 125).

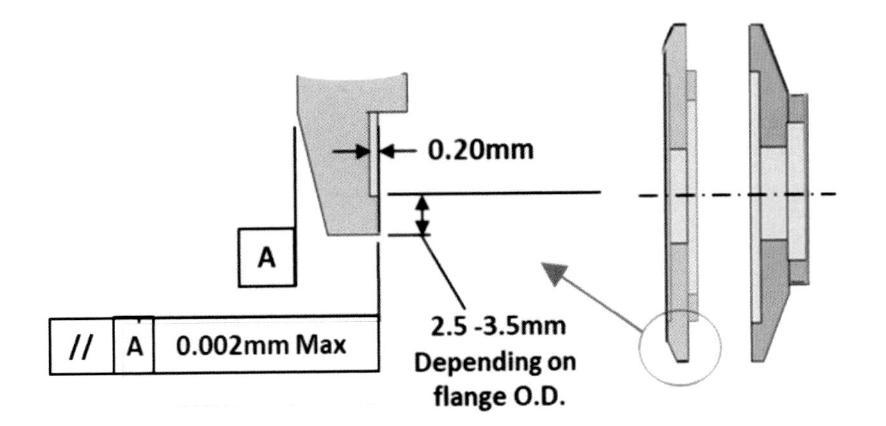

Figure 125. – Generic important dimensions at the flange blade clamping area

The parallelism of the blade mounting surface to the back side of the back flange is very important and is determining the axial runout of the blade and the kerf width.

Concentricity of the blade step to the inside diameter: –

The concentricity of all diameters to the center accurate hole are important to minimize vibration. However, the blade step is the most important diameter as it will affect the radial run-out of the blade (See fig. 126).

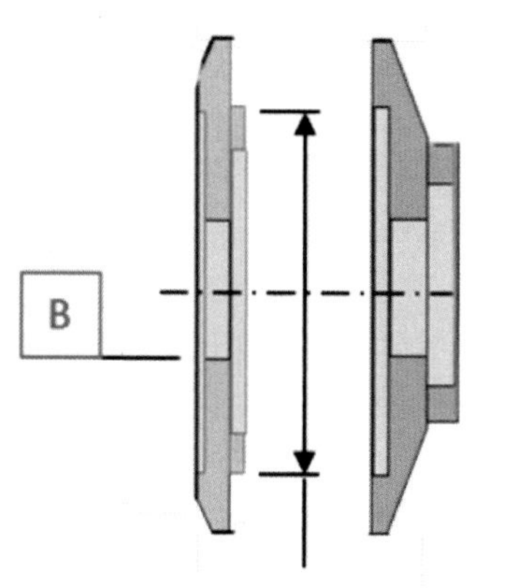

Overall concentricity to B – 0.003mm

Figure 126. – Blade step concentricity to the flange inner diameter

Perpendicularity of blade mounting surface to the inner diameter: –

Will affect the axial run-out and the kerf width (see fig. 127).

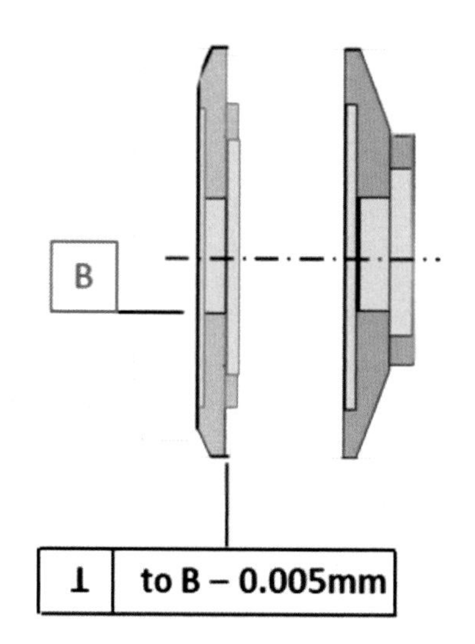

Figure 127. – The importance of blade clamping surface to flange I.D.

The above flange sketches are showing the blade step on the back flange however this is just a generic sketch, some flanges have the step on the front flange, but the mentioned important accuracies are the same. Following in fig. 128 are surfaces with better surface finish, marked in Red.

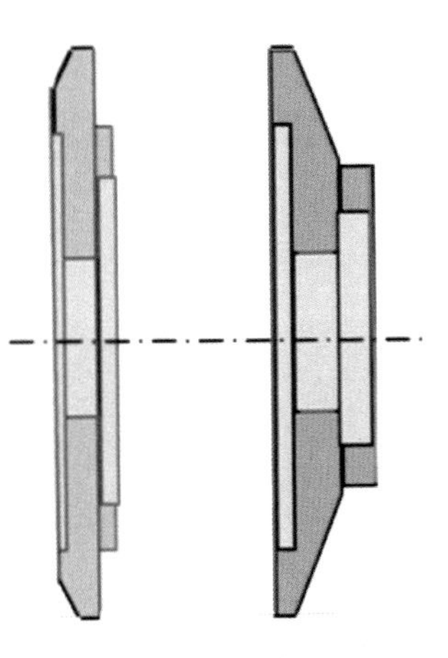

Figure 128 – Important flange surfaces with special attention to the surface finish

One other important parameter in any flange is the flange grip. The flange is holding a delicate dicing blade and a poor flange grip can cause difficulties when moving the flange to the saw resulting in blade breakage. The grip needs to be "friendly" to the fingers, not too small, not too rounded. A sharp edge would be better to eliminate slipping.

14.4 - SPECIAL FLANGES

During the years many special flanges were designed to meet special requirements in the market. The main requested special flanges are gang type flanges both in 2" and 4" geometries. This requirement is not a daily request. However, it is a challenge from time to time. Before getting in some real difficult cases, below is a table describing the advantages and disadvantages of using a gang set-up:

Parameter	Single blade	Gang set-up
Production Throughput	Normal / medium	High
Complexity	Low	High
Flexibility	Short	Low
Set-up time	Short	Long
Accuracy • Index • Cut depth • Kerf width	 High High Normal	 Lower Poor More difficult to control
Break recovery	Fast	Slow
Cost	Standard	Very high

A gang set-up requires engineering time to first understand the application spec and then to design all parts. This is adding to the cost of a first-time order and most first-time orders are for a single set-up. If more set-ups are ordered later, the design cost is eliminated. Let's take the easiest case of using two blades. Any two blades set-up requires a wider flange step, a spacer between the two blades and a special cooling nozzle to cool both blades (see fig. 127).

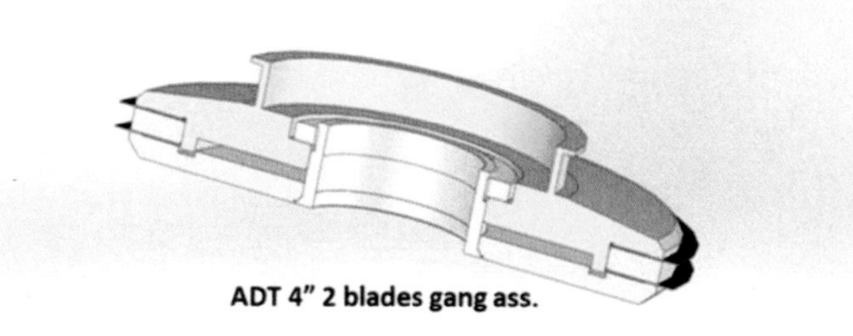

Figure 127. – Dual blades in a special flange and spacer design

It looks that a spacer with the index dimension will perform on the substrate with the exact dimension, however this is not the case. As an example, let's use two blades of .010" thick and a spacer width of .040" (see fig. 128).

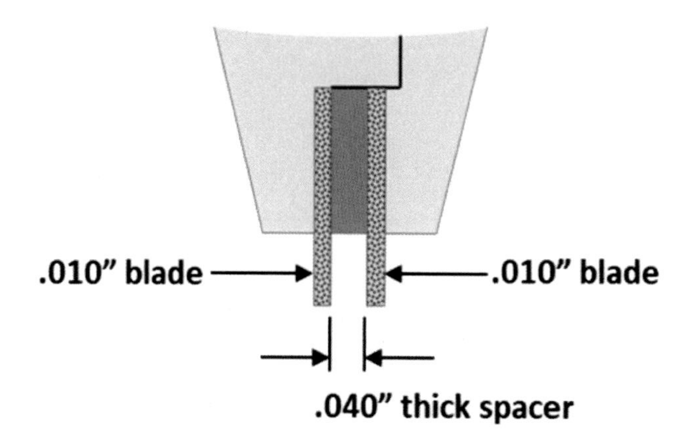

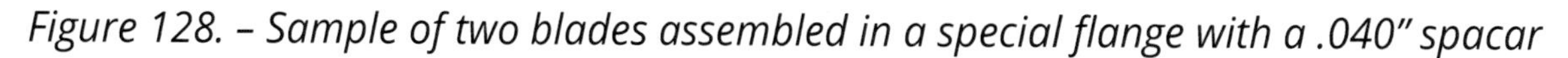

Figure 128. – Sample of two blades assembled in a special flange with a .040" spacar

Being lucky, the .040" gap between the blades may be accomplished in spec. However, in most cases this is not the case. In order to achieve the right gap/index, a torque meter needs to be used to tighten the flange and a special torque needs to be developed. Any gang flange needs to be clamped using a flange nut. The problem to be controlled is eliminating overtightening the flange that can deflect the blades which will result in the wrong gap/index and wrong kerf width (see fig. 129).

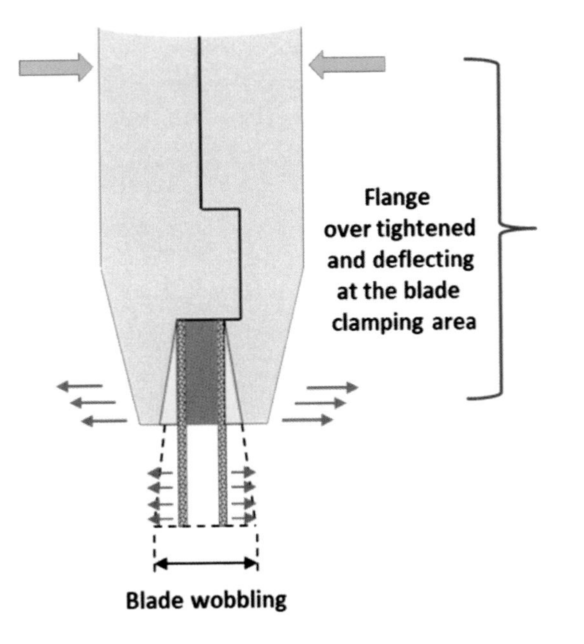

Figure 129. – Over tightening causing poor blade clamping and blade wobbling

Some of the gang flange assemblies are way more difficult with more than two blades, sometimes up to eight blades. The above roles for using two blades are the same with more blades. However, there are more spacers to handle more issues with cooling the blades and in many cases there is a need to change the cooling block covering the gang assembly and the wheel-mount. Another part needed is a set of tooling to handle the assembly of the blades and spacers especially if the blades are very thin. There is also a need to have a tool to easily mount the assembly to the spindle. Needless to say, the weight of a gang assembly is an issue and can cause vibration. Let's review some of the different requirements:

- **Flange material**: – Will affect vibration, so a titanium material is preferred mainly on 2" spindles. On 4" spindles, titanium is an option. However, a SS back flange and an aluminum front flange are mostly used. In any case, the standard flange tolerances are used regardless of the material.
- **Spacer's material and geometry:** There are a few options, stainless steel, titanium, aluminum with anodized coating, WC tungsten carbide, and hard alumina. A major concern in any decision regarding what spacer's material to use is cost and availability/delivery time. The following are the advantages and disadvantages of each option:

- **Stainless steel –** the most common spacer material, easy to machine, relatively heavy, fast delivery, reasonable cost.
- **Titanium -** More difficult to machine, longer delivery, lighter, more expensive, softer than stainless steel, tend more to get nicks and scratches.
- **Aluminum with anodized coating –** Easy to machine, fast delivery, lighter, softer than stainless steel, tend more to get nicks and scratches, less expensive.
- **WC – Tungsten Carbide –** Long delivery for raw material, more difficult to machine, heavier on weight, hard and maintains longer life, more expensive.
- **Ceramic –** Longer delivery for raw material, more difficult to machine, light on weight, longer life, more expensive. Probably the best choice but more expensive and longer delivery.
- **Geometry of spaces –** It depends on the spacer outside and inside diameters and the spacer width. In some cases, if the spacer is not too thin, side releases help to better clamp the blades to maintain the accuracy of the application. The side releases should be similar to the side release on the flanges (see figs. 125 and 130).

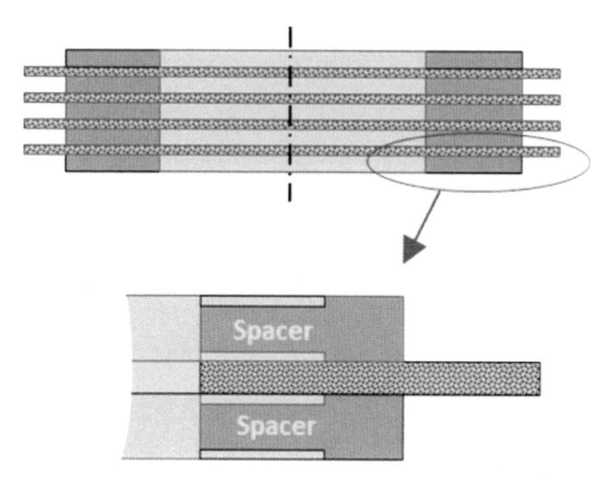

Figure 130. – Standard gang assembly cross section

Following is an example of a 2" SS gang set with 5x SS spacers followed with a few assembly tooling (see figs. 131 and 132)

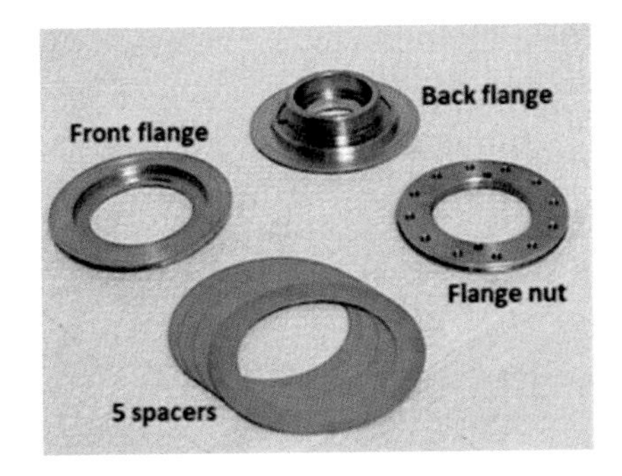

Figure 131. – 2″ 5x blades flange design

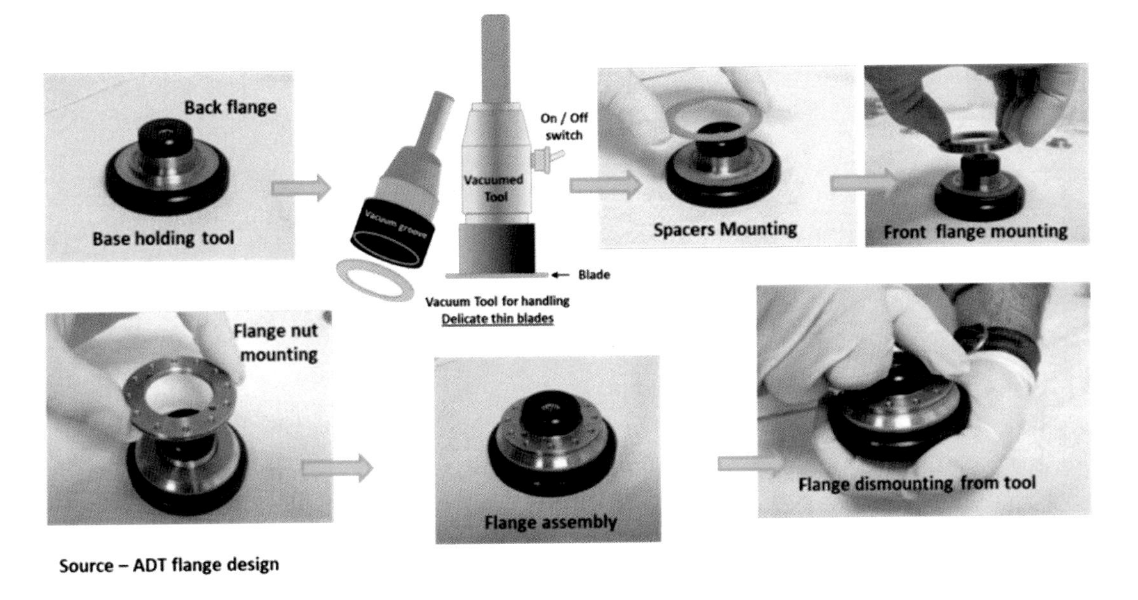

Figure 132. – 2″ flange gang mounting instructions using special designed tools

Coolant design of the multi-Blades requires changes and can be done in a few different designs. A 2-blade gang is relatively easy. It also depends on the index; if the index is small a wide oval cooling nozzle can perform well (see fig. 133).

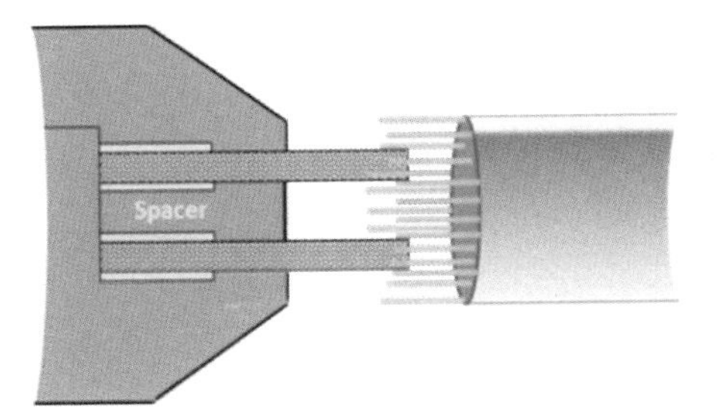

Figure 133. – A wide cooling nozzle dia. used for a 2x blade gang assembly

For a wider gang using more than 2 blades other cooling nozzle designs need to be made. Following are a few photos showing a few multi blade gang set-ups using different cooling systems. See fig. 134 of a few ADT designs in the past.

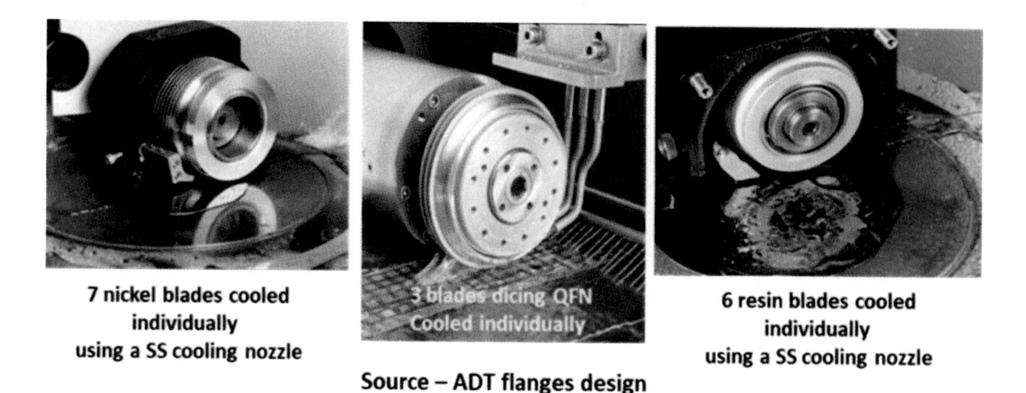

Figure 134. – Spec. cooling nozzles used for gang assemblies with more than 2x blades

14.5 - HCF -HIGH COOLING FLANGE

A unique flange option that was aimed at solving coolant issues mainly on thick and loading applications was designed by K&S/ADT. The basic principle of the high cooling flange design is to "shoot" the coolant water into the flange. Due to the centrifugal forces at higher spindle RPM, the coolant is being spread via side grooves on both the back and front flange sides and then on both sides of the blade exposure. This high radial water pressure is penetrating much better into deep and narrow cuts. Using a standard, perfectly aligned cooling nozzle on a thick substrate is way more difficult for the coolant to penetrate into the bottom of the kerf. (see fig. 135).

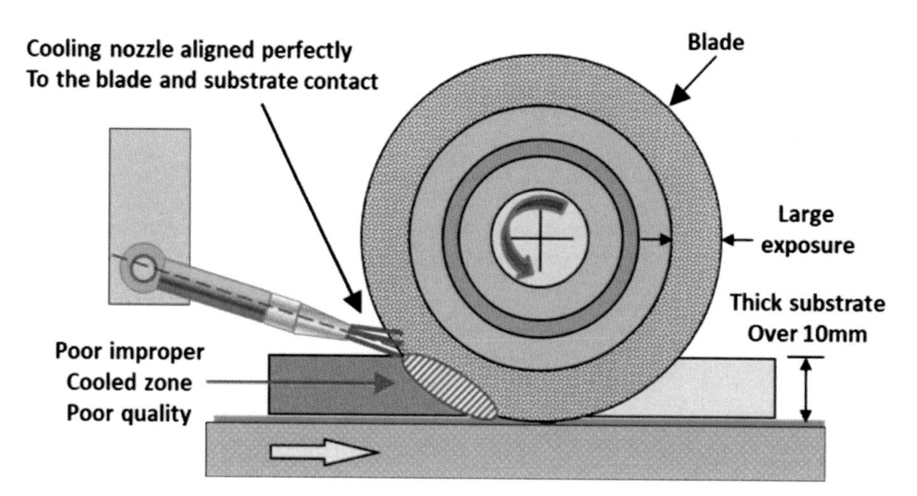

Figure 135. – A deep cut with poor coolant

Most HCF applications are on 4" saws. The main reason is the fact that most heavy and thick substrates are diced on 4" saws so the main advantage to use a high cooling flange is on 4" saws, however 2" and 3" O.D. systems did perform well with HCF flange sets.

The below sketch is a generic view of the HCF operation principle: (see fig. 136).

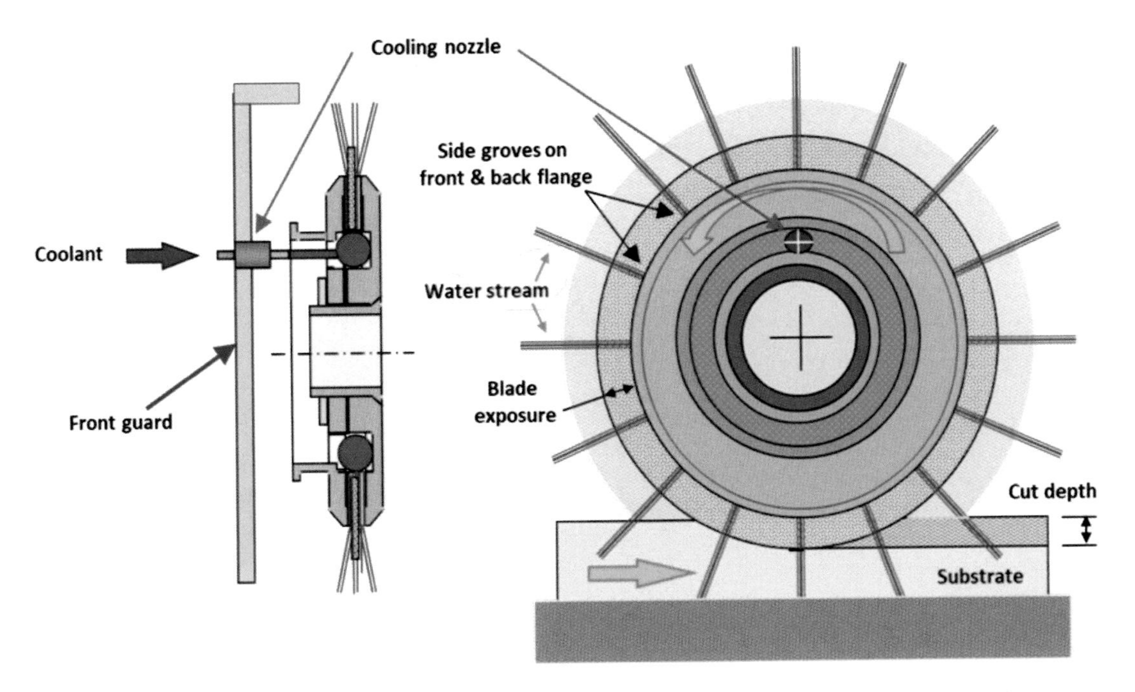

Figure 136. – Front and cross section of a typical High cooling flange design

The HCF requires some changes on the dicing saw. A special coolant line is required with a small jet aiming a stream of water into the flange. Originally the HCF was designed to be used with a special side grooved resin blade (SPG) which was also developed by K&S/ADT. The side grooves on the SPG blade need to match with the grooves on both the front and back side of the HCF. However, the orientation is not critical, and standard blades without the special groves can be used as well. The SPG resin blades will be discussed in the application section. The HCF is an accurate and expensive machined part. However, a few customers in the market have adapted it, and some are running high-volume production lines of way over 10–20 saws.

Following are a few photos showing the ADT HCF flange (see fig. 137).

Figure 137. – High cooling flange design showing the water in and out channels

14.6 - SPECIAL FLANGE SET-UP FOR EXTREME LARGE EXPOSURES

There are applications requiring the use of a relatively thin blade with a very large exposure to dice between high components mounted on different substrates. A standard blade with a very large exposure will deflect, perform with non-straight cuts, and eventually break. To solve this problem, a special flange with a wider blade step is used to accommodate the thin blade, and two WC tungsten carbide thin spacers to side support the blade and at the same time to be able to dice between the high components (see fig. 138).

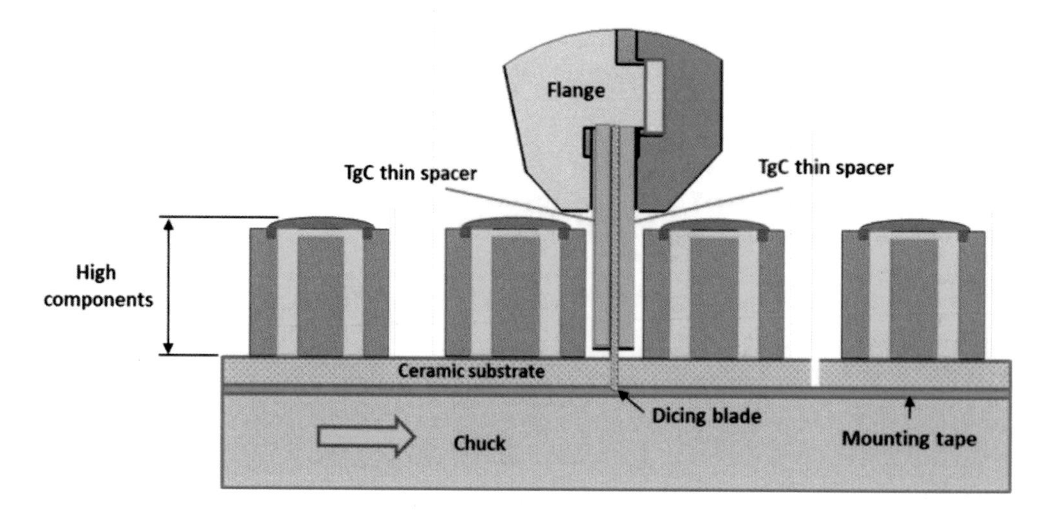

Figure 138. – Thin TgC spacers clamping the blade on a large blade exposure

A few words on the standard flanges used in the market:

14.7 - STANDARD 2" FLANGES

The principle and layout of flanges on the wheel-mount were discussed already. Following is a summary review: There are basically two types of flanges. One popular flange is a SS snap type flange that is mounted on a cylindrical spindle mount. The flange itself does not require a nut, the back, and the front with a blade in between are clamped together by an O-ring forcing the front part to be clamped to the back flange. The flange assembly is disassembled from the O-ring pressure using a three-pin fixture. The flange assembly is clamped to the spindle using the spindle nut (see fig. 139).

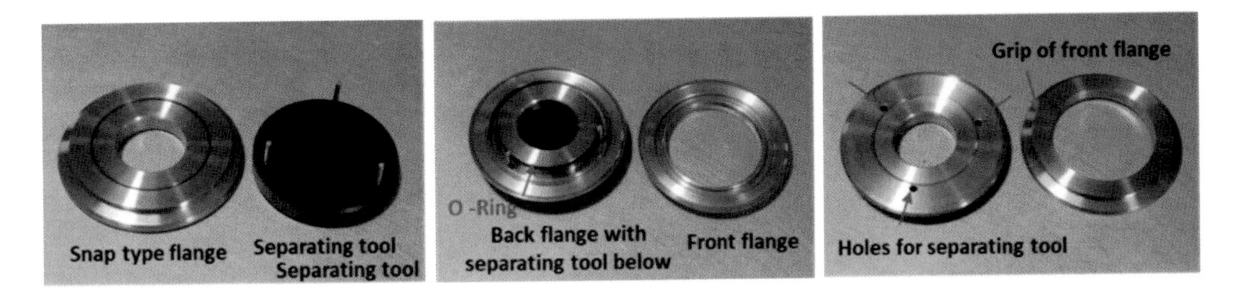

Figure 139. – 2" flange clamped by a snap push over an – O – ring

The other 2" flange is the nut type flange with two mounting options. One is mounted directly on a cylindrical spindle and one directly on the spindle cone. Following is the standard cylindrical mount (see fig. 140).

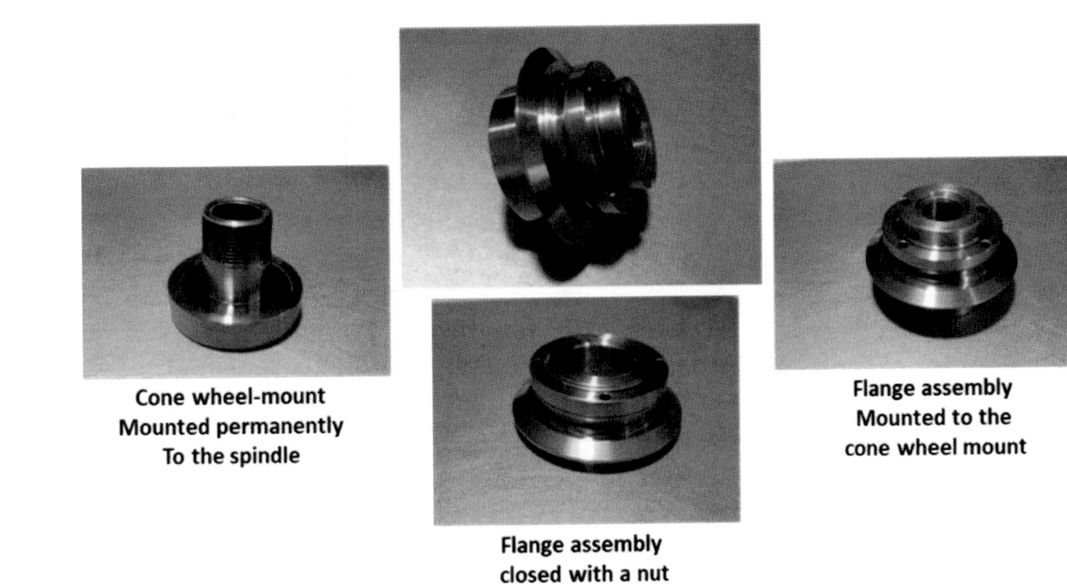

Figure 140. – 2" flange assemblies using a nut for blade clamping

Following is the flange set-up with the back flange mounted permanently to the cone spindle wheel mount (see fig. 141).

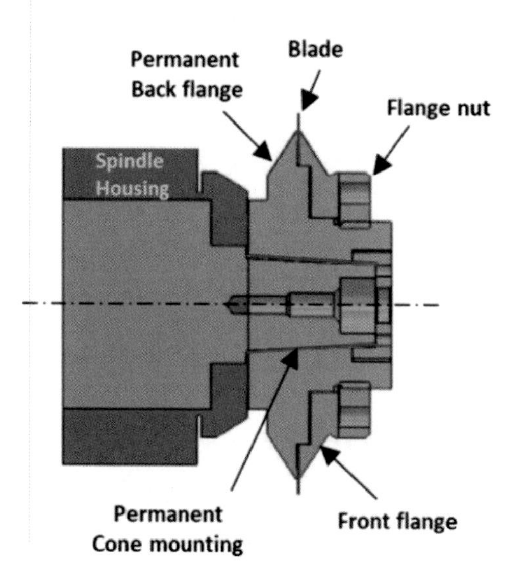

Figure 141. – 2" flange mounted directly to the spindle, back flange is mounted permanently

14.8 - STANDARD 4" FLANGES

There are two concepts however the design is similar. The traditional design concept is mounting the flange on a cylindrical spindle shaft using a SS bushing as the center part of the entire flange parts (see fig. 142).

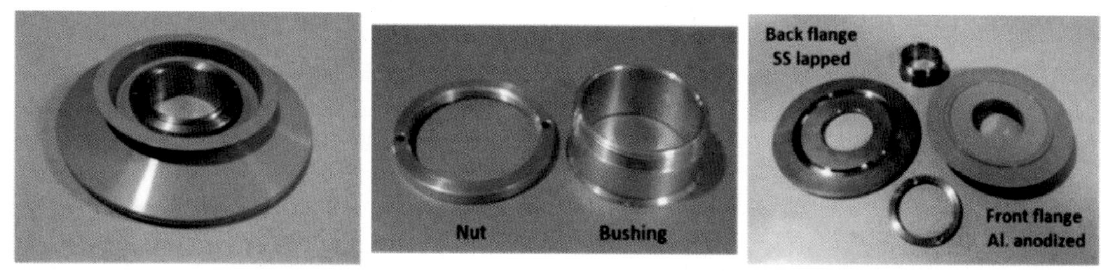

Source - ADT

Figure 142. – 4" flange using the bushing design

The main advantage of using the bushing design is the accuracy of the back flange. The final manufacturing process of the back flange is a dual sided lapping process that results in 0.001–0.002 mm flatness and parallelism. This is important to maintain the best possible axial runout. The other advantage is the ability to replace only the back SS flange, which is a cost saver in the long run. The disadvantages of this design are the more parts of handling and assembling. The other issue is the blade step on the front aluminum flange that requires more attention in the blade assembly process. The other design option is eliminating the busing by making the back side flange with the bushing in one part. The blade step is also on the back flange which for some customers is easier to handle.

This design cannot be lapped so accuracy on the back flange is more difficult to achieve. The cost of this flange is a bit higher as there is way more machining time to machine the back flange with the bushing in one part (see fig. 143).

Figure 143. – 4" flange design with the back flange and the bushing in one part

14.9 - FLANGE BALANCING

General:

Any round object that rotates at any given rpm causes some vibrations due to asymmetric mass distribution. When using higher rpm, the vibration phenomenon is accelerated. The vibration issue is critical on any dicing system, as parts are rotating at relatively high rpm. The main reason for vibrations on a dicing system is multiple parts assembled on the spindle:

- The spindle itself
- The wheel mount
- Flange assembly with blade

Although the spindle and wheel mount are balanced, the flange and blade assembly are the main reason for more vibrations. In most applications the existing vibrations of a flange & blade assembly can be tolerated without any special balancing requirements. However, for some delicate applications an extra dynamic balancing is required to achieve better cut quality. Following are some applications requiring perfect balancing in order to achieve better cut quality requirements:

- Ultrasound sensors dicing PZT for the medical industry.
- Magnetic head for the storage industry
- Some optical applications

Flanges can be supplied already dynamically balanced. This is a major help as it minimizes the vibrations on the saw. However, as the flange is mounted at a random orientation on the spindle shaft + there is a clearance tolerance on the flange I.D. and the blade I.D/ that is mounted in the flange, so all affects the balancing and is critical at high spindle speeds. Therefore, dynamic balancing on the saw is needed. Balancing the flange by the flange vendors is done on special balancing systems. The system is measuring the heavier mass and location on the flange. This extra mass needs to be machined off by either drilling a shallow hole at the heavy location or by grinding a small portion out of the heavier location. This operation is very delicate and requires special care. See fig.144 of a generic dynamic balancing set-up.

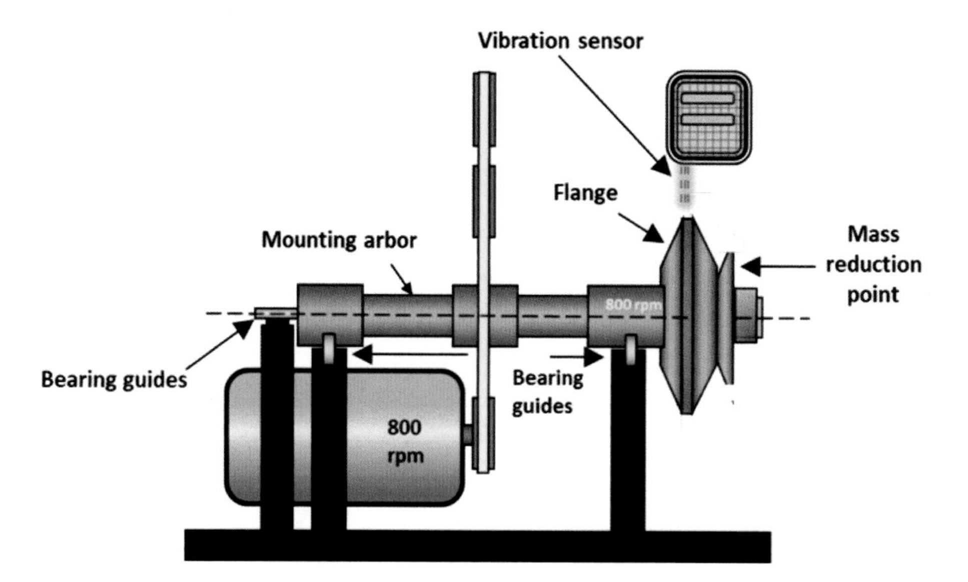

Figure 144. - Flange dynamic balancing set-up

There are balancing systems that have automatic drilling systems to drill out the exact amount of mass at the right location. See more information at the JP-Balancer site with a video of the system. The site location is at the reference section. See the below sketch Fig. 145 of a back side flange with the small, drilled holes after the dynamic balancing process.

Figure 145. – A back flange after balancing by drilling shallow holes to take off some mass

The below photo is showing the same mass machining off done in a manual mode by small grinding off the extra mass at the right location. This operation is done a few times till the right balancing is achieved (see fig. 146).

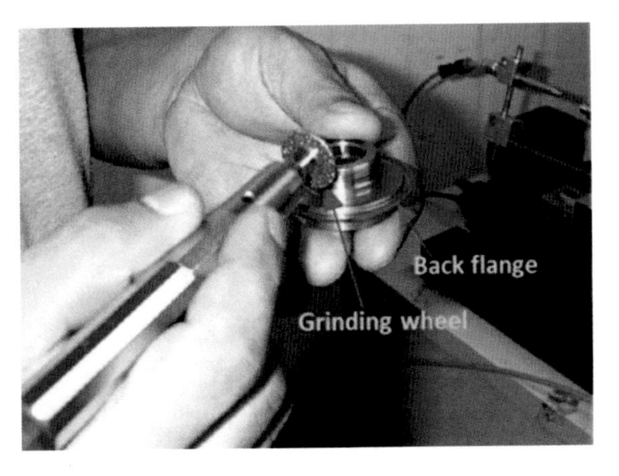

Figure 146. – Manual grinding some mass on a 2" flange

The main important issue is how to dynamically balance the flange and blade on the saw. This process is done at many customers houses where vibration is an issue and quality requirements is a must.

14.10 - PRINCIPLE OF DYNAMIC BALANCING ON DICING SAWS

Dynamically balancing a flange/blade assembly needs to be performed on the saw. A balanced flange out of the saw may help but is not a perfect solution as the flange and the blade can be mounted in different orientations and will result in different unbalanced characteristics. The dynamically balanced process is well-known and is a common practice on balancing car wheels and many other industrial products. A sensor is monitoring the vibrations and calculating the mass to be added at a specific location in order to perfect the mass distribution while the wheel is rotating. This process is repeated till the wheel is in the vibration tolerance. The correction process of dynamical balancing the flange set on the saw is the opposite of the dynamical balancing of the flange offline. In the offline process, the correction is done by machining off material mass from the flange, while on the saw, the correction is done by adding mass on the flange. One of the common systems for dynamic balancing of flange sets is the Schenck system. See the attachment link in the reference section. In order to perform a dynamic balancing process, 8–20 set screws (depending on the flange diameter) are needed on the flange front grip or on the spindle nut. For the purpose of this example, let's use a flange set with 12 setscrews (see fig. 147). Those set screws are designed to add mass at the right location. Aside from the balancing system, a box with small set screws with many different weights is needed, ready to be added per the unbalancing weight requirements.

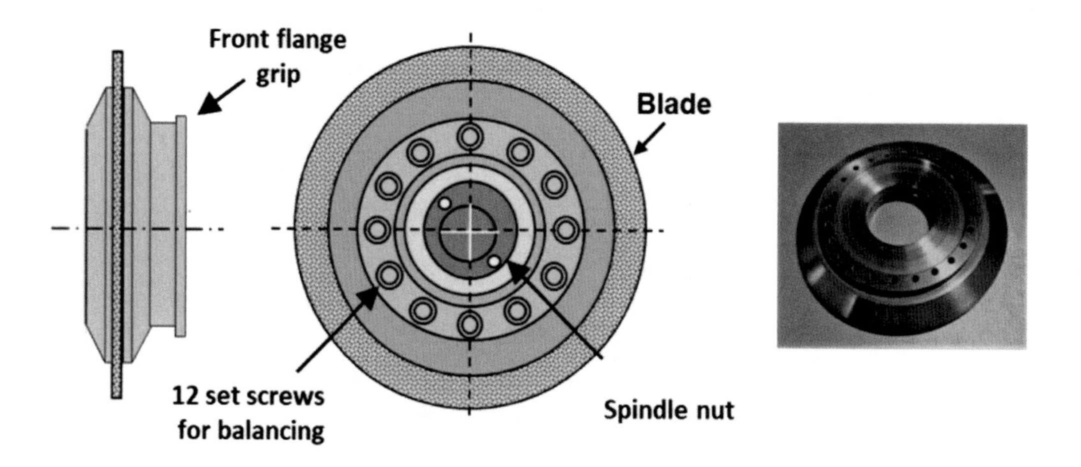

Figure 147. – Front flange with 12 set-screws for adding mass

An acceleration sensor is mounted to the spindle housing area close to the wheel mount and an RPM sensor to the chuck area. Both the acceleration and the RPM sensors are connected to a microprocessor (see fig. 148).

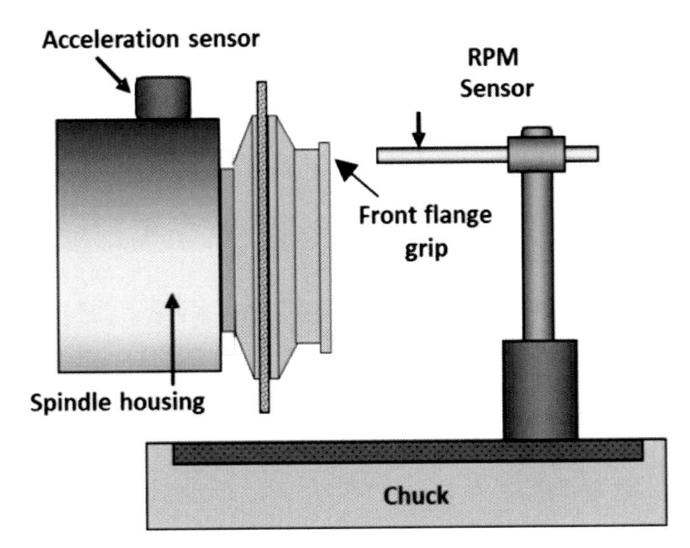

Figure 148. - Side view of balancing sensor facing the front flange

On the front flange where the set screws are located, at 12 o'clock a bright line is marked. This is needed for the RPM sensor (see fig. 149).

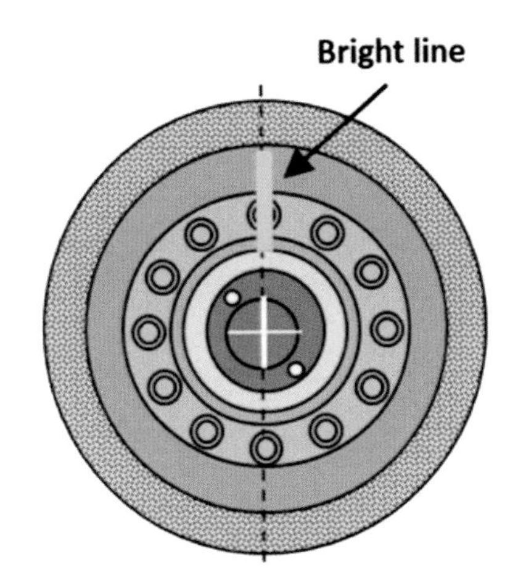

Figure 149. - Front flange 12 oclock marking line

After setting up all the sensors on the saw, the spindle is being rotated to the same production RPM and the microprocessor is monitoring the vibrations in mm/sec^2. The vibration reading is digitally displayed or printed depending on the system options. The instrument is also displaying the mass (weight) needed for correction of the vibration and the location the mass needs to be added. The location can be either set screw no. or the angle from the bright reference mark on the front flange set. For example, hole no.5 or 120°. - (360: 12 = 30 x 4 = 120°). See the below sketch fig. 150.

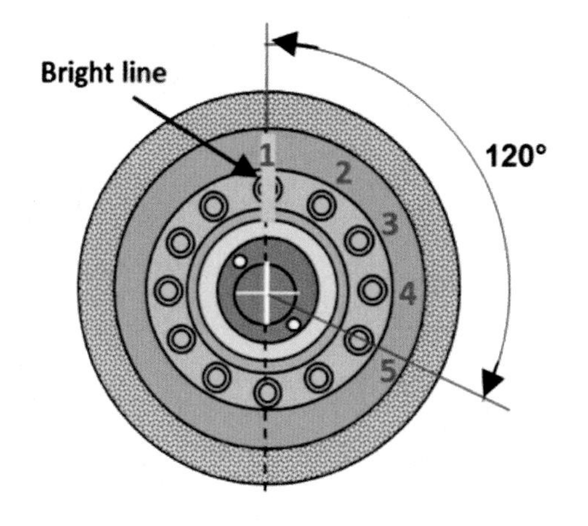

Figure 150. – An example of location to add mass

There are two options for mass distribution correction:

- A set screw with a specific weight is being weighed on an accurate scale with an accuracy of milligrams and is screwed at the right location. The set screws are taken from the set of screws, including many small nylons set screws with different lengths and each with a different mass (weight).
- Another option is to use a thin metal sheet with an adhesive on one side. A small piece of the metal sheet is cut with scissors being weighed on an accurate scale with an accuracy of milligrams. The small metal sheet piece is then glued to the front flange (see fig 151)

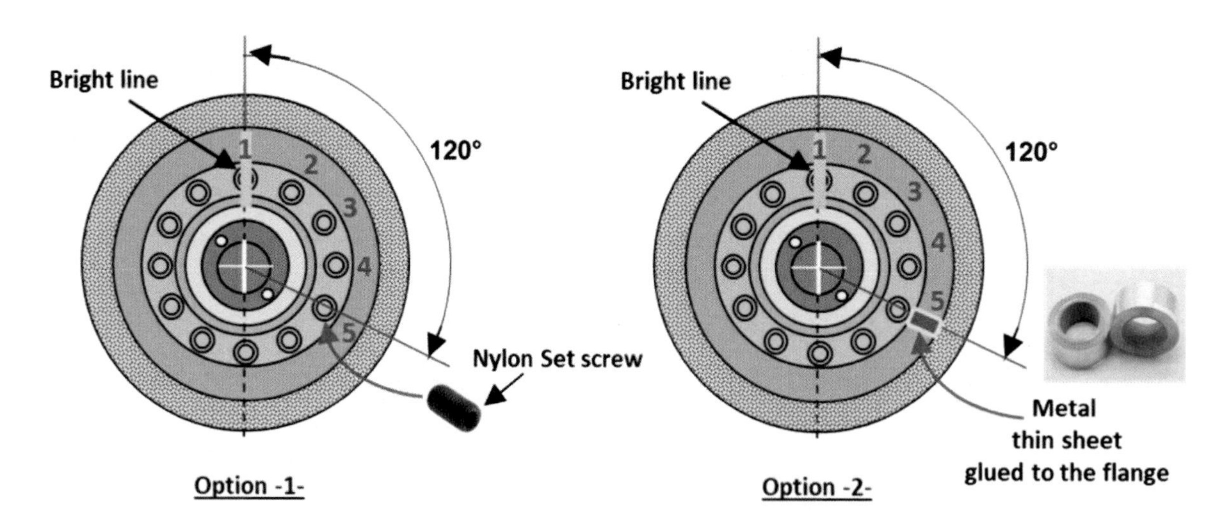

Figure 151. – Adding mass using a set-screw or a thin metal sheet

After the above process, a second spindle rotation is conducted to the production RPM and a second vibration check is monitored. This process is repeated till the flange set is meeting the vibration goal. A normal acceptable vibration figure is in the range of 0.02 mm / sec^2 but this figure may change per each application requirement.

14.11 - FLANGE INSPECTION AND MAINTENANCE

This is an issue some customers are ignoring till they face a quality issue during production. New flanges are delivered after a tight QC inspection done by the dicing vendors, however all accurate mechanical machined parts wear during time, especially in the type of environment the flanges are running. Following is a generic sketch with highlighted areas, showing the critical points and areas that are important to inspect (see fig. 152).

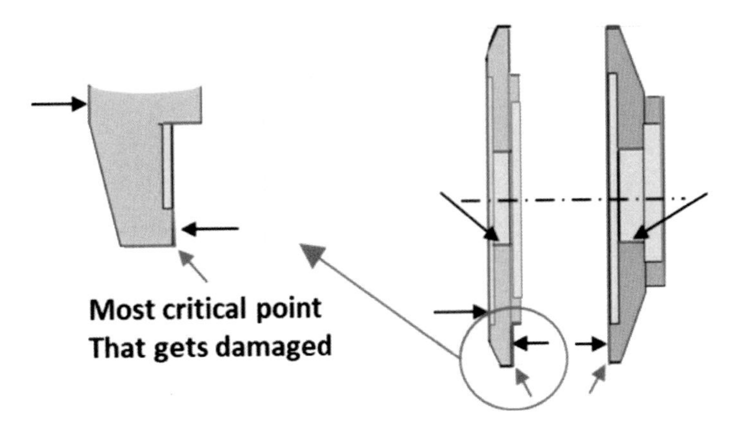

Figure 152. - Critical points and areas to inspect

A good practice is to establish a weekly or by weekly inspection depending on the production volume of the flanges. The above highlighted areas should be checked under a zoom microscope to check for nicks and deep scratches. The I.D. of the flange should also be inspected as during time the I.D. can wear and the fit to the spindle may get loose causing vibrations due to unbalancing. Following are a few examples of possible rejects observed under the microscope (see fig. 153).

Figure 153. – Examples of flange damaged areas

Small and minor scratches and nicks can be fixed by fine lapping using a fine Arkansas stone. (see fig. 154).

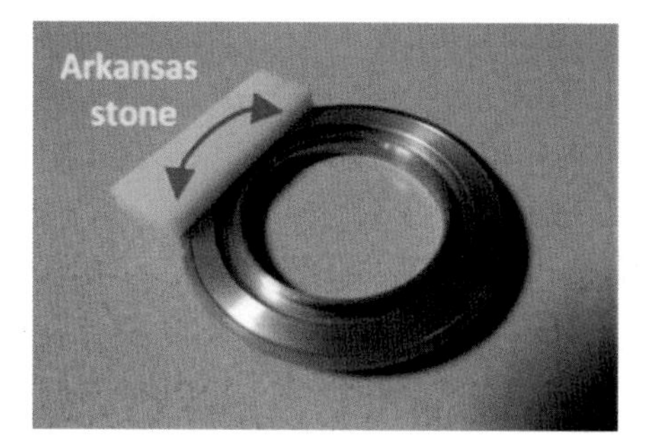

Figure 154. – A fine Arkansas stone lapping off minor scratches or nicks

CHAPTER 15

SUBSTRATE MOUNTING METHODS.

15.1 - SUBSTRATE MOUNTING METHODS

Mounting substrates on the saw chuck is a major important parameter in the dicing process. It is important to have a firm mounting without any movements of the substrate during the dicing process. Mounting is also an economic issue to consider mainly in a mass production environment. Automation today is also a major factor dictating what mounting and dismounting method to use. Following are the different mounting methods the industry is using:

- Standard Blue tape (PVC)
- UV high tackiness tape
- DDAF tape (Dicing Die Attach Film)
- Wax type mounting
- Wax and adhesives on carriers
- Other glues
- Tapeless
- Mechanical mounting
- Magnetic mounting
- Freezing mounting

15.2 - TAPE MOUNTING IN GENERAL

In the past, mounting directly on the chuck was used. However, this is not an option anymore, mainly for customers using porous-type vacuum chucks. Tape mounting is probably the No. 1 industry mounting option. Following is the history of tape mounting: Years ago, the car industry had a lot of shiny metal parts, starting with fancy bumpers and many other parts that were all nickel-plated and required protection from scratches. Even today, there are many metal parts, like kitchen sinks and others, that require this type of protection. All those parts were protected using low tackiness tapes. Those companies making the different protecting tapes with low tackiness properties were asked to develop the mounting tapes with better tackiness properties to fit the microelectronic dicing industry needs. There are many tape type manufactures making agriculture tapes for agriculture needs and other simple tape products. From experience, meeting the dicing mounting tape requirements is not an easy task and only the companies that spend a lot of R&D efforts are today's players in our dicing market. The need for automation leads the industry to find a process enabling the substrates to move from one production station to the next one without disassembling the substrates from station to station. The tape mounting was an excellent solution that is used today in most dicing houses.

15.3 - STANDARD LOW TACKINESS "BLUE TAPE" (PVC)

This is the most popular mounting method used today in the market. The Dicing Tape is a flexible PVC in most cases with synthetic acrylic adhesive bonded to one side with a clear PET (Polyethylene) to protect the adhesive side. It is tough, has high tear strength and elongation. Some customers are performing the tape mounting manually on a flat table however the majority are using a professional tape mounter that performs much better and eliminates air bubbles between the tape and the substrate (see fig. 152).

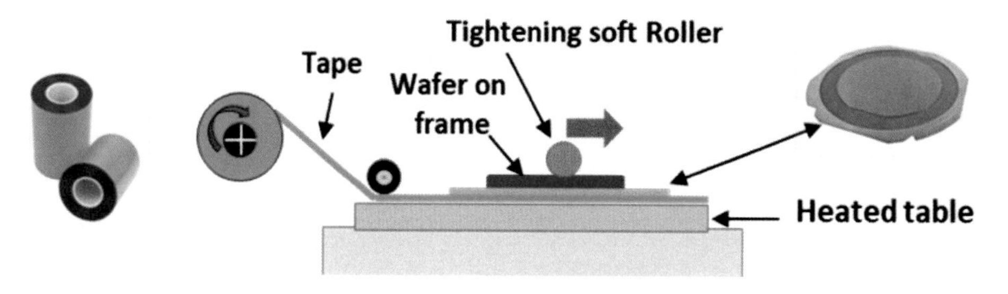

Figure 152. - Tape mounter side view

There are a few types of different tackiness and tape thicknesses. The adhesion characteristics (Tackiness) of the most common tapes are 215–315 gr/25 mm. One of the popular tapes is the low tackiness x 0.075 mm (.003") thick used for wafers with relatively larger die size where the tackiness is enough to hold the die and easily enough for the die bonders or pick-and-place equipment to pick the diced dies. The next group of the "Blue tape" is a medium tackiness designed for dicing silicon wafers with smaller die size. The adhesive layer is slightly greater than the Low tack to have a firmer tackiness but still easy enough to remove the die on the next die bonder or pick & place process. Using the tape requires a special frame that the tape is mounted on. The main tape frames used are SS frames.

However, plastic frames with the same geometry are also used. The frames have special geometry groves/slots to fit tight on special chuck mounting pins (see fig. 153).

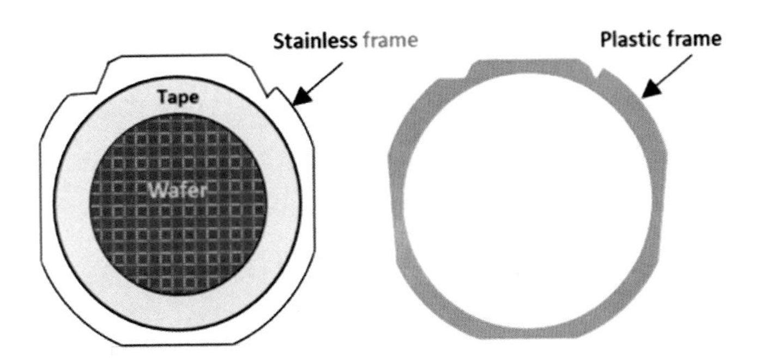

Figure 153. – Stainless steel and plastic mounting frames

In the olden days and still today at some customers the ring type tape mounting is used mainly on older saws and on non-automation dicing processes (see fig. 154).

Figure 154. – Plastic ring type tape mounting

This method principle is based on stretching the tape by pushing an outer ring on top of an inner ring while the tape is in between the rings. This process is done on designating jigs (see fig. 155).

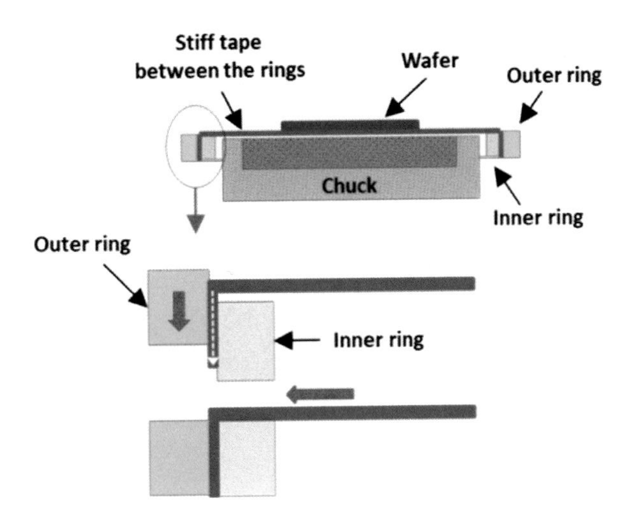

Figure 155. – Ring type tape mounting process

The majority of the used tape is the blue PVC tape made by a few vendors. The most popular thickness is .003" (0.075 mm) which most silicon dicing houses are using and is economical to use. There are other tape thicknesses for other applications. The tape is heated during the mounting process on the wafer mounter however, extra heating/curing to about 70ºC for 10–20 minutes is done by a lot of customers to perfect the adhesion. The timing and time need to be optimized per application. This extra curing can be done manually on hot plates for small operations or in designated ovens on high volume mass production products. Much more will be discussed on the application side of using substrates mainly silicon tape mounted.

15.4 - UV TAPE

UV tape is widely used today especially in applications with a small die size of 2mm x 2mm & smaller. It is also used in applications performing with relatively higher loading during dicing like Ceramics, Sapphire, BGA, QFN and others where thicker blades are used. UV tape is a high tackiness tape with different tackiness options and tape thicknesses to fit different applications. It has a high tackiness when mounting the wafer and a low stickiness after exposing to UV irradiation light. See fig. 156 describing the UV tape mechanism.

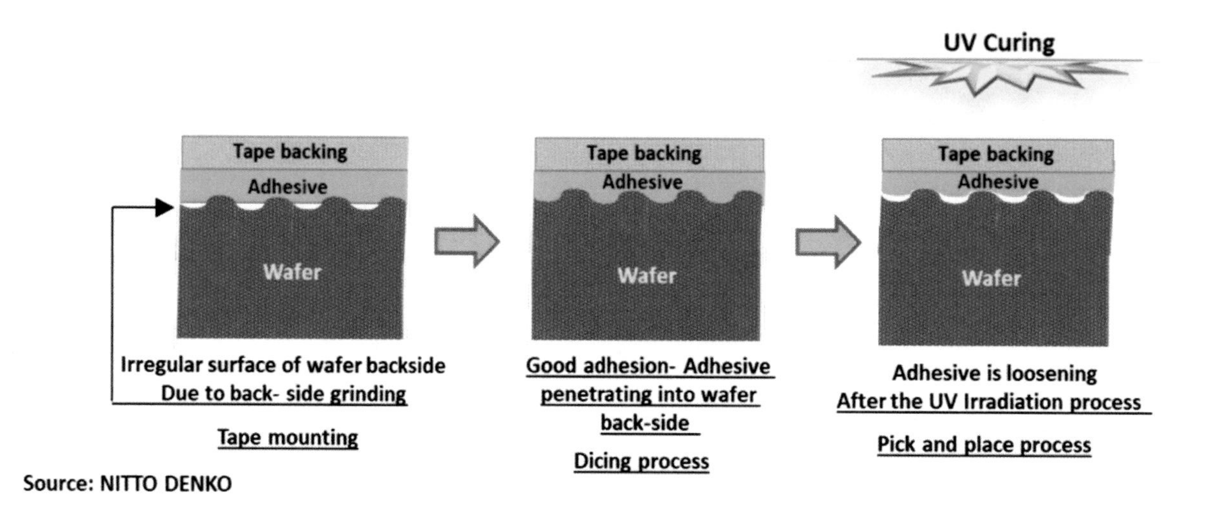

Figure 156. – UV mounting and dismounting process mechanism

Handling and mounting UV tape is similar to the non-UV tape. Mainly using SS frames on wafer mounting systems. However, dismounting the diced wafer requires a wafer curing system. There are many brands in the market making those systems. It is a simple fast process of exposing the diced wafer to UV (Ultraviolet) light for a short time that needs to be optimized per application. The UV wavelength required for curing/releasing a UV diced wafer is a continuous 250–360 nm (Nanometers). Most curing systems in the market go up to 360 nm wavelength. Below is a generic UV tape sketch with the UV curing characteristics (see fig. 157).

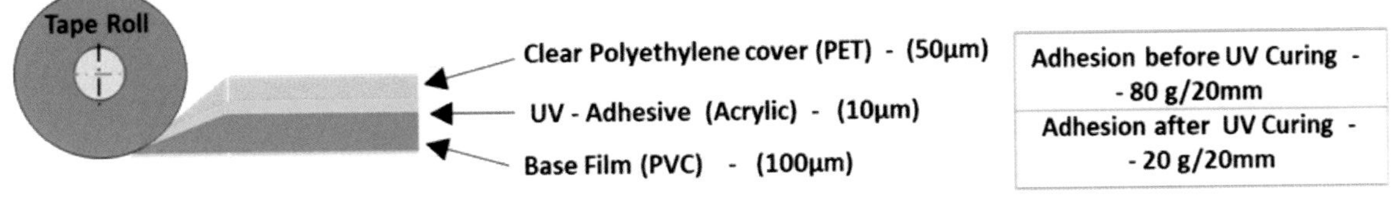

Figure 157. – UV tape characteristics

15.5 DDF – A COMBINATION OF DICING TAPE & DIE ATTACH FILMS

The Die attach film – DAF is mounted on top of the Dicing UV tape and both are mounted at the same time on the wafer (see fig. 158).

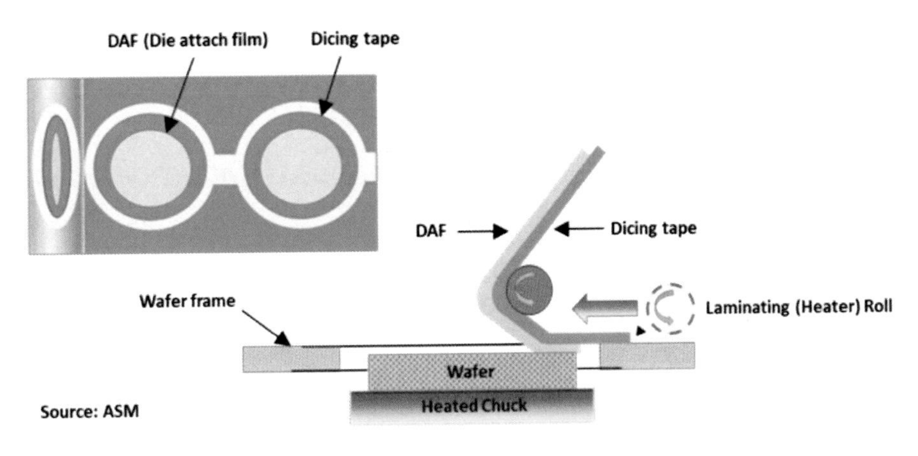

Figure 158. – DDF cross section process

This product method was designed to help the die bonding process after the dicing process and the tape expansion process of separating the dies (see figs. 159 and 160).

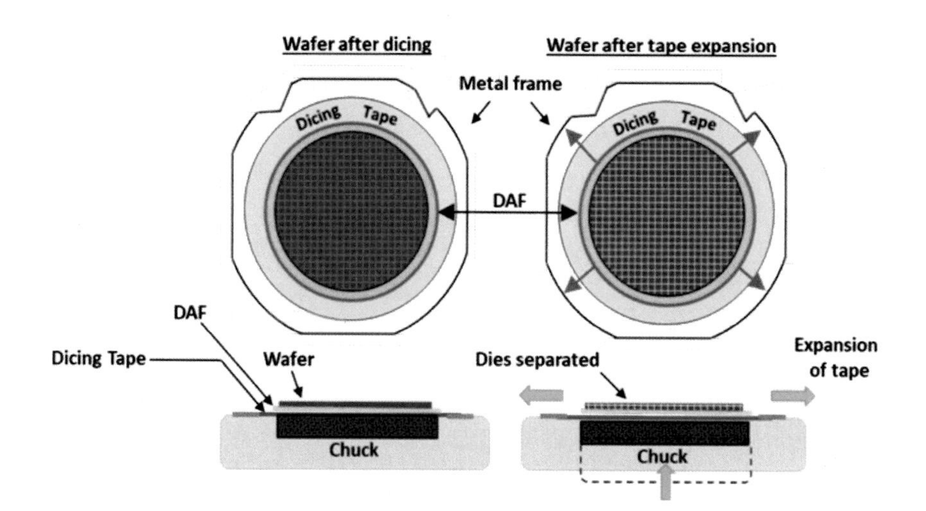

Figure 159. – Dies separated after dicing by expansion

Appearance of Dicing/Die-Bonding Double Functional Tape

Source: Showa Denko Materials

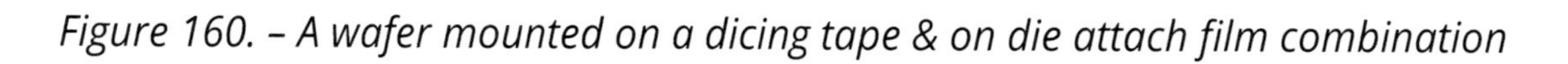

Figure 160. – A wafer mounted on a dicing tape & on die attach film combination

The expansion process to separate the diced dies is critical to maintain good yields during the die bonding process. A non-expansion tape after dicing will result in die contact during the die bonding process (see figs. 161 and 162).

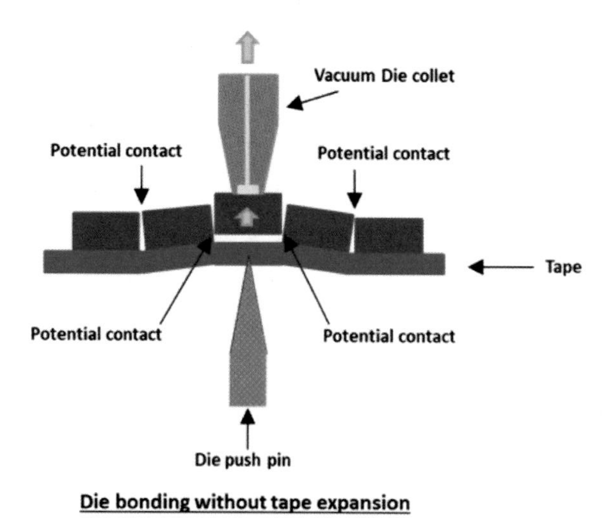

Figure 161

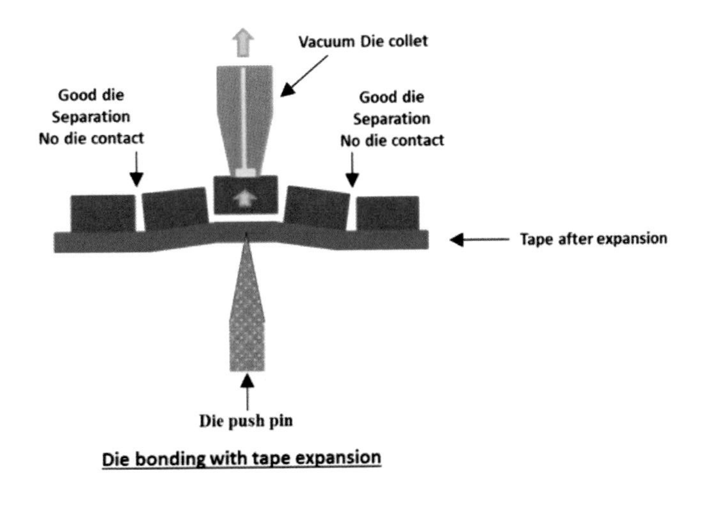

Figure 162

Figs.161 & 162. – Die bonding pick up process on non-expansion and on expansion tape

Using the Die attach tape unable customers to bond multiple die designs on the same lead-frame substrate. This is done after wire bonding the first die to the lead-frame and then mounting the second top die. This process can be done multiple times (see fig. 163).

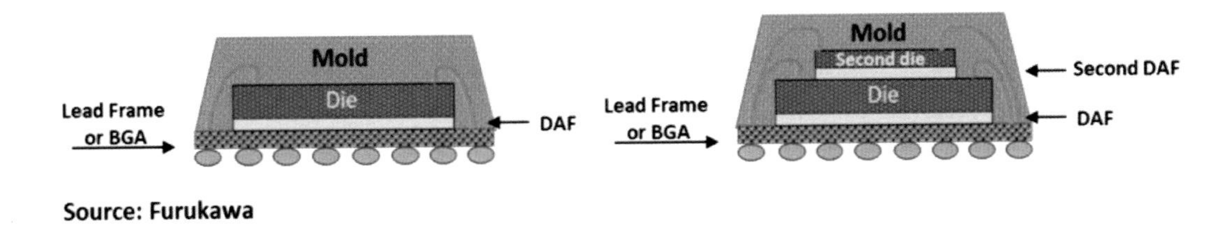

Figure 163. – Multiple bonds on the same substrate using the die attach film

15.6 - WAX MOUNTING

There are many types of wax products on the market. Some are organic, and some are synthetic. For this review, let's just call them wax. Wax is an excellent mounting method for relatively higher dicing loads. This method is aimed at thicker substrates where tape mounting is not an option as the tackiness of the tape will not hold the diced dies/device in place. Wax is also used on small parts that cannot be mounted on tape or mechanical mounting is not possible. The main advantage is the ability to cover the substrates on the sides and on the top of the substrate, which helps to minimize side movements of the dies. Mounting substrates using wax requires a base substrate that can be mounted on a chuck by using standard tape. The idea is to dice through the substrate into the base material. The base mounting substrate can be of many different materials. The idea is to use a material that will not load-up the blade. A very popular base substrate is glass. However, many other base substrates are used in the market like sintered lava, unfired ceramics, ferrites, and others. Fig. 164 shows a dicing cross section set-up.

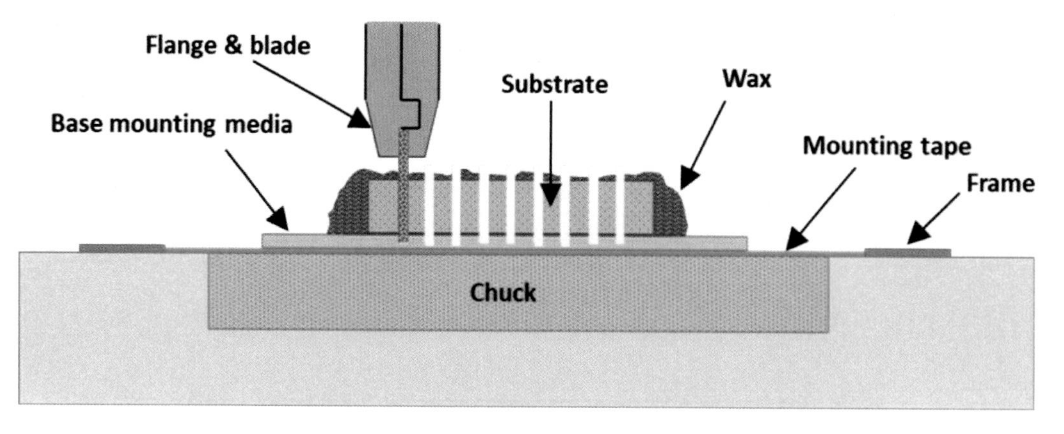

Figure 164. – Dicing a substrate mounted using wax

One important parameter to take into consideration is the dismounting of the diced dies. The dismounting is an additional process but is a must. The improved quality "pays" for the extra handling time. Wax materials are available in molded bricks, flakes, lumps, powders, chips, and others. They vary in melting point (The average melting point is in the range of 60ºC –70ºC), flash point, specific gravity, hardness, brittleness, Surface characteristics (like dry, sticky, oily), flexible and elastic characteristics.

Main advantages of using wax: – The ability to cut deep into the base material which eliminates the lip affect created by the blade edge. This improves BSC and will be discussed in the dressing section (see fig. 165)

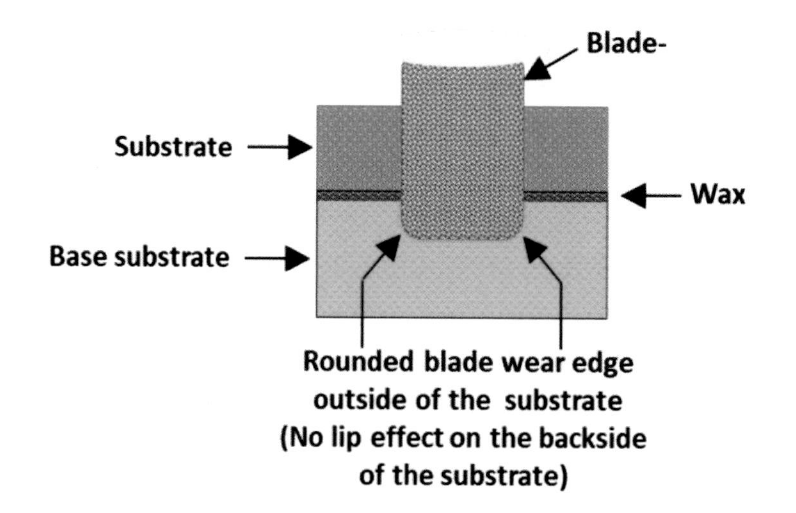

Figure 165. – Improved cut quality on wax mounted substrates

The ability to mount non flat substrates which in many cases is the only mounting option (see fig.166).

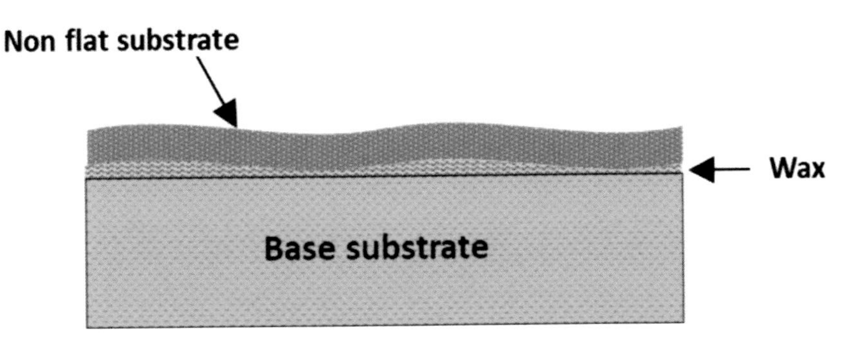

Figure 166. – Wax mounted of non-flat substrates

Wax mounting is also used on very thin and brittle substrates requiring high quality performance that cannot be achieved by using tape mounting due to die movement during the dicing. A few good vendors for wax and other mounting materials are available in the market. They do have a lot of good information regarding the different products, how to use the products and how to separate the diced dies and clean them. Below is a general product photo. More specific information can be found in the reference (see fig. 167).

Source: NIKKA DENKO

Figure 167. – Different wax type mountinn products

There are more good sources for wax type products used in the microelectronic dicing market. See in the reference section. Another good product used already for many years is the Crystalbond. It is for holding small and difficult parts. The product has a low melting point of 70ºC–75ºC and can be dissolved and cleaned with a supplied solvent or with Acetone. The product has a good coverage property to protect the substrate from movement during the dicing. See a sample photo of a diced substrate using a small index with a well-covered Crystalbond to eliminate any row movement (see fig. 168).

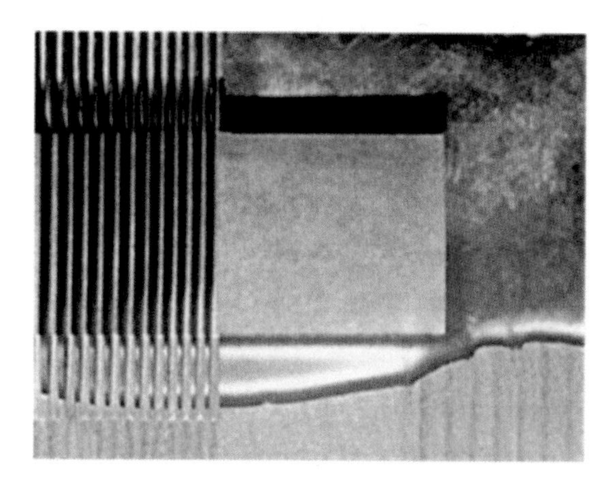

Source: AREMCO

Figure 168. – Dicing a thick substrate mounted using Crystalbond

The below photos are showing a variety of other similar products with different characteristics for different applications (see fig. 169).

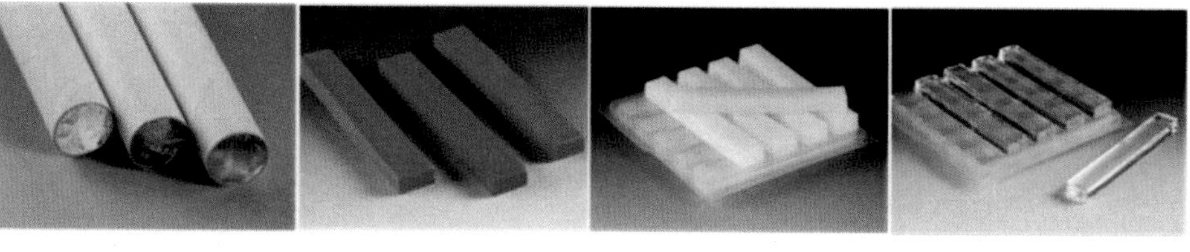

Source: AREMCO

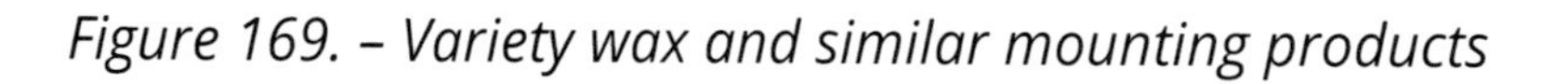

Figure 169. – Variety wax and similar mounting products

Some customers are chopping the Crystalbond into flakes in order to ease the spreading process on a hot plate with a base substrate before mounting it on the substrate to be diced. In the reference is a site info with a detailed walk-through of the mounting and dismounting process of the different products.

15.7 – WAFER MOUNT / ADHESIVES ON CARRIERS

Adhesive mounting carriers is a method of using Thermoplastic film adhesive and (EVA – Polyvinyl Acetate) polymers with good adhesive properties. The wafer mount is coming in round forms with different diameter sizes or in squares up to 10 x 10 inch that can be cut to small sizes to fit smaller substrates. The carriers are heat activated on hot plates or in ovens with a weight mounted on the substrate to eliminate air bobble and to perform an accurate even mounting. This type of mounting is for relatively low production load however it is less "messy" process compared to the standard Wax or Crystalbond mounting (see fig. 170).

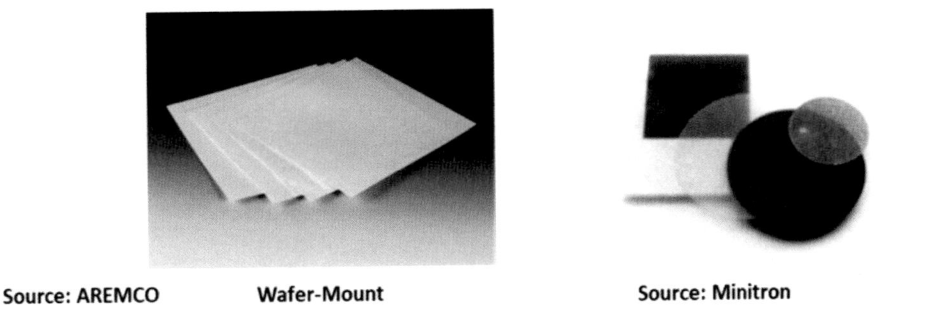

Figure 170. – Wafer mounting bases

Detailed technical mounting and dismounting instructions are well described in the vendor's sites. See in the reference section.

15.8 - TAPELESS MOUNTING

Tapeless is a major mounting process used today in mass production products like BGA, QFN and similar. It is used in fully automated integrated dicing systems with pick and place quality inspection and unloading of the diced substrates. It is a complex and very accurate process including many steps in one singulation system. Each individual device is clamped in a rubber vacuum cavity nest. See the bellow sketch of the tapeless concept, fig. 171.

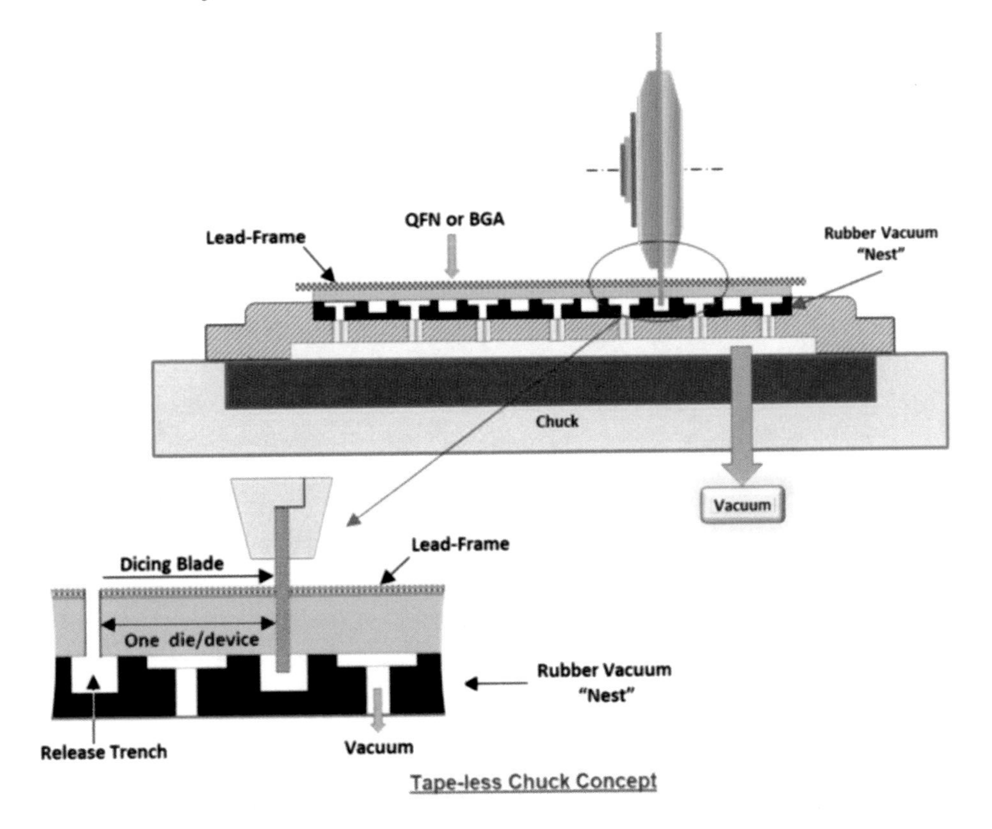

Figure 171. – Cross section of a tapless dicing process

For every device geometry a special chuck with a different rubber nest is needed (see fig. 172).

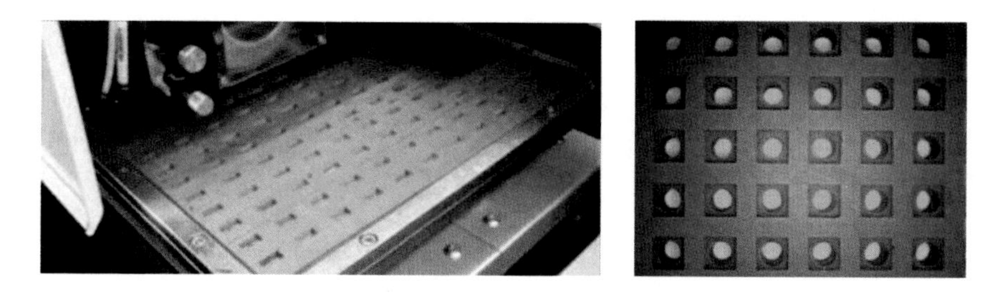

Figure 172. – Rubber nest mounting designs

The need to change the tapeless chuck assembly for every device geometry is time-consuming. However, the loading, vision alignment, dicing, cleaning, inspection, and pick-and-place unloading to a waffle tray is fully automated and saves a lot of time. A few manufacturers of singulation systems are integrating a dicing system made by a leading dicing manufacturer and some are using their own dicing system. Below is a market leader of the singulation systems made by HANMI. The system has a DISCO dual dicing spindle saw integrated in their system. It is a fully automated system with two separate dicing chuck tables, a high-performance vision system for aligning and dismounting the diced devices, using two dicing spindles, cleaning, pick-and-place after dicing, inspection, and down-loading to a waffle tray (see fig. 173).

Source: HANMI

Figure 173 – The HANMI fully automated singulation system

More information about the HANMI system is in the reference section.

Another unique singulation system made by Besi – Semiconductor is a fully automated system with all the elements to handle BGA, QFN & BCB substrates from mounting to down-loading the diced devices. It includes vacuum mounting, aligning, sawing, cleaning & drying, vision inspection, pick and place for sorting and offloading to a waffle jig (see fig. 174).

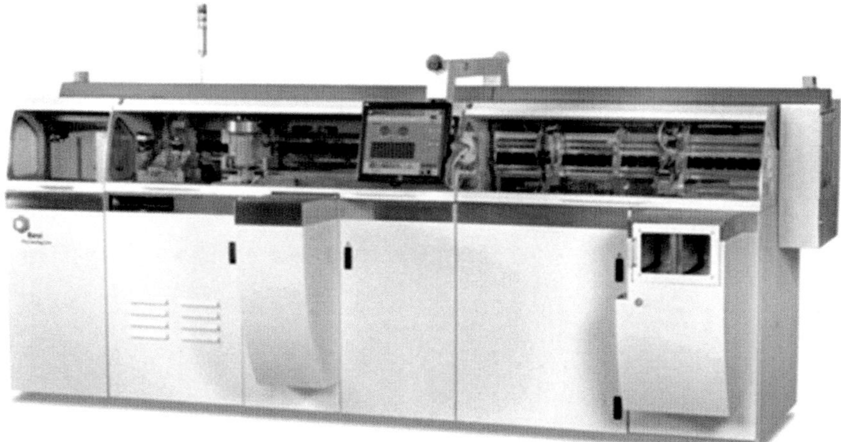

Source: BESI Semiconductor

Figure 174. – The BESI Semiconductor fully automated singulation system

The Besi dicing system is a unique system, very different from all other Singulation systems. It is an inhouse made dicing system with 3" (70–80 mm) blade O.D. for longer blade life. The vacuumed substrate is mounted above the spindle and the blade is dicing the substrate below (see fig. 175).

Source: BESI Semiconductor

Figure 175. – Dicing the vacuumed clamped substrate above the spindle

More information leading to the Besi system is in the reference section.

15.9 - OTHER LIQUID TYPE GLUE MOUNTING

Some applications of very delicate and brittle substrates, mainly on optic-type substrates but not only, require different mounting of high-flow glue. It is not a popular method seen in the dicing industry, but for some delicate unique applications, it is a good option. High flow UV glue is also used for this type of application. In some cases of delicate substrates with thin die openings and thin walls, the high flow glue needs to penetrate those open gaps in order to act as a support and to eliminate breakage during the dicing process. Mounting the substrates using glue requires using a base substrate to hold the actual substrate. Needless to say, any of those fine high-flow glues need to have a good process to release the diced parts and to perfectly clean them after dicing. Fig. 176 shows a generic process.

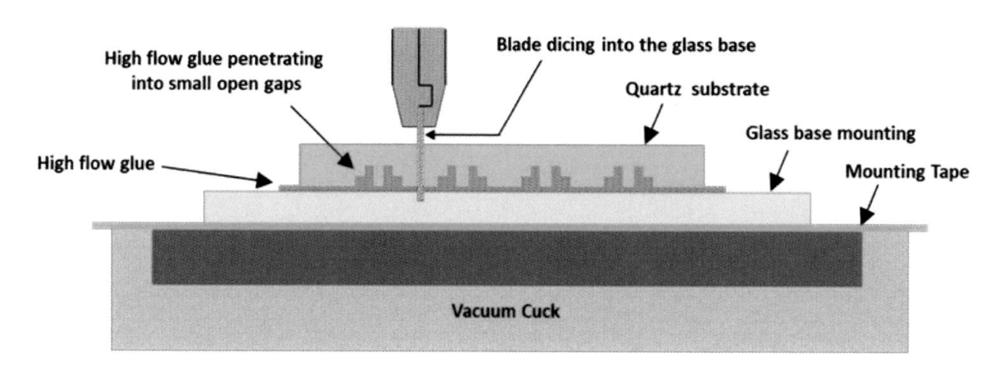

Figure 176. – A delicate Quartz geometry substrate glued on glass using high flow UV glue.

15.10 - MECHANICAL MOUNTING

Mechanical mounting is used on special applications where other mounting methods are not possible or on special applications customers are preferring not to use tape mounting or wax / glue type mounting that require messy handling. Below are a few practical examples from the marketplace. The first example is dicing straight carbide bars with extreme accuracy tolerances on straightness, parallelism, and perpendicularity. Needless to say, the mounting jig needs to maintain an accuracy of 0.001mm on all dimensions. (see Fig. 177).

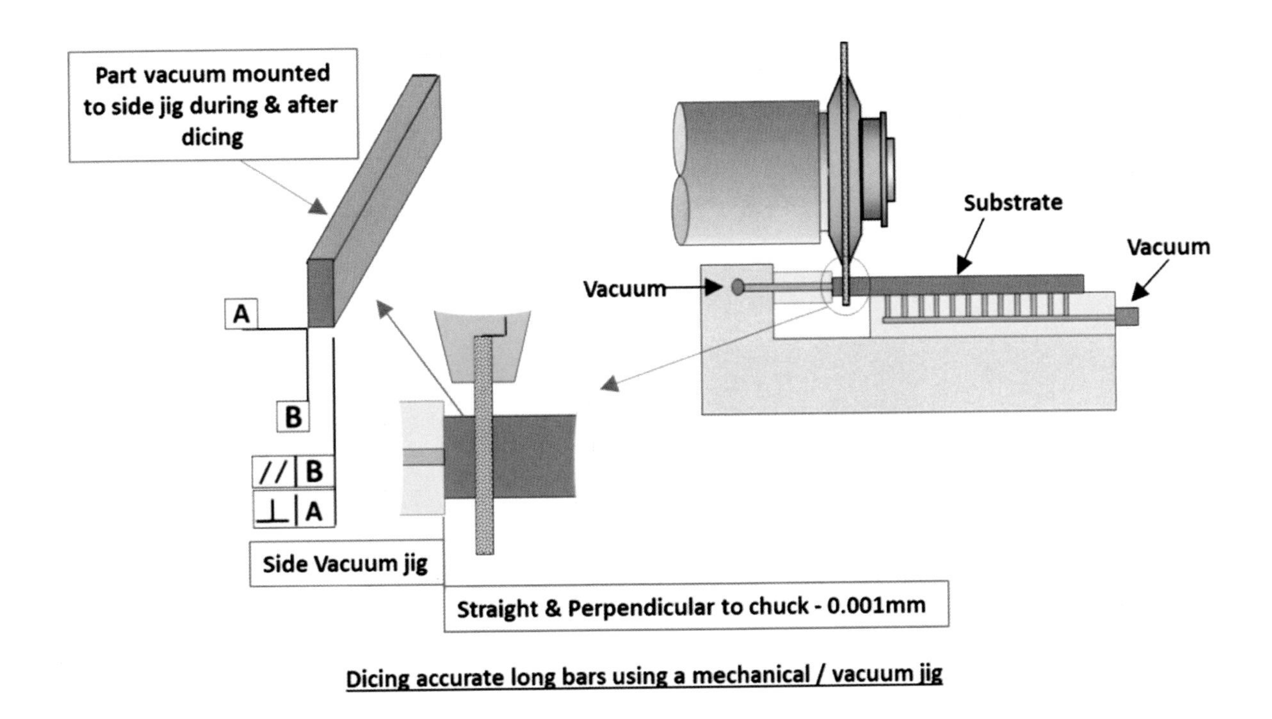

Figure 177. – Dicing thin Carbide bars mounted on a unique accurate side vacuum jig

Another example is a request to trim a square hard Alumina substrate on four sides without using any mounting media like tape or any wax/glue media. This is in order to save the mounting materials and the mounting time required prior to the dicing. The idea is to mount the ceramic alumina by using only a press button to vacuum activate the mounting jig (see figs. 178 and 179).

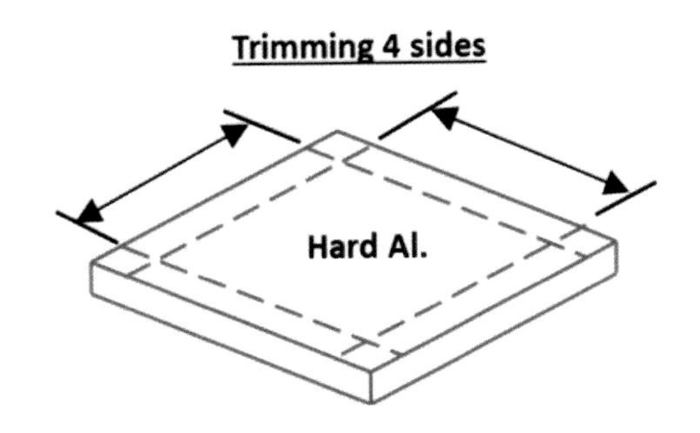

Figure 178. – Ceramic substrate to be trimmed on all sides

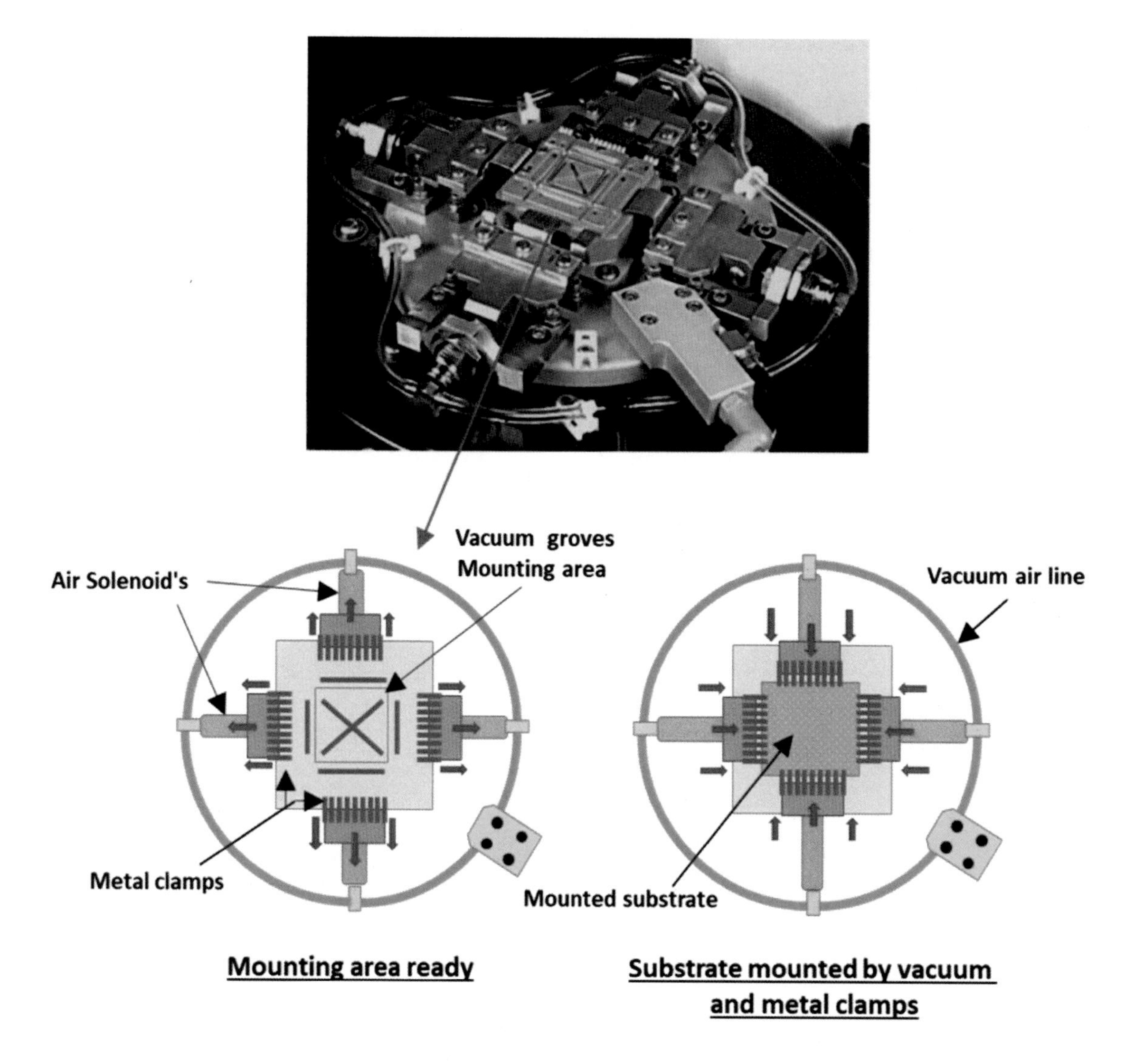

Figure 179 – A mechanical mounting jig for clamping a single ceramic substrate

A unique application was developed of dicing fiber optic products to a very accurate length and a perfect perpendicularity of the edge surface to the fiber length (see fig. 180).

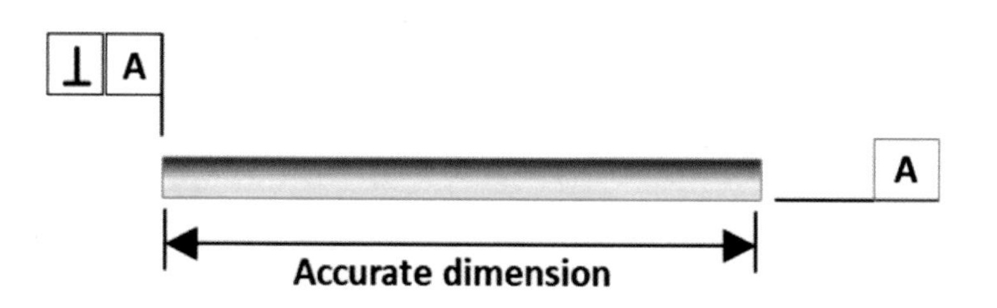

Figure 180. – Fiber optic to be diced with an accurate edge geometry

A glass plate is grooved using a – V – blade to fit the fibers to be exposed about .010" (0.250mm) above the glass and parallel to the side glass plate. The fibers and the top plate are waxed down to the –V – groves. (see Fig. 181).

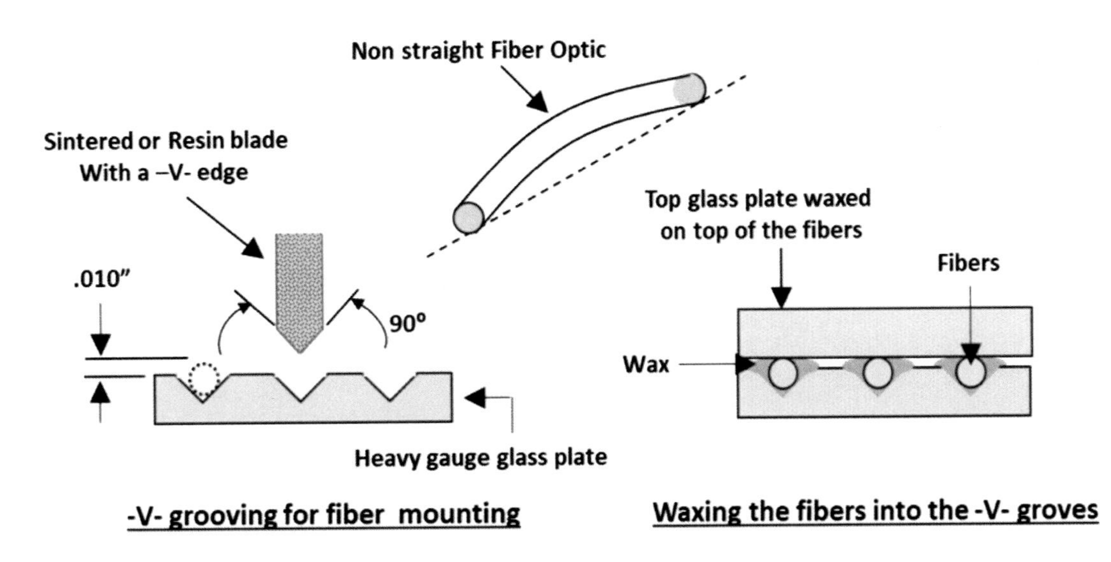

Figure 181. – A special glass mounting jig with – V – groves to wax down the fiber optic rods

The glass "sandwich" jig is aligned on the saw chuck on the –Y – axis and two cuts through the top glass plate and the fibers are performed per the below sketch. (see fig. 182).

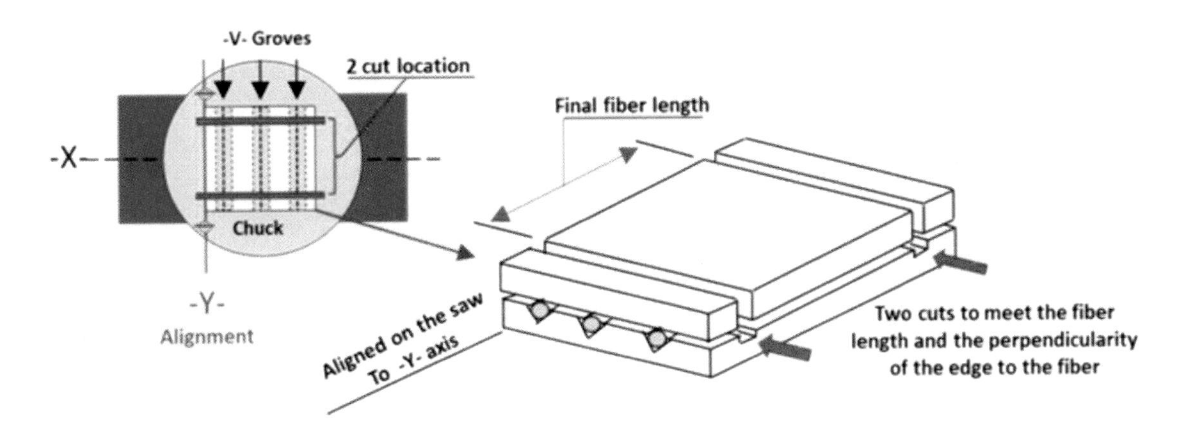

Figure 182. – A mechanical – V – grove mounting jig design to clamp down the fiber optic rods

The reason to align the substrate jig on the –Y – axis is to maintain the perpendicularity of the fiber edge to the fiber. A similar application of dicing ceramic pipes to a specific length was performed on a mechanical fixture without using any wax or glue. The pipes are mounted in metal – V – groves and clamped down by a mechanical clamp.

The blade is cutting through the ceramic pipes into a release grove. This fixture setup is used many times after the dicing is completed (see fig. 183).

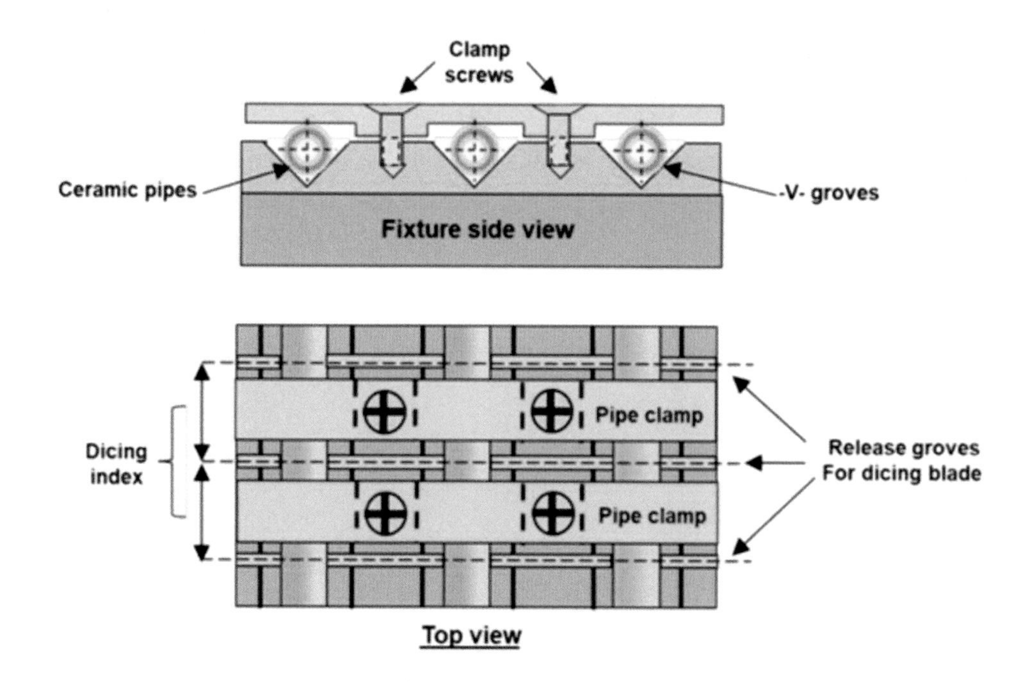

Figure 183. – A mechanical jig for dicing fiber optic rods & Ceramic pipes to an accurate length

15.11 - MAGNETIC CHUCK MOUNTING

Magnetic chuck mounting is the most common mounting method in surface grinding. There are manual activated magnet chucks and electrical activated chucks. The purpose in grinding is to strongly hold down magnetic metallic parts without any movements during the grinding process (see fig. 184).

Figure 184. – Different Magnetic chuck geometries

Magnetic chucks are not popular in the dicing market, however for some applications it was and still is an option. The way the magnetic chucks are used in dicing is mounting on the magnetic chucks the actual jigs for mounting the production devices to be diced.

15.12 - FREEZING MOUNTING

This is a unique method used in the past in the mechanical watch industry. It is not a practical method in the dicing industry however I have seen it in a demo, and it works well. In order to cover the many different mounting methods, the freezing option should only be mentioned. In principle, a macromolecular freezing agent whose freezing point is higher than the water is used in a special freezing system. This option is still used in different metal machining processes of mounting complicated thin geometries that any other mounting is not practical. It is good to know that this option exists. For more information on freezing mounting, see a Pdf site in the reference section.

CHAPTER 16

DRESSING REVIEW

Dressing is one of the most important issues in any dicing process, so this review is very important, interesting and a long one.

16.1 - DRESSING MECHANISM

To better understand the need for dressing and the dressing mechanism, we need to go back to the machining mechanism of the dicing blades. The basic dicing process is a machining operation of many small diamonds penetrating at high velocity into a mounted solid substrate. The ability of the diamonds to penetrate freely into the substrate depends on how well they are exposed. Below in fig. 185 is a sketch of well-exposed diamonds and non-exposed diamonds.

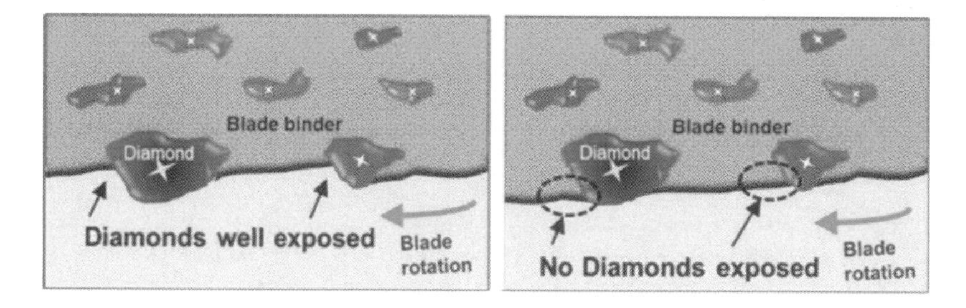

Figure 185. – Well exposed diamonds and poorly exposed diamonds

Below is a sketch demonstrating the performance of a well-exposed single diamond compared to the performance of a poorly exposed diamond on each blade rotation (see fig. 186).

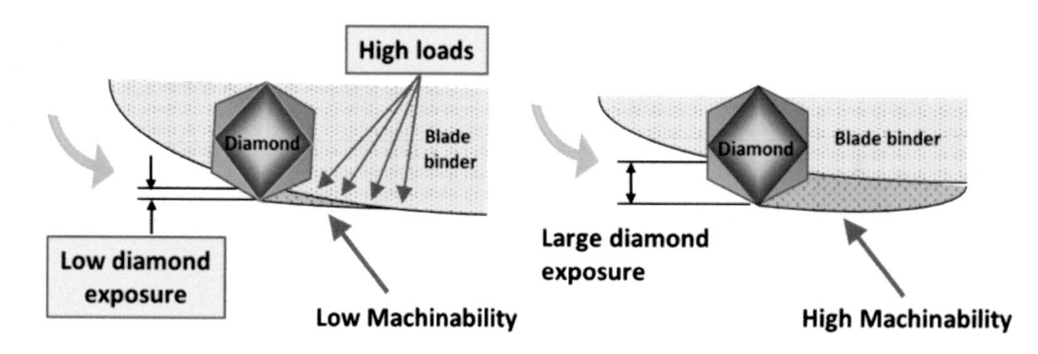

Figure 186. – Machinability of well exposed diamonds compared to a poor exposed diamonds

A well-exposed diamond will penetrate freely with minimum loads into the substrate. A poor exposed diamond will create high loading, high temperatures, and poor cut quality. The blade dressing starts at the manufacturing blade vendor site but needs to be dressed in a production mode during the first installation of a new blade and later during the dicing process.

16.2 - DRESSING CONVENTIONAL GRINDING WHEELS

Before discussing the many different dressing options on diamond dicing blades let's discuss how dressing is done on conventional grinding wheels made of aluminum oxide, silicon carbide and on diamond grinding wheels. This is important information and will help to understand the dressing process on diamond dicing blades. Dressing grinding wheels is a process developed a long time ago prior to dressing diamond dicing blades. Dressing is a common process in the grinding industry. It is used on new grinding wheels to straighten the edge and, in some cases, straighten the wheel sides. It also opens well the abrasive by grinding off the wheel binder and metal powder residue after grinding. This process is also perfecting the run-out of the grinding wheel edge to the spindle. The process is performed by contacting and moving a single diamond point on the grinding wheel edge. The diamond is usually a natural diamond 1–3 carats in size, depending on the grinding wheel diameter and width. The diamond is embedded and exposed in a metal holder (see fig. 187).

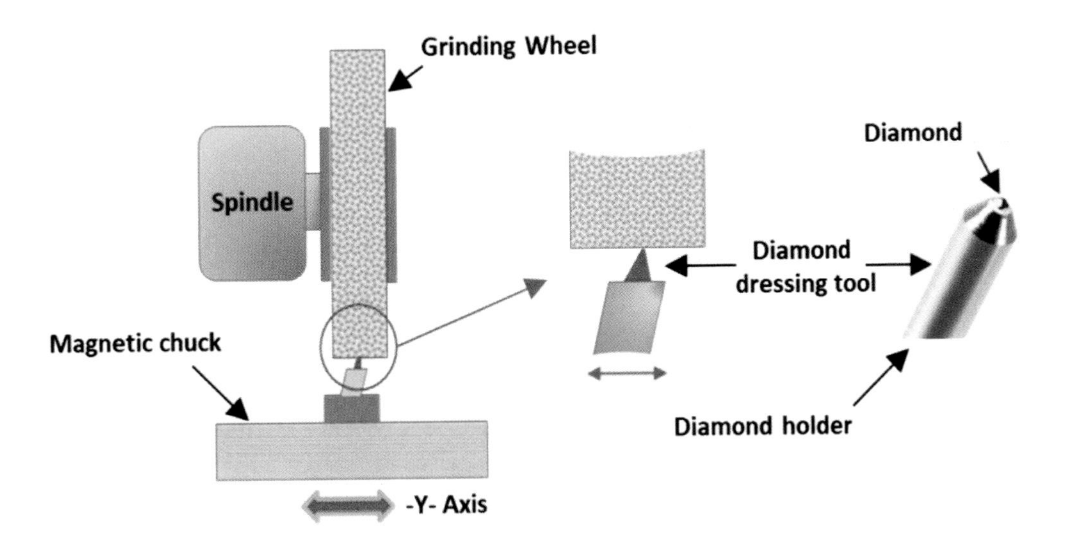

Figure 187. – Dressing process of grinding wheels

The above sketch demonstrates the principle of dressing a grinding wheel by moving a mounted diamond tool below the grinding wheel using the machine - Y - axis back and forward. There are more sophisticated dressing tools mounted above the grinding wheel using an accurate slide holding a similar diamond dressing tool that can be moved across the grinding wheel (see fig. 188).

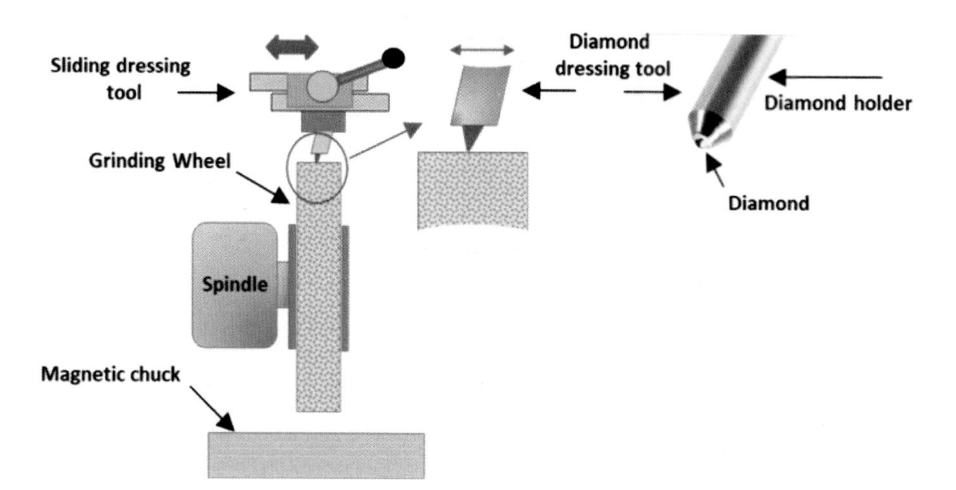

Figure 188. - Dressing the grinding wheel by moving the diamond above the grinding wheel

16.3 - DRESSING DIAMOND GRINDING WHEELS AND DIAMOND DICING BLADES

Dressing diamond grinding wheels requires a different setup of using a silicon carbide grinding wheel. It can be done offline on a grinding/dressing machine or on the grinding machine using a small grinding/dressing jig. Using a single diamond point tool on diamond grinding wheels will grind the sharp diamond tip immediately when in contact with the grinding wheel. Instead, a relatively soft Sil. Car. grinding wheel is oscillating over the diamond grinding wheel and gradually machining off the binder holding the diamonds. This process involves gradually losing loose diamonds, exposing new diamonds, and flattening the diamond grinding wheel edge (see fig. 189).

The process of dressing diamond grinding wheels is very similar to dressing diamond dicing blades done initially at the dicing blade vendor.

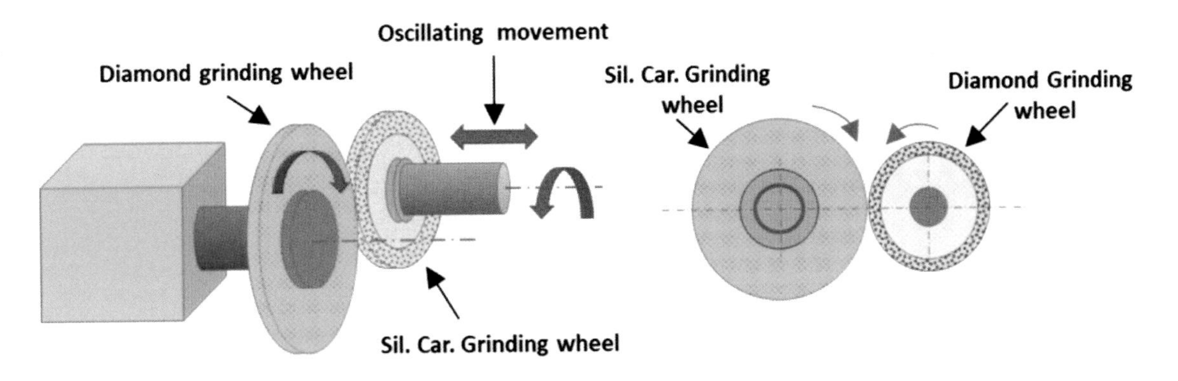

Figure 189. - Dressing process of diamond grinding wheels

16.4 - DRESSING DIAMOND DICING BLADES IN THE MICROELECTRONIC INDUSTRY

Following is a short review of the initial dressing process in the microelectronic dicing industry during the mid-70s when dicing semiconductor silicon wafers was done using annular blades before introducing the hub blades. The silicon wafers at that time were 2–4 inch (50–100 mm) in diameter x .010"–.030" (0.254–0.762 mm) thick. The wafers were mounted directly on ring type vacuum chucks and diced only partly, 60%–80% deep (see fig. 190).

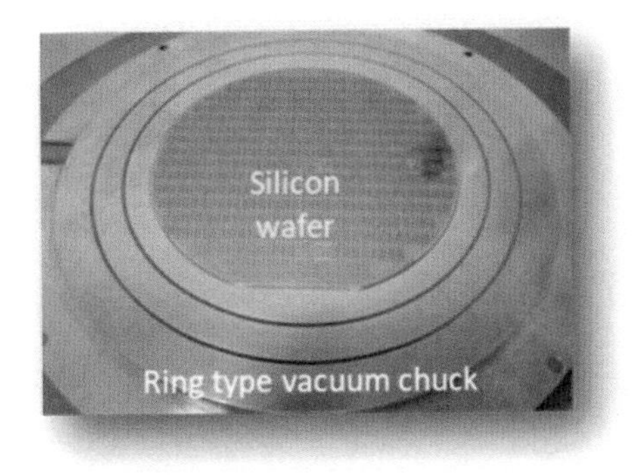

Figure 190. - Silicon Wafer mounted directly on the vacuum chuck

This process was before automation was introduced. The partly diced wafers were broken to separate the dies by using a rubber roller over the top side of the wafer. (see fig. 191). (Same as fig. 2)

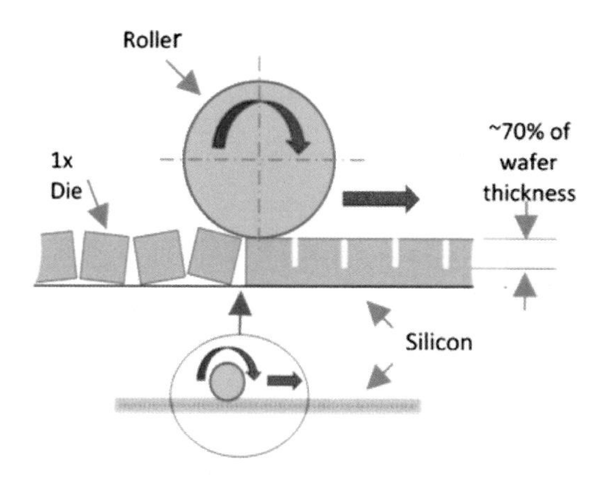

Figure 191 – Rolling over the partly diced wafer breaking the wafer into individual dies

At that time, the annular dicing blades used for dicing silicon wafers were of the nickel electroformed type with the following parameters:

Diamond grit - 4-6mic.

Diameter - 50-58mm

Blade I.D - 40.00mm

Thickness - .008"-.014" (0.200 - 0.355mm)

The blade edge of some vendors was edge grounded and for other vendors not grounded. In any case, the blade edge had diamonds that were not well exposed, and the edge runout to the spindle needed to be perfected. The dressing media to perform the dressing was a Bakelite board with 600 mesh silicon carbide media (see. fig. 192).

Figure 192. - Bakelite dressing media with 600 mesh Sil. Car. grit

The blade was zero chuck calibrated and then diced into the dress board. Every customer developed its own procedure. Following is a procedure that I used to recommend customers for dressing thin annular nickel blades:

Use 30K RPM spindle speed

Index – 3x blade thickness

10x cuts .002" deep at 6" / sec.

10x cuts .002" deeper than production depth at .2"/sec.

10x cuts .002" deeper than production depth at .5"/sec.

Continue on a blank Sil. wafer or on a production wafer:

Index – 4x blade thicknesses

10x cuts .001" deeper than production depth at .1"/sec.

Continue with .2"/sec steps 10-20x cuts depending on

cut quality up to production speed.

Continue with .2"/sec steps 10–20x cuts depending on cut quality up to production speed.

The production speed at that time was up to 6"–7"/sec, dicing 60%–80% of the wafer thickness. The above dressing were only recommendations and could easily be modified per customer quality requirements and test results on specific wafers. Blade dressing on silicon applications was gradually reduced and optimized when automation was "born" using tape mounting and, in addition, when the hub blade was introduced. The semiconductor silicon market was continuously pushing to improve cut quality, better UPH and minimize the dressing cycle. The dicing saws software was also optimized, making it easier to perform and optimize an automatic dressing cycle. The hub blade manufacturing process was forced to be optimized to perfect the accuracy of radial and axial runout on the dicing spindles.

In addition, the diamonds on good hub blades are optimized to be well exposed on the aluminum side which eliminates the need to perform a long and time-consuming dressing (see fig. 193).

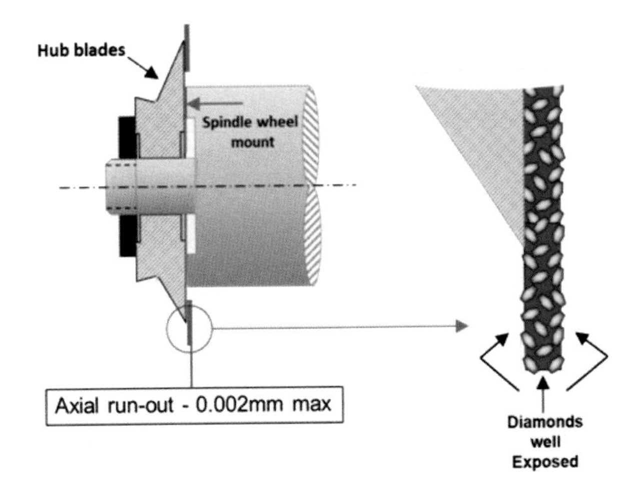

Figure 193. – A new well exposed hub blade

16.5 - HUB BLADES DRESSING

Hub blades today are basically ready to go out of the package and do not require the relatively long dressing board process. The dressing process for hub blades is performed on blank silicon wafers or on production wafers. The methods used in the market are called override or pre-cut. The process is automatically starting at low feed rates and gradually going up reaching production speed. The other used process is also starting at low feed rates and at a shallow cut depth and gradually reaching both production feed rate and cut depth. Both options can be optimized to meet cut quality specs and to shorten cycle time. Dressing dicing blades is a process done not only for silicon wafers.

General:
In today's microelectronic market we face many different materials that are diced using different blade binders and different blade geometries, they all require different dressing processes. Before covering the different dressing methods and options lets discuss the different manufacturing dressing processes done during the final manufacturing of the different blade binders. In general, different blade binders require different edge treatments to meet the different applications in the marketplace. The idea is to supply customers with blades that will perform well with minimum initial dressing. Let's cover the factory different binders and initial factory treatment/edge grinding.

16.6 - NICKEL ANNULAR DRESSING

When discussing dressing of nickel blades, we need to divide them into thick blades and thin blades. In general nickel blade binders with diamonds embedded need to be O.D. grinded in a wet process. This process is done on cylindrical grinding machines. Thick blades are mounted on an accurate mandrel in batch quantities after the blade surface was lapped to be perfectly parallel. This is important to maintain a good and solid batch of blades clamped together without any gaps between the blades. It is also important to minimize any burs on the blade edge. Silicon Carbide grinding wheels are used for the grinding with mesh sizes of 280 to 500 depending on the blades diamond grits size. (see fig. 194).

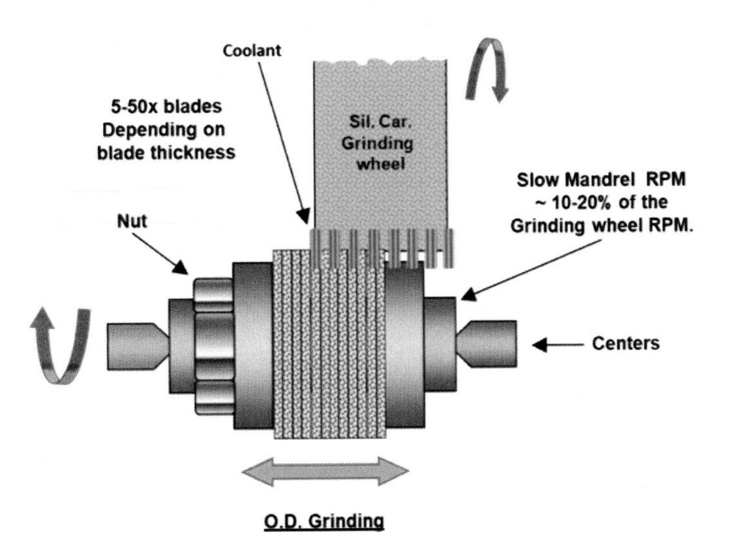

Figure 194. – Wet O.D. grinding of blades mounted on an accurate mandrel

On most grinding machines the table holding the mandrel with the blades is oscillating during the contact with the grinding wheel. The grinding process is a slow process of entering only a few microns on each pass after passing the blade stack, either on both sides of the mandrel or on one side, depending on diamond grit size or the final requested finish. The blade's mandrel RPM is relatively slow, about 20 percent of the grinding wheel RPM. The reason for the low blade RPM is the low machinability of the blades. The blades with the diamond particles are way harder than the grinding wheel and a fast RPM of the blades will wear very fast the grinding wheel with a very low wear on the blades. These phenomena in generally are similar with all blade binders/matrices (see fig. 195).

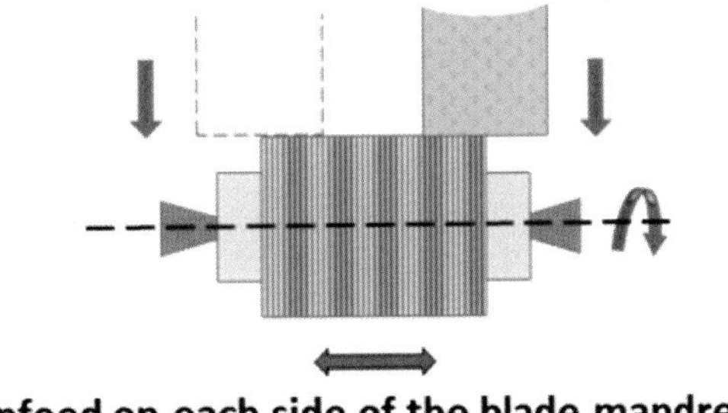

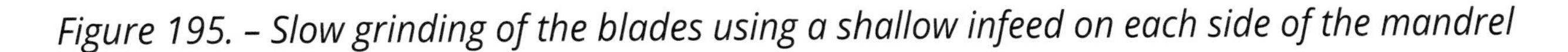
Figure 195. – Slow grinding of the blades using a shallow infeed on each side of the mandrel

The main goals for grinding the blade's edge are to perfect the concentricity of the blade I.D. to the blade O.D. and to achieve a nice flat edge with the diamonds exposed. The diamonds will need to be better exposed which is done later by the customer on the dicing saw (see fig. 196).

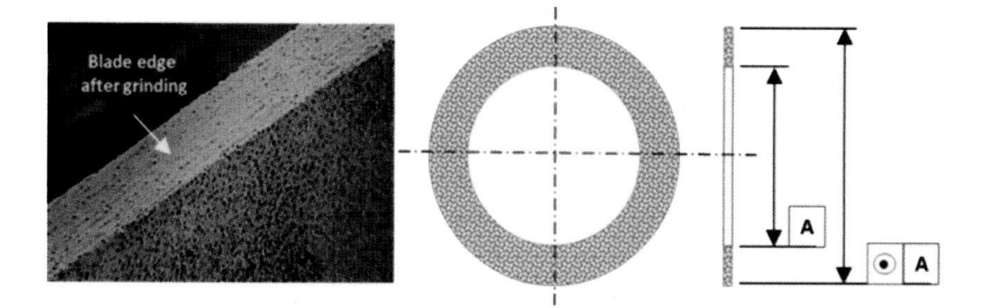

Figure 196. – SEM of a nickel blade edge after O.D. grinding

Another option to aggressively expose the diamonds on nickel blades is to perform a chemical electrical etching which is also called "electropolishing." This process is aggressive and needs to be closely controlled. It is performed in a strong acid tank connected to a power supply but in the opposite power connection so instead of plating on, the process is chemically electrically etching some of the nickel binder to better expose the diamonds. Needless to say, this process is done only at the blade vendor house (see fig. 197).

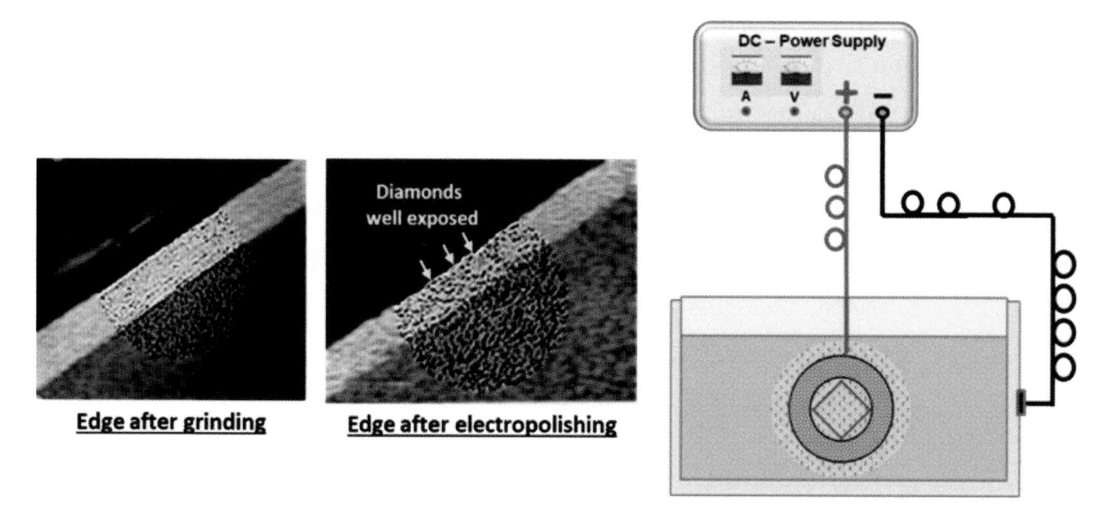

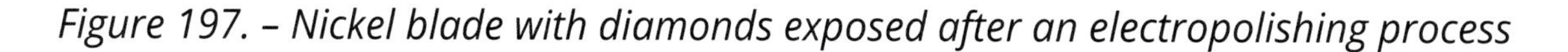

Figure 197. – Nickel blade with diamonds exposed after an electropolishing process

O.D. Grinding thin blades .0008"–.0025" (0.020 mm–0.640 mm) can theoretically be done only at the vendor's house. The blades are also mounted on an accurate mandrel. However, they need to be mounted with a soft spacer between the blades. The spacers need to be of a soft enough material to back up the diamonds and eliminate any diamond indention to the next mounted blade. In addition, it needs to be made of a material that will not overload the grinding wheel. Diamonds that are pushed into the next mounted blade will cause an indention that will reject the blades. The grinding itself is a very delicate process using a fine mesh silicon carbide grinding wheel and the process requires a very small infeed (see figs. 198 and 199).

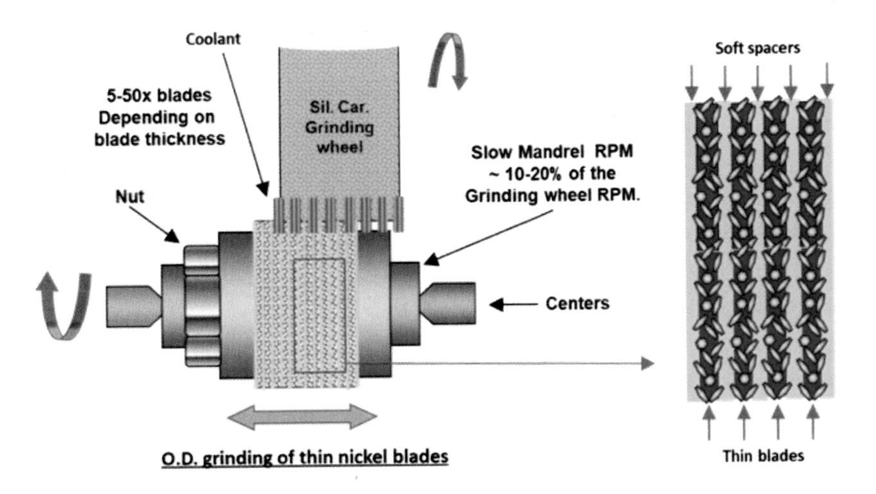

Figure 198. – Special mounting of thin nickel blades on a mandrel with spacers

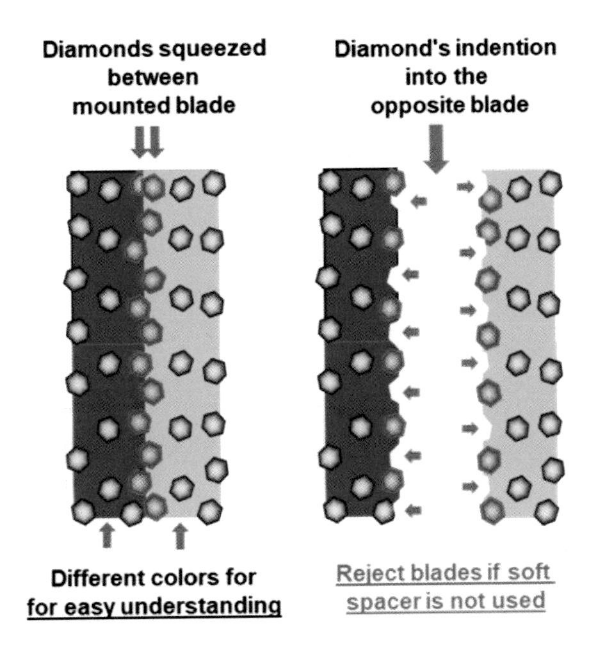

Figure 199. – The damage effect of grinding thin blades without soft spacers between the Blades.

16.7 - METAL SINTERED BLADE DRESSING

Dressing the sintered blades at the blade manufacturing vendors is similar to the nickel blades. After lapping the blade side surfaces to be flat and parallel, they are mounted on an accurate mandrel and wet grinded using Silicon Carbide grinding wheels. The sintered blades are softer than nickel blades, however still hard and the grinding process is also a slow process. The goal is the same goal to perform a good concentricity of the O.D. to the I.D. and a flat edge at the right blade O.D. size (see fig. 200).

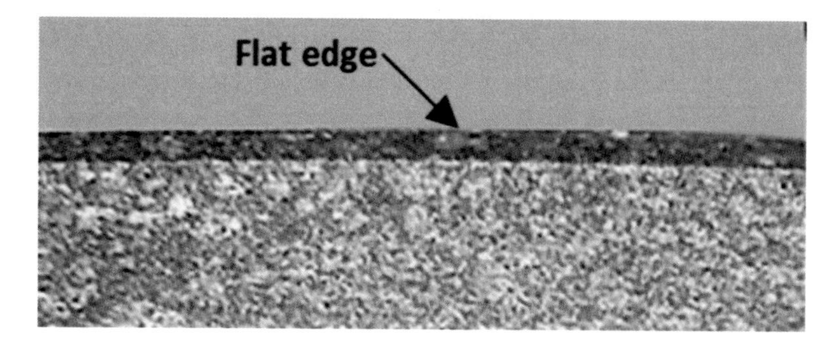

Figure 200. – Edge grinding of Metal Sintered blades

Depending on the application, some sintered blades are being used in a production mode without any dressing and some are going through a dressing mode usually developed by the customer.

16.8 - VITRIFIED AND RESINOID BLADE DRESSING

VITRIFIED:

Vitrified is a softer bond compared to the nickel and metal sintered blade but harder than resinoid blades. The edge grinding is a wet process, a similar process of grinding the metal sintered blades.

RESINOID:

Resinoid blades are way softer than both nickel and metal sintered blades. This leads to a much easier grinding process. The edge grinding process of resin blades is in most cases a dry grinding process done on dressing machines or on cylindrical grinding machines. The grinding process is also done using Silicon Carbide grinding wheels and the blades are mounted in batches of 10–50 blades. The process is much faster as the infeed of the grinding wheel is much higher, and the oscillating of the blade mandrel is also faster. The blade edge becomes flat but not as flat and even like the nickel and metal sintered blades. This is mainly because the resin blade bond is not as dense. Resinoid blades with finer diamond grit that are much denser, perform with a flatter edge after the grinding (see fig. 201).

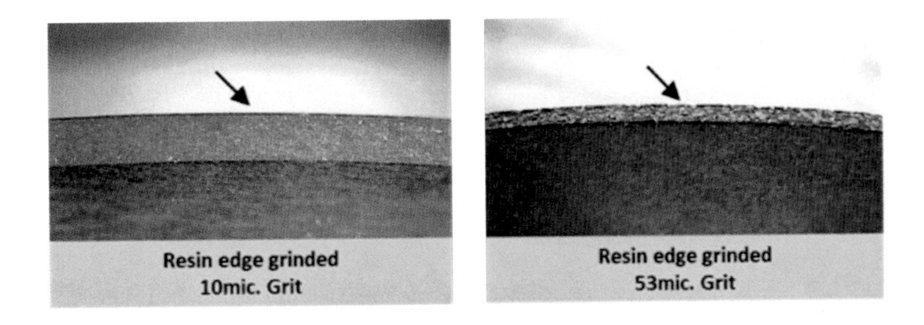

Figure 201. – Edge grinding of resin blades

As mentioned, there are many dressing machine options used for O.D. grinding resin blades. The most popular method is using a designated dry universal dressing machine that can be used also for profile grinding of the blade edge (see fig. 202).

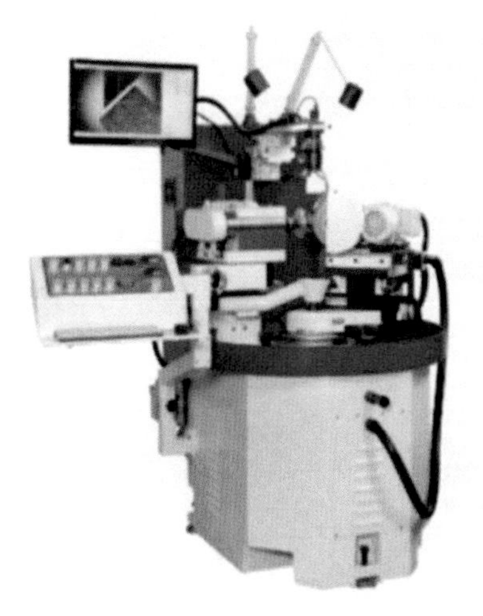

Source: FARMAN MACHINERY

Figure 202. – O.D. dressing machine

In most cases, the designated dressing machines are manually operated, which makes them easy to use and very universal to change from one application to another. In most cases, the principle of those dressing machines is to use two spindles mounted on accurate slides, one for mounting the grinding wheel and one for mounting the blades or diamond grinding wheels. In addition, the grinding wheel slide and spindle are mounted on a swivel arm so the grinding wheel can be rotated to any angle for profile grinding. The grinding wheel feed at the end of the oscillating movement can be accurately manually operated using a leadscrew with a micrometer scale. The edge of the dicing blade is visually inspected by an optic system. There are available optic systems that can measure delicate and accurate geometries. Fig. 203 shows the principle of a dressing machine.

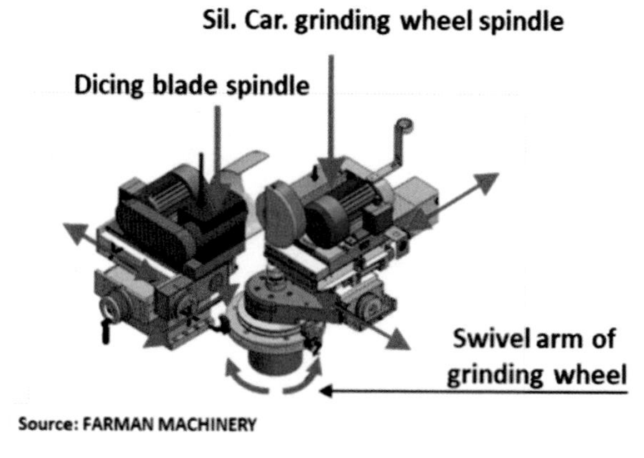

Figure 203. – Principle of a universal dressing machine using 5 axis

There are CNC dressing machines used for complicated edge geometries in high production volumes. It all depends on the edge requirements and quantities. The standard regular straight edge is a relatively simple process of mounting a large batch of blades on a mandrel and can easily be handled on a standard manual dressing system using a good visual optical system. One important parameter on any dressing diamond blades or diamond grinding wheels is the rotation relation between the Si. Car. wheel and the diamond tool. In grinding the rotation of the grinding wheel should be against the metal workpiece. In dressing it should be the opposite (see fig. 204).

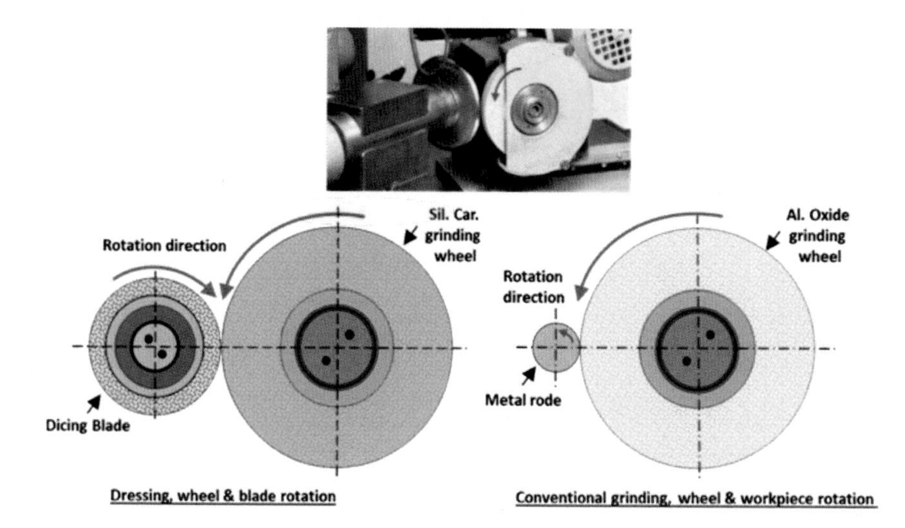

Figure 204. – Principle of O.D. grinding grinding wheels and diamond dicing blades

In conventional grinding, the idea of rotating one against the other is needed to create a machining action. In dressing such a machining action will damage the diamond's sharpness. In dressing, the idea is to loosen the binder, let loose diamonds wear out, gradually expose new diamonds, and at the same time flatten the edge surface. This is the reason for the different rotations between grinding wheels and diamond dicing blades. Another edge grinding that is performed by the blade manufacturing vendors and sometimes at the dicing users when using high production volumes is profile edge grinding. Below are a few of the popular geometries used in the market (see fig. 205).

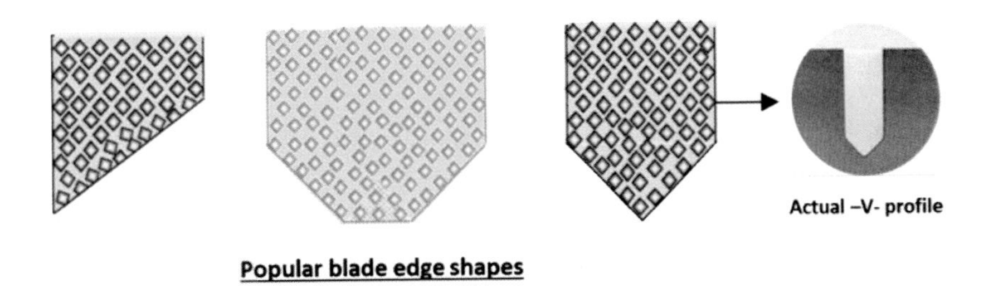

Figure 205. – A few blade edge geometries made on a universal dressing machine

The blade edge grinding profile process is done on universal dressing machines by swiveling the grinding wheel to the right angle and oscillating with the Sil Car. grinding wheel over the blade edge (see fig. 206).

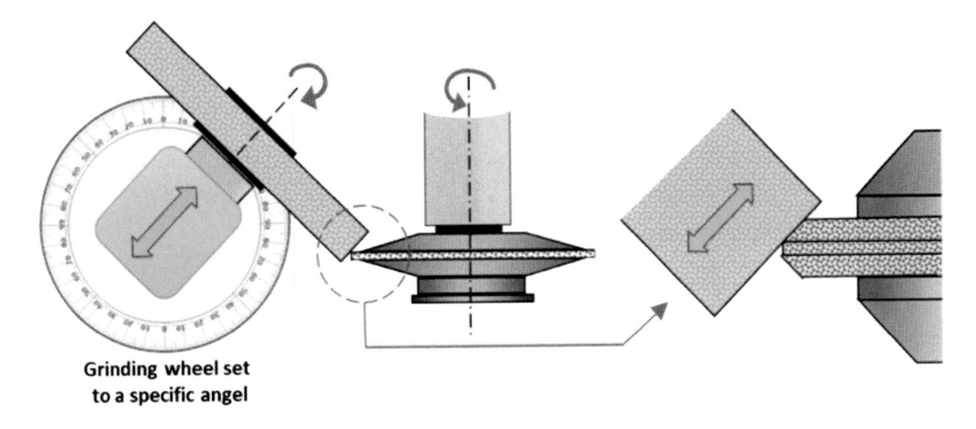

Figure 206. – Blade edge – V – grinding principle

A unique application of dicing thick alumina, PZT, and similar materials to create high "towers" (see fig. 207) requires the need to open the side surfaces of resin blades to minimize loading. The idea is to lap and wear only a small amount of the resin sides binder to expose the side diamonds. Lapped resin blades are helping to perfect the cut's perpendicularity. Non-lapped blades in this type of application may result in high loading, slanted cuts and "tower" breakage.

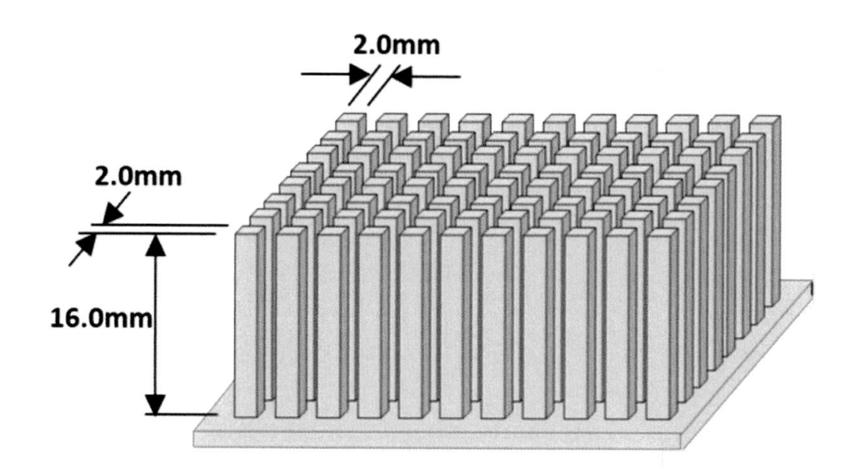

Figure 207. – Deep dicing x narrow indexing using a side lapped resin blade

The process is done by a min. surface lapping process on dual side lapping machines and is done by the blade vendors (see fig. 208).

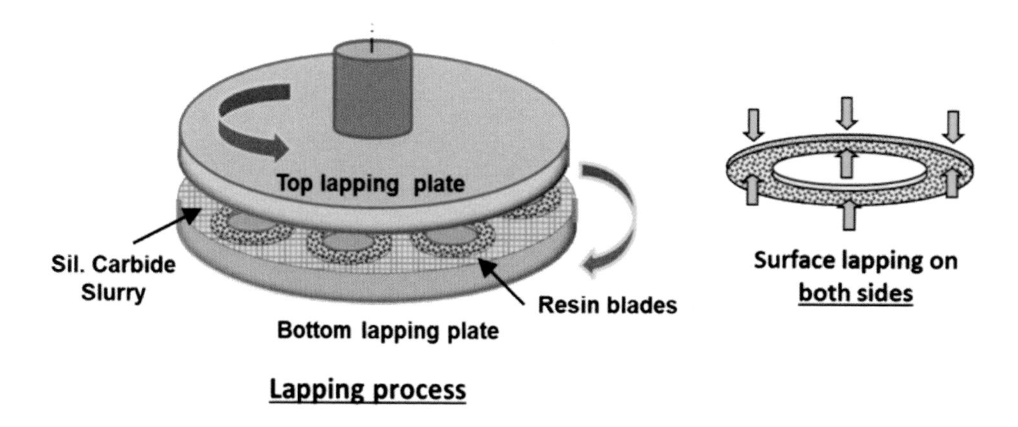

Figure 208. – Fine lapping set-up of lapping resin blades

The above is a generic explanation of the principal dressing process done by lapping. Let's now concentrate on the dressing processes performed by the customers on their dicing saws. The dressing for the different blade binders is quite different because of a few elements: the bond hardness, the diamond type, the diamond grit size, the diamond concentration and the blade thickness. However, the blade bond is a major factor. Following is a discussion on the mechanical behavior of the major different bonds:

16.9 - SELF RE-SHARPENING/ DRESSING OF RESIN BLADES

Resin bond is the softest bond used in the dicing industry. Some are called resin self-re-sharpening blades. During the dicing, there is continuous wear on the blade edge, resin and diamonds are wearing out continuously and new sharp diamonds are being exposed. This continues wear characteristics requires min. extra dressing and performs with good cut quality. The disadvantage is higher blade wear (see fig. 209).

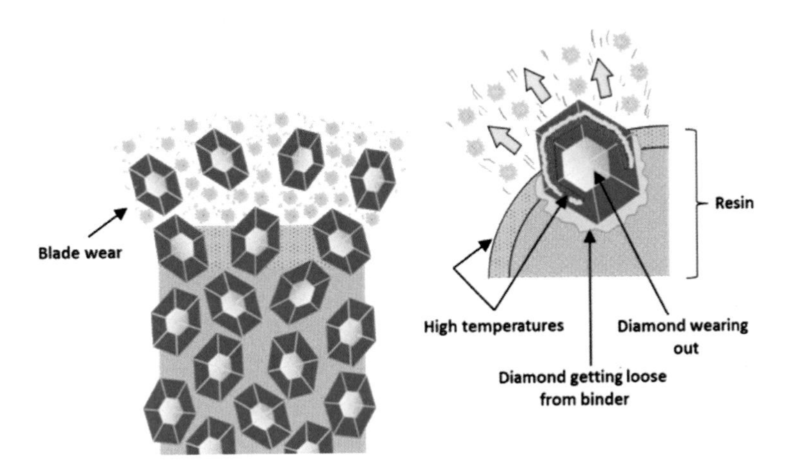

Figure 209. – Self resharpening of diamond resin blades

Some customers do dress on site resin blade for the following main reasons:

Perfect the concentricity of the blade edge to the spindle. (see fig. 210).

- Releasing loose diamonds and exposing new sharp diamonds
- Maintaining a flat edge. One recommended process is to lower the blade RPM in order to machine / shape the blade edge more aggressively.
- On most applications the above is not needed however for some demanding applications this is a routine process on new blades and during production.

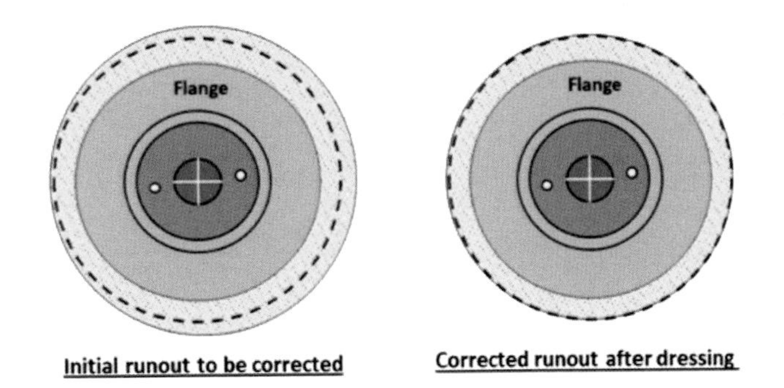

Figure 210. – Perfecting the blade edge concentricity to the spindle

16.10 - THICKER METAL NICKEL & SINTERED BLADE DRESSING

Dressing thin nickel blades were covered already using the 600 mesh Bakelite Sil. Car. dress board. Thicker nickel and Metal sintered blades have very different hardness characteristics. In addition, the diamond type is harder and will maintain the sharp edge much longer compared to the friable diamonds used in resin blades. Below is a sketch showing the mechanical properties of a metal-type blade edge (see fig. 211).

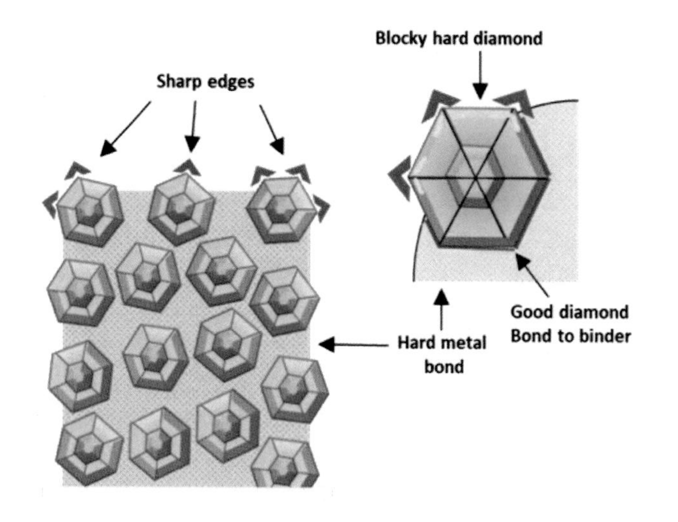

Figure 211. - Metal bond (Nic. & Sint.) blades maintaining the diamond sharpness longer

The mechanical bond characteristics of each diamond to the metal binder are way stronger compared to the embedded diamonds in a resin binder. This is one of the reasons for using harder and blockier diamonds. Wearing a diamond out of the binder takes a much longer time during the dicing and dressing when new sharp diamonds need to be exposed. Dressing online during dicing is cleaning residue between the diamonds and exposing new sharp diamond edges (see fig. 212).

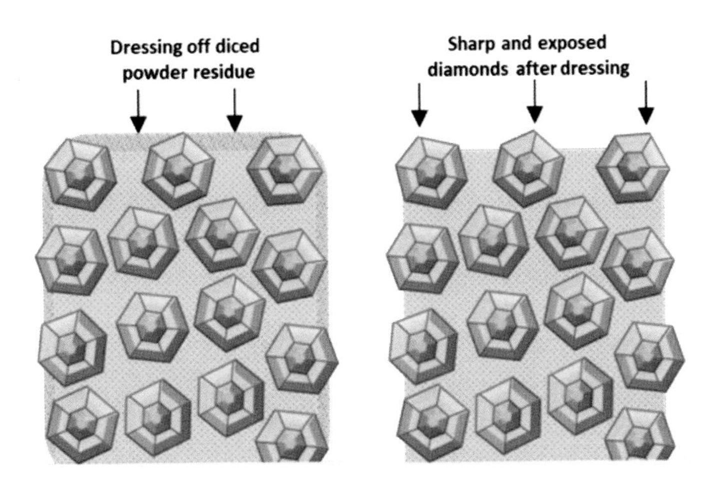

Figure 212. – Dressing off powder residue and exposing well the diamonds

Losing the edge geometry is difficult to correct on the dicing saw. In this case, an offline dressing / edge grinding is needed and will be viewed in one of the applications. Following are a few dressing methods used on the dicing saws:

The conventional dressing using a dressing media mounted on the saw chuck. In most cases the dressing is an inhouse process. (see fig. 213).

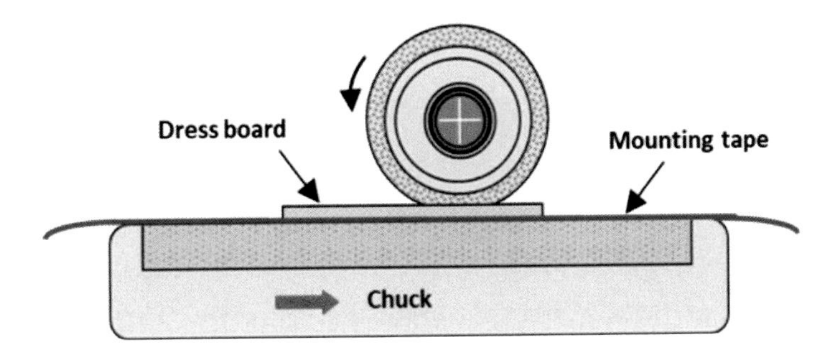

Figure 213. – Conventional dressing process on a chuck mounted dress board

Some demanding applications require a more frequent dressing to lower loads and maintain tight quality specs. For those applications, an online dressing is applied by adding a special jig holding a dress media besides the substrate. The idea is that the blade is passing the dress media after every cut on the production substrate (see fig. 214).

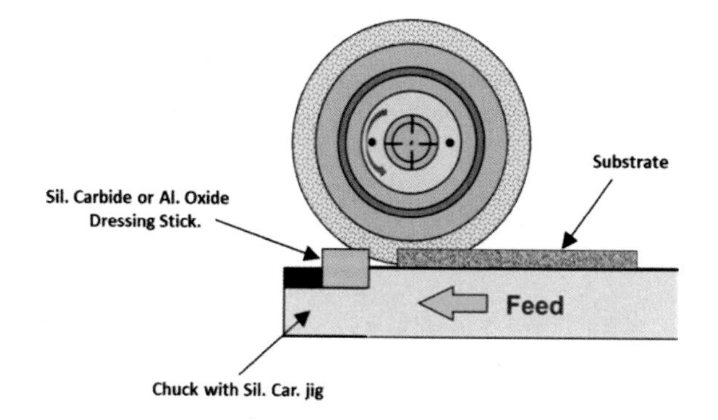

Figure 214. - Dressing Online

The other option of dressing the blade during the dicing process without interfering with the dicing by dismounting the substrate from the chuck is using a side dress station which is a standard option on most new dicing saws (see fig. 215).

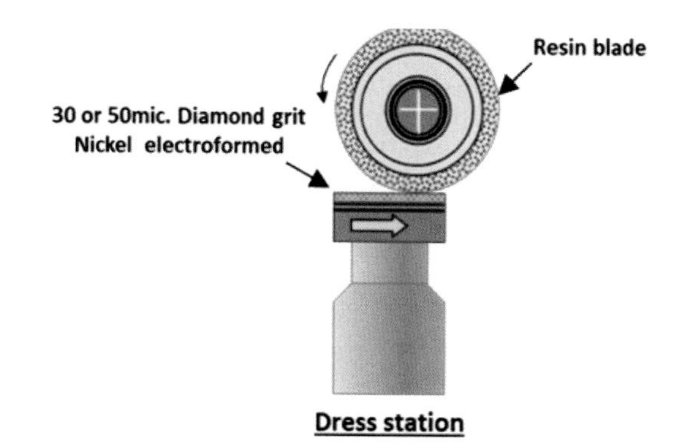

Figure 215. - Dressing of line on the side dress station

The idea is to program the saw to stop the dicing after a certain number of cuts and move to the dress station to perform a certain no. of cuts into the dress board. After the dressing, the blade performs a height sense on the noncontact height to compensate for the radial wear. The blade continues to dice at the point the dicing process was stopped. More on online dressing will be discussed in the applications section.

There are a few dress media that can be used on the dress station. The normal most popular media are Sil. Car. and Al. Oxide with different mesh sizes and hardness characteristics. The mesh size depends on the blade type, mainly the diamond grit size. The above dress media and process are designed mainly to clean the blade edge from the powder residue created during the dicing and to well expose the diamonds. A special process is required if the blade edge geometry needs to be corrected/flattened. Correcting the flat blade edge using the dress station on the saw can be done only on resin blades as the resin bond is relatively soft and can be relatively easily ground down using a much harder dressing media of a nickel or metal sintered board with diamonds embedded in the dress plate. Below are sample photos (see figs. 216 and 217).

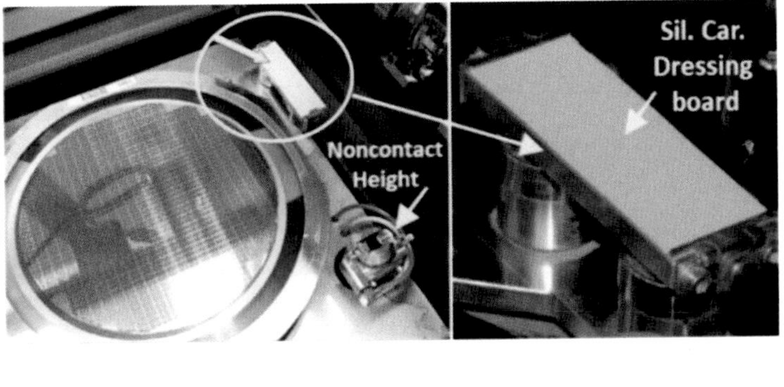

Dress station beside the mounting chuck

Source: Advanced Dicing Technologies

Figure 216. – Sil. Car. dressing media on the dress station for cleaning powder residue

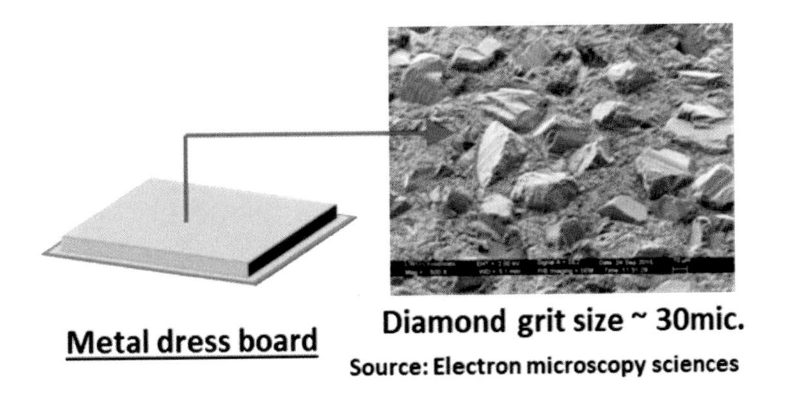

Figure 217. – Nickel or metal sintered diamond plate mounted on the dress station

The dressing process needs to be done in a slow and controlled mode. The cut depth into the diamond board needs to be very shallow with a relatively slow feed rate of about 5mm/sec. For a more aggressive action, it is recommended to use a lower spindle speed, lower than the production spindle speed. A few indexes need to be performed till the blade edge meets the spec. All parameters need to be optimized depending on the diamond grit, the blade binder, and the blade thickness. Below is a sketch showing in a generic mode of the dressing process (see fig. 218).

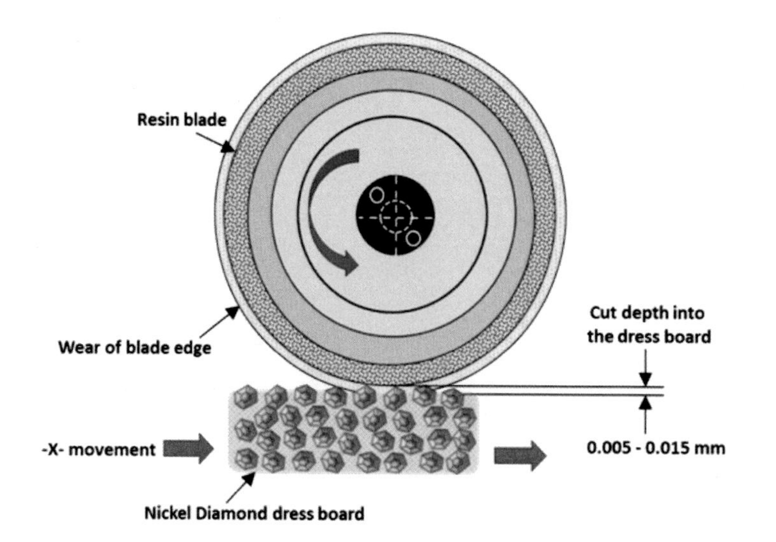

Figure 218. – Resin blade on metal / diamond dress media to correct the blade edge flatness

Below are two real cases examples of flattening .020" thick resin blade, 2" Dia. & 4" Dia. that lost the blade edge geometry: See Fig. no. 219 & 220

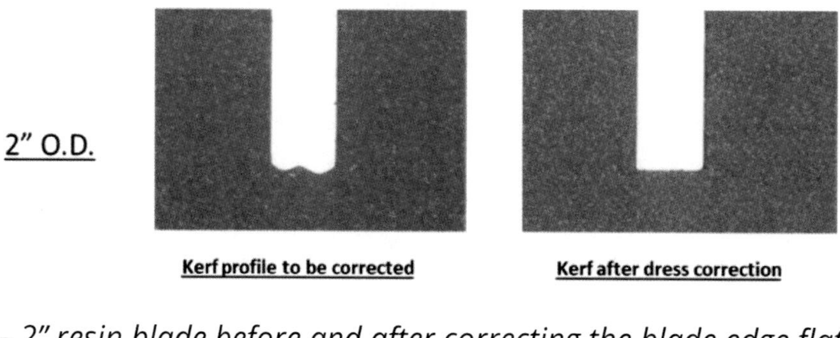

Figure 219. – 2" resin blade before and after correcting the blade edge flatness

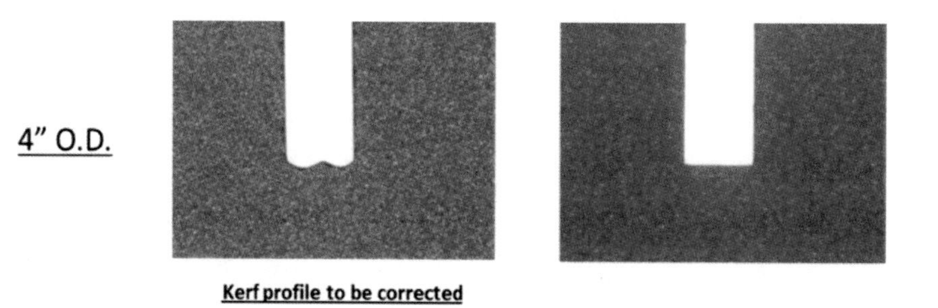

Figure 220. – 4" resin blade before and after correcting the blade edge flatness

Profile dress correction process on 2" O.D. blades:

Blade parameters – Resin bond – 58mm O.D. x .020" (0.50mm) thick x 45mic.grit.

Dressing media – Nickel bond x 30mic. diamond grit

Spindle speed – 20Krpm

Cut depth into the dress media – 0.015mm

Index – .040" (1mm)

Feed rate – 5mm/sec

No. of passes to achieve a flat edge – 5x

Profile dress correction process on 4" O.D. blades:

Blade parameters – Resin bond – 4.6" O.D. x .020" (0.50mm) thick x 45mic.grit.

Dressing media – Nickel bond x 30mic. diamond grit

Spindle speed – 8Krpm

Cut depth into the dress media – 0.015mm

Index – .040" (1mm)

Feed rate – 5mm/sec

No. of passes to achieve a flat edge– 5x

A major problem in most dicing applications is getting the blade edge rounded. When the blade edge is getting rounded creating in the diced substrate a lip that can cause back side chipping and in some cases causing the device geometry to get out of spec. The main issue is when mounting the substrate on tape which is minimizing the cut depth causing the lip effect (see fig. 221).

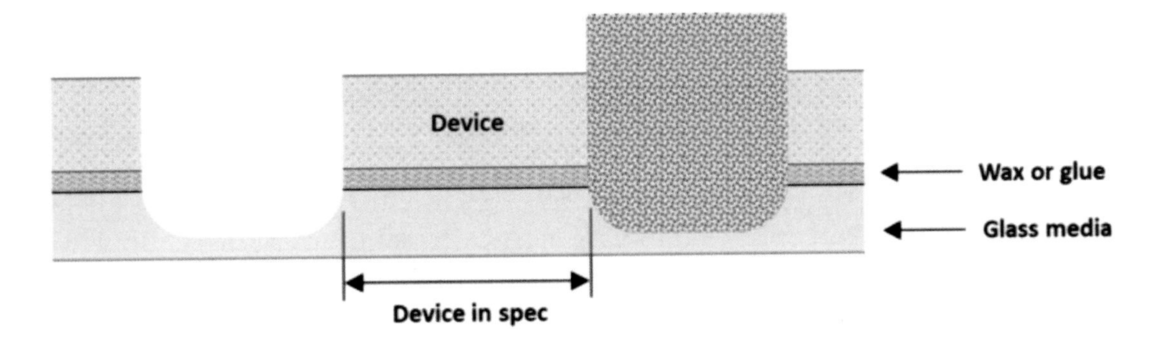

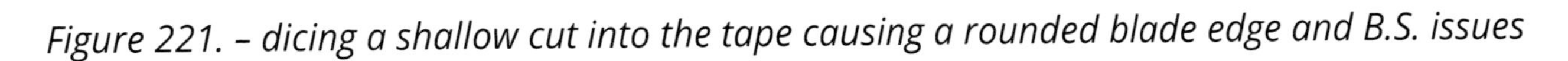

Figure 221. – dicing a shallow cut into the tape causing a rounded blade edge and B.S. issues

The above illustrates the importance of maintaining the blade edge and the dressing process to fix the blade edge. When mounting the substrate on a mounting media like glass or similar using wax or glue, the edge of the blade becomes less critical as the lip effect can be eliminated by cutting deeper into the media below the actual device. (see fig. 222).

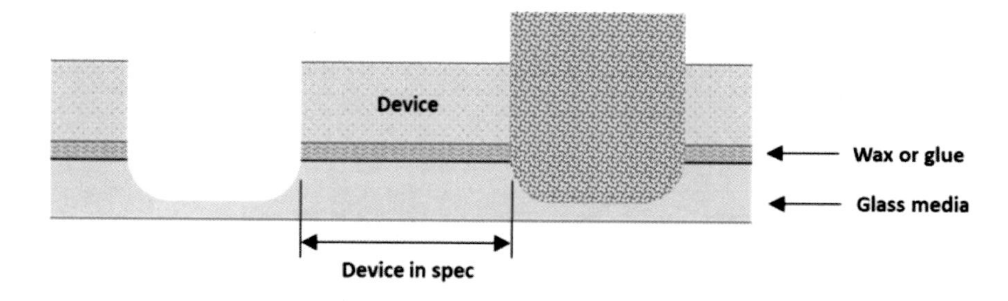

Figure 222. – Blade rounded edge inside a thick mounting media = better cut quality

More blade edge issues will be discussed in the application section.

16.11 - DRESS MEDIA'S

As discussed, many dress media are available with different grit types, different mesh sizes, different bonds, and different geometries to fit the application and the saw mounting options. The following are the main dressing types:

- Silicon Carbide in a Vitrified bond with 350 – 2000 Mesh size
- Silicon Carbide in a Bakelite bond 600 – 2000 Mesh size
- Aluminum Oxide in a Vitrified bond – 400 – 2000 Mesh size
- Diamonds in a nickel electroformed bond – 30 – 50-micron grit size
- Diamonds in a metal sinter bond – 15 – 75-micron grit size

Most dressing vendors can change the abrasive grit size, the bond hardness and the geometry per customers' requests. Following is a photo of different dress boards. (see fig. 223).

Figure 223. – Different dress medias geometries

16.12 - DRESSING SUMMARY

All that was discussed is leading to the fact that dressing is a major important factor in any dicing application. The main available dressing options were discussed. However, every customer is developing and optimizing the dressing process that fits best the application they run. The blade is a major factor in any application; however, the right optimized dressing process is also a key important factor to maintain any application spec in the marketplace.

CHAPTER 17

OTHER NEWER SUB-STRATE SEPARATING TECHNOLOGY TECHNICS.

This technical article is aimed at the dicing process using diamond dicing blades, however before reviewing the conventional dicing applications, it should be noticed that other technologies are used in the marketplace. There are four major technologies but only three are being practically used. The four technology options are:

- Diamond scribe & Break
- Laser scribing / Stealth dicing & Laser ablation
- Plasma Etching
- Water jet cutting. (Not really used)

(see figs. 224 and 225).

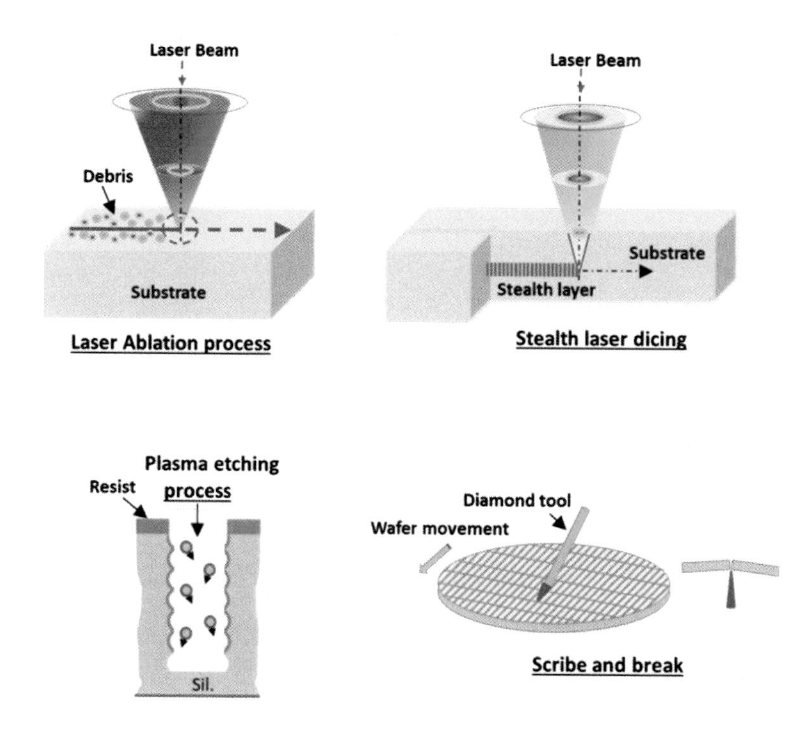

Figure 224. – Generic sketched of a few non diamond dicing technologies

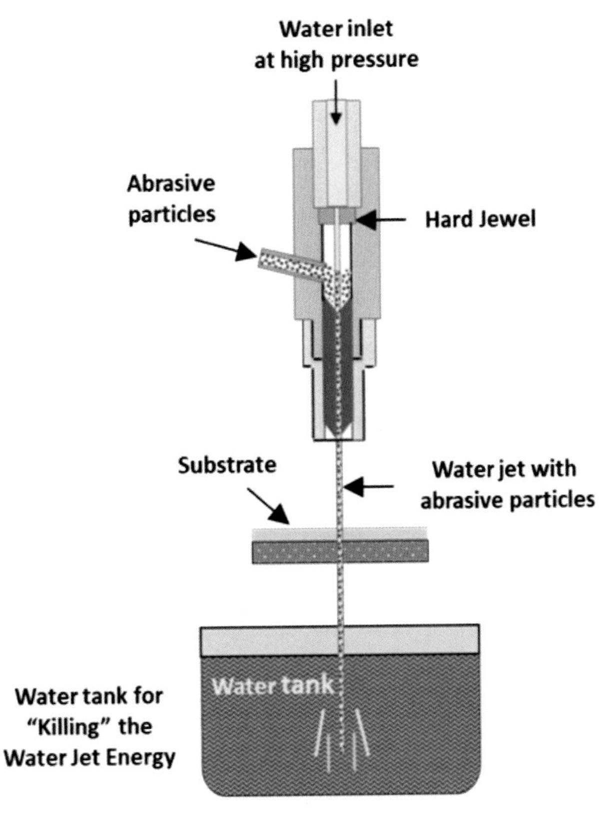

Figure 225. – Water jet dicing principle

17.1 - DIAMOND SCRIBE & BREAK

The diamond scribe and break process was used in the semiconductor industry for over 50 years, way before the diamond dicing process was adapted which is the major process today. However, a few applications like thin discrete semiconductor III-V (Chips and wafers for LED and Laser diodes) are preferred to be singulated using the scribe and break process. Quality, yields, and cost of the diamond dicing process are less effective on GaAs devices and GaN-LED Sapphire substrates due to the following reasons:

- Sapphire wafers & III-V materials are more difficult to dice due to their brittleness and the tendency to chip easily.
- Sapphire and III-V semiconductor wafers are more expensive compared to silicon wafers and therefore the streets are narrower to gain more "real-estate" devices. Narrower streets can be 0.020mm wide and thinner x 0.350 mm x 0.350 mm die size and smaller. Those tiny, small dimensions are very difficult to be diced using the diamond dicing process.

- On some GaAs wafers, the scribe lines are not aligned to the cleavage plane which makes the quality of the die separation much better compared to the diamond scribing process.

17.1.1 - THE SCRIBE AND BREAK PROCESS

The scribe and break process is using a crystal cleavage plan.

The process is accomplished by creating a stress line along the wafer streets and then fracturing the wafer along the stress line. The stress line is created by a scribe line performed using a natural diamond tool with a special geometry. A few diamond geometries are used, each diamond tool has a few sharp scribe points that can be used after the previous point is not sharp enough. There are a few process options depending on the wafer material to be processed. See a few options below in fig. 226.

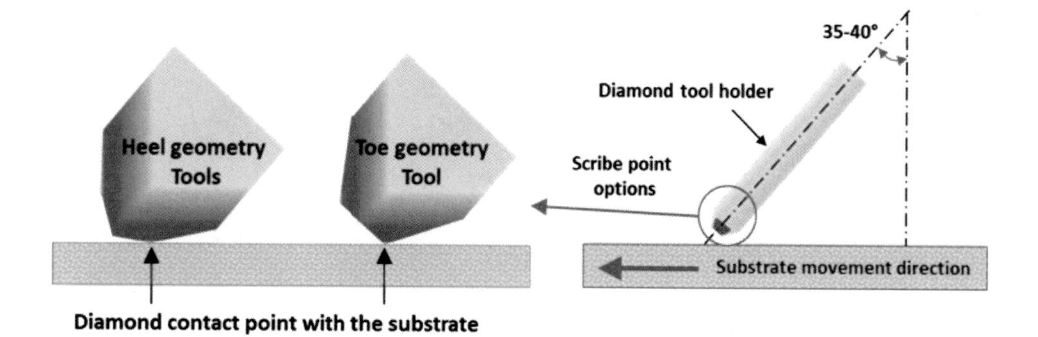

Figure 226. – Diamond scribing using a Toe and Heel geometry

Below are different diamond tip geometries including different process adjustments and how they perform during the scribing process (see fig. 227).

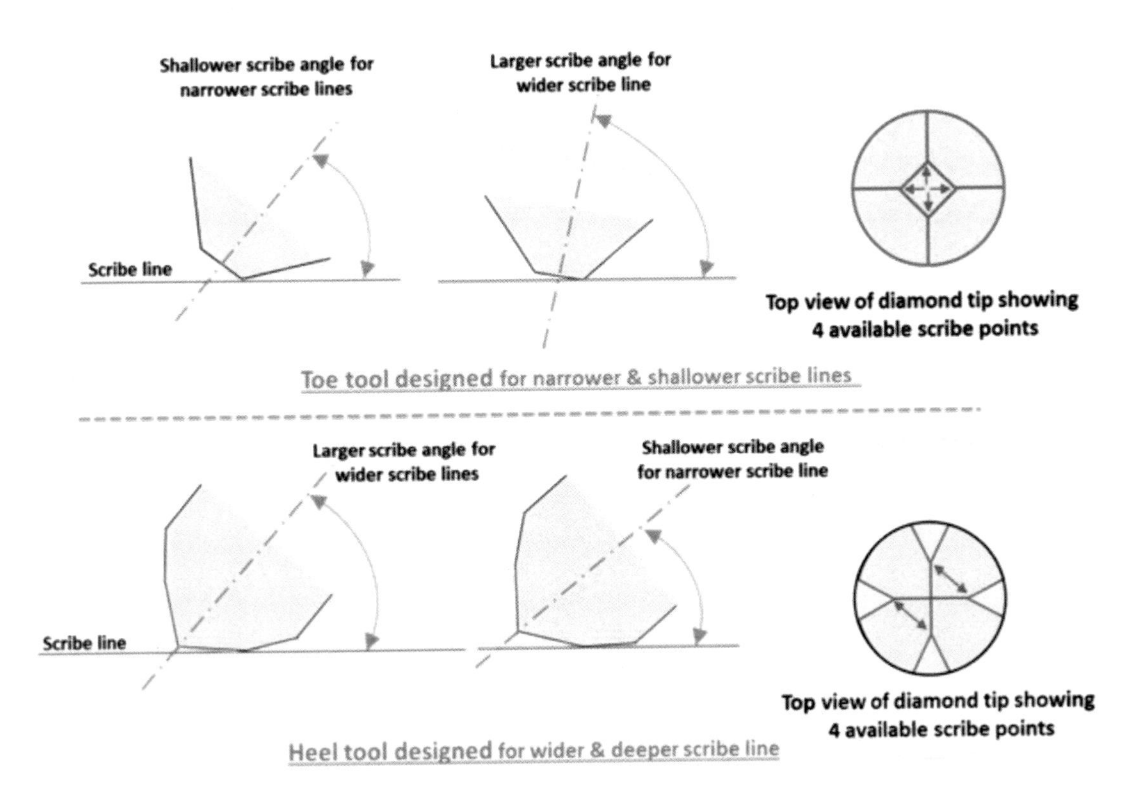

Figure 227. – Different diamond scribing processes used with different tool geometries

The diamond scribe tool needs to be perfectly sharp to create a good, concentrated stress effect, it is way more important than the depth of the scribe line. The typical groove geometry of a scribe line has a – V – shape, 3-5µm deep x less than 5µm wide (see fig. 228).

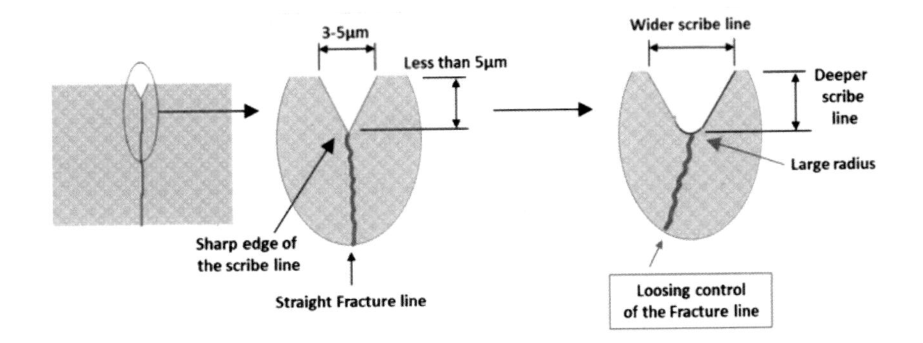

Figure 228. – The effect of different – V – shape geometries on the break line

17.1.2 - BREAKING METHODS

There are a few methods for breaking the scribed line. In general, it is a three-point bending process with some modifications per the scribe machine vendor. Fig. 229 shows one major breaking process.

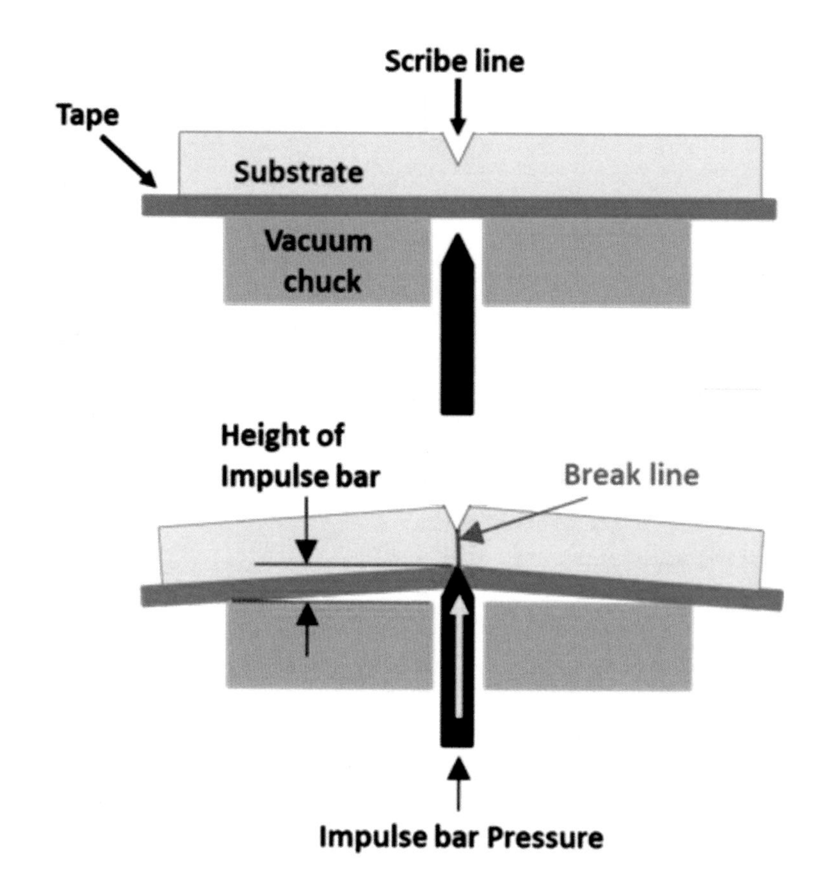

Figure 229. – Scribe breaking process

Diamond scribing is a process requiring skilled knowledge however once it is optimized it performs well in a production mode. Following are a few important parameters required to perform well in a production process: Scribe tool angle, Scribe force, Impulse tool height and pressure.

17.2 - LASER SCRIBE

During the 70s, the laser scribe process was a potential threat to the diamond dicing process mainly on dicing hard alumina microelectronic substrates. At that time this did not happen mainly because of the quality performed by the laser technology at that time. Below are SEM photos showing the different quality between laser scribe of hard alumina substrate compared to the same material diced using a diamond dicing blade process (see fig. 230).

Hard Alumina diced using a Diamond resin blade

Hard Alumina scribed using a Laser bean

Figure 230. – Old laser process compare to a diamond resinoid dicing process

The threat did not happen even though the laser process was performing with a much higher feed rate. Eventually, the old laser scribe process was used and still is used for rough trimming of blank ceramic substrates in many different geometries where edge quality is not an issue. The internet is loaded with companies offering hard alumina substrates in many different geometries. In today's world, the use of the laser scribe process is very different with some specific difficult applications where the standard diamond dicing process performs with real difficulties and sometimes it is even impossible to perform. The main reason to develop new laser technologies for separating wafers into small dies while maintaining good quality is the new products, mainly thin silicon substrates, low K silicon wafers and probably the no. one application, MEMS substrates. A few laser technologies are used in the semiconductor industry. Let's discuss a few major technologies performing in the semiconductor industry:

17.2.1 - LASER ABLATION

Laser ablation creates a groove / Kerf while melting/vaporizing the wafer top surface layers away using a laser beam at a wavelength that causes the material to absorb it. This process is causing thermal damage at the edge of the machined material and in addition debris on the wafer surface. To minimize any of these effects, a protective film layer must first be formed prior to the laser process to prevent contamination and another extra washing process after the laser process. There are two laser processes developed over the years, the long pulse laser beam, which is an aggressive process with quality issues and the short pulse laser beam, which is the one used today with much better-quality performance. Below is a generic sketch showing the differences (see fig. 231).

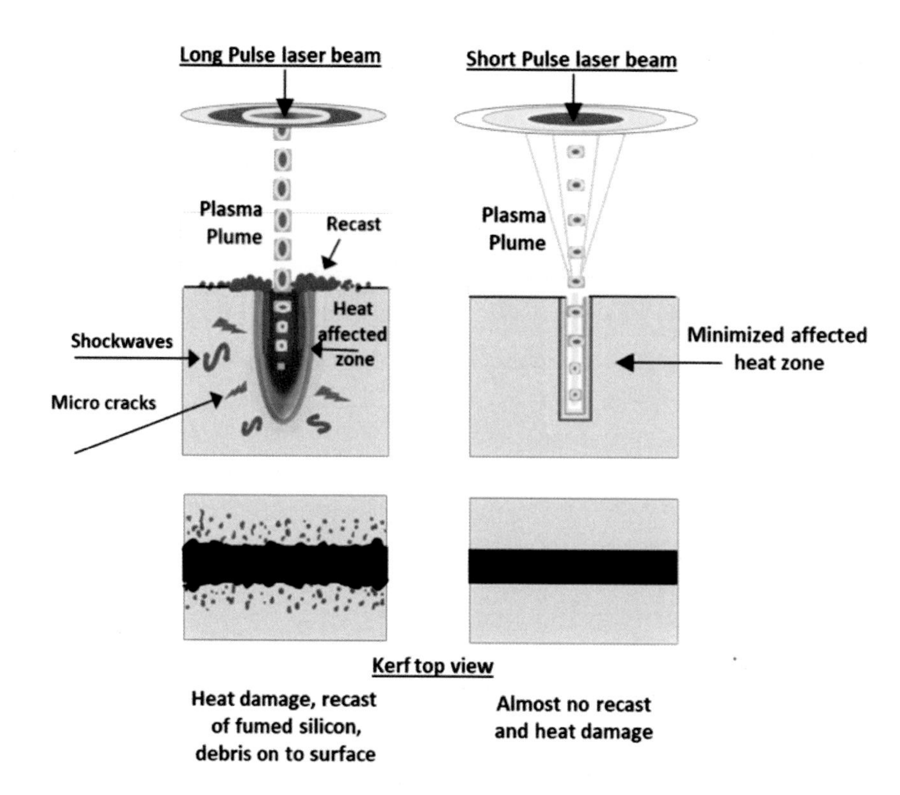

Figure 231. Quality Differences between Long Pulse Laser and the short Pulse laser bean

By adjusting the laser depth, three types of processes can be performed: - Grooving, Full cut, and Scribing.

17.2.2 - GROOVING

This process removes thin layers of hard metallization in the wafer streets ready to be completed by a conventional diamond blade dicing process. This process improves cut quality by minimizing chipping and layers peeling on the wafer. This process is used for low-K film wafers, coating of aluminum nitride (AlN) in wafers streets and many others.

17.2.3 - FULL CUT

This process is mainly used for thin silicon and GaAs wafers less than 0.200mm thick. A good cut quality can be achieved, something that is more difficult using the standard diamond dicing process.

17.2.4 - SCRIBING

The scribing process creates a thin groove in the wafer streets. The wafer is then separated into individual dies by applying external stress to break all the scribe lines. One major application using this process is sapphire. There are a few laser dicing vendors available in the market. Those systems are very robust and accurate. Fig. 232 shows two systems available for the ablation laser process.

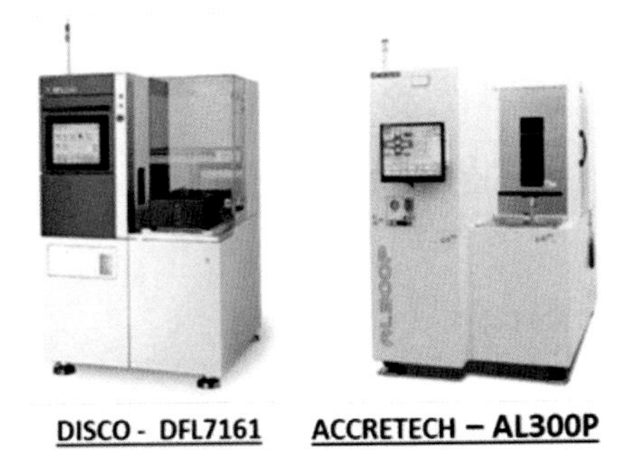

Figure 232. – Popular Ablation laser systems in the market

To eliminate any surface damage to the wafers, the wafers are coated with a water-soluble protective film using a spinner prior to the laser process. This eliminates any debris particles to adhere to the wafer surface. After the laser process, a cleaning process is performed to remove the protective water-soluble film with the debris from the wafer surface. Below is a sketch showing the differences between using the water-soluble film prior to the laser process and without it (see fig. 233).

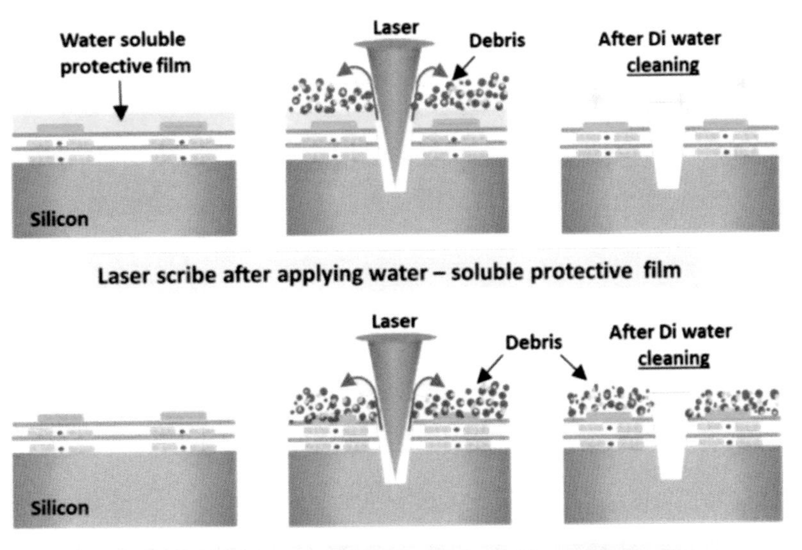

Figure 233. – Quality difference between using soluble film and without it

One popular application for laser ablation is dicing Low-K wafers, which will be discussed in the silicon application section.

17.3 - STEALTH LASER DICING

Stealth Laser Dicing is a unique amazing technology enabling singulating new difficult, very thin, and delicate semiconductor substrates. Stealth dicing is a totally dry process which is a major advantage, with no need for wafer top side coating and no need of DI cleaning. A laser beam is focused to transmit a wavelength to the center of the wafer. This process is performing internal stress layers inside the wafer called "SD layers" (Stealth Laser) performing starting points for later die separation due to internal tensile stresses (see fig. 234).

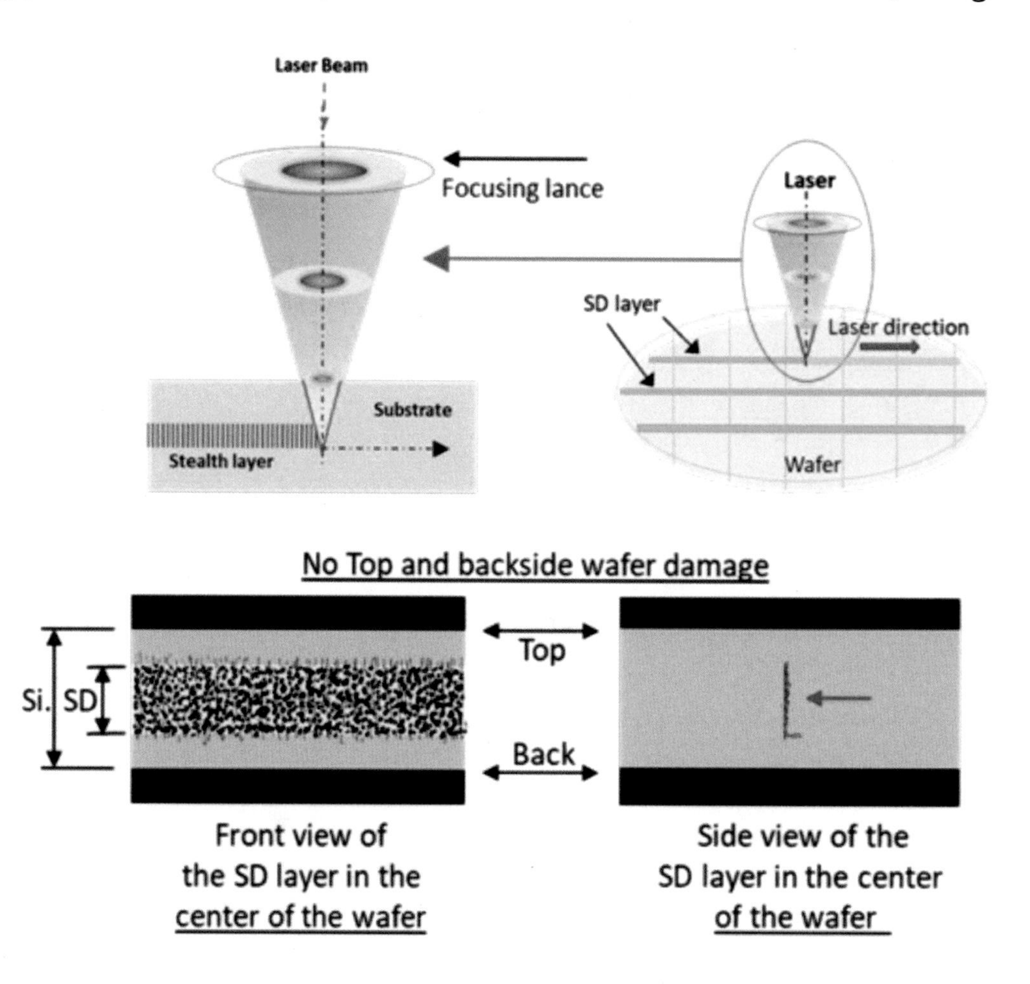

Figure 234. – Stealth Laser Dicing performing internal stresses

The condensed wavelength in the center of the wafer forms a mechanical damage layer (stress layer) in a localized point near the laser focus area. This center stress area is later mechanically separated by expanding the wafer. Fig. 235 shows the laser indexing on the wafer and the internal SD which is the stress layer.

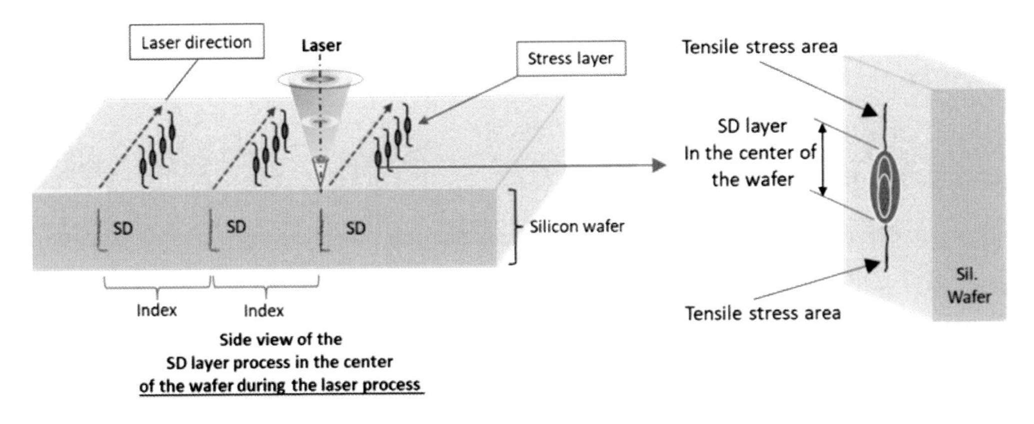

Figure 235. – Stealth Laser indexing and cross section of the internal stress

Separating the individual dies after the SD laser process is done on the tape-mounted wafer while expanding the tape. This is working on the internal stress/SD area in the center of the wafer by causing a continuous micro breakage that is separating the individual dies (see fig. 236).

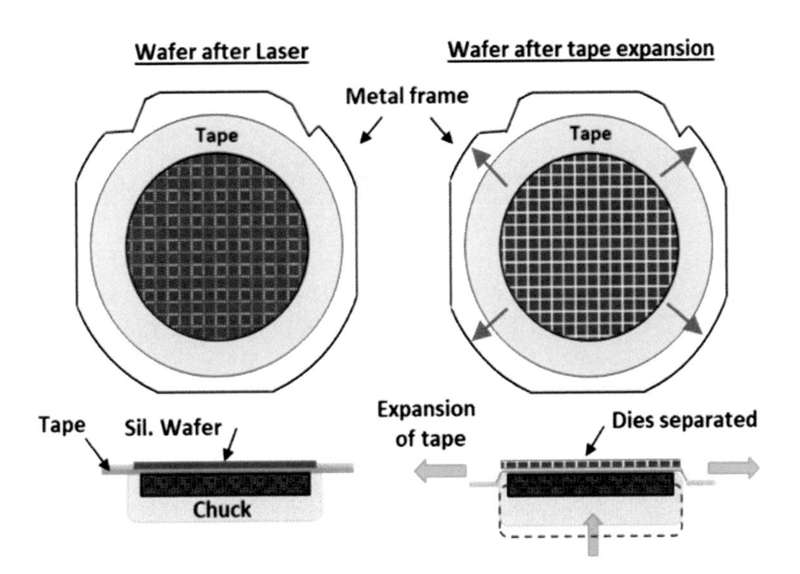

Figure 236. – Die separation after the SD process and the wafer expansion process

The SD process has a few major advantages:

- Very thin wafers can be processed., 0.050 mm and less. Will be discussed in the silicon application.
- No need for special spin protective coating prior to the laser process.
- A dry process so no surface contaminants like in traditional diamond dicing requiring aggressive washing, which is a great solution for delicate products like MEMS and very thin substrates.

- No heat to the wafer surface during the laser process.
- High yields/no die loss because there is no cutting material loss like in conventional diamond dicing. During the design stage, it is possible to increase the number of dies on the wafer.
- Special irregular shape geometries can be processed, something not possible with standard diamond blade dicing (see fig. 237).

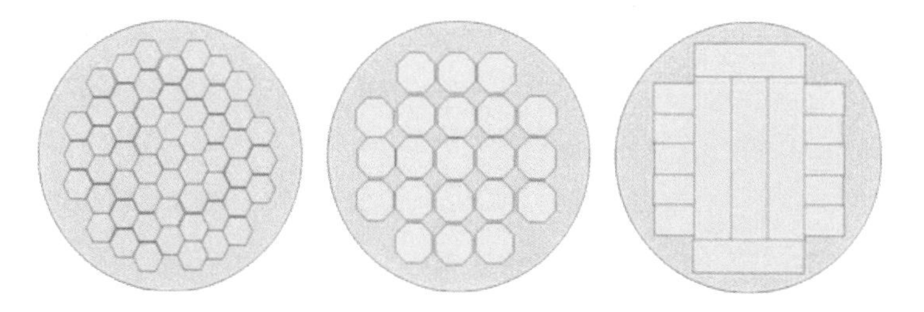

Figure 237. – Special irregular geometries made by the SD process

The following materials can be processed using the SD laser technology:
Si, Glass, Sapphire, SIC, GaAs, $LiTaO_3$ / $LiNbO_3$, Crystals.

Some specific SD applications will be discussed in the application section. Below are market-leading SD systems supporting Φ300 mm wafers (see fig. 238).

Figure 238. – Main Industry SD dicing leading systems

Below are a few samples photos laser diced using the SD laser system and a compared poor diamond dicing of a thin silicon wafer (see fig. 239, 240, and 241):

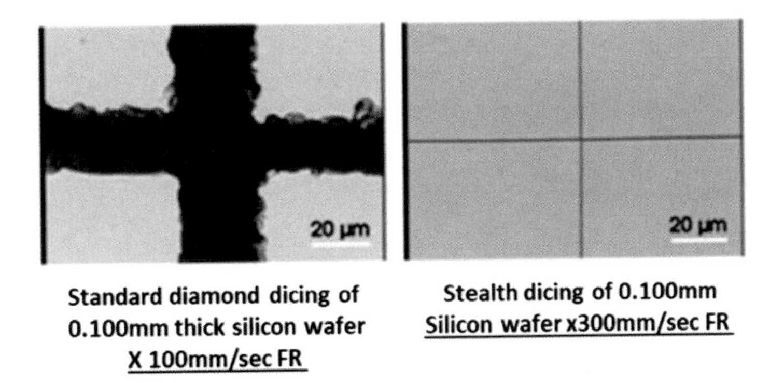

Figure 239. – Stealth dicing quality compare to diamond dicing – Reference by DISCO

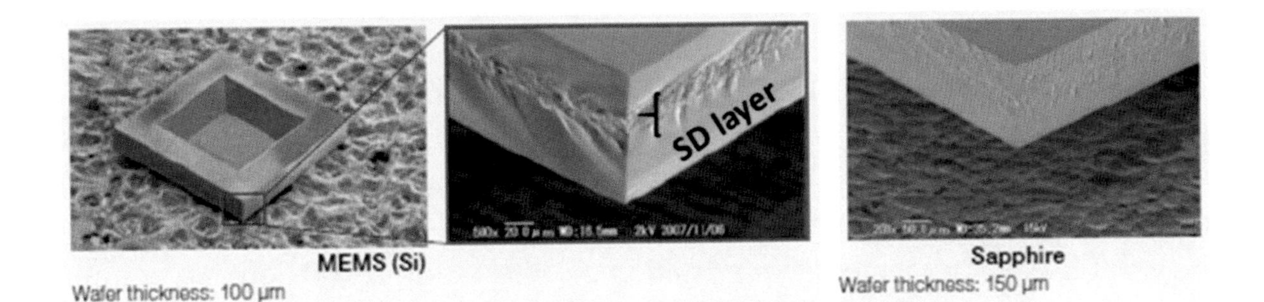

Figure 240 – MEMS and Sapphire wafers processed by the SD process – Reference of DISCO

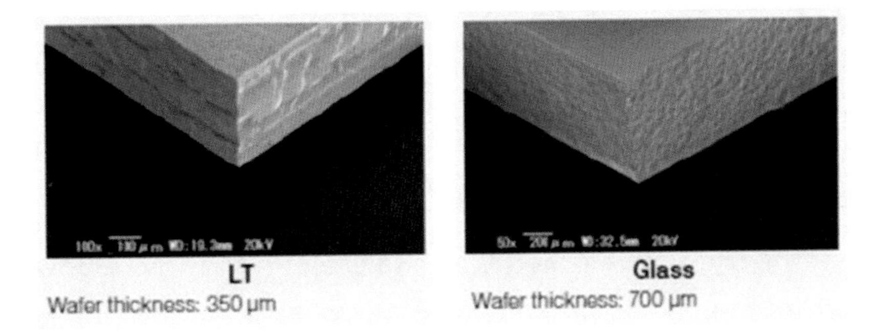

Figure 241 – Lithium T. and Glass wafers processed by the SD process – Reference of DISCO

Improvements to the SD process were made over the years by correcting the spherical Aberration which is improving the laser focus properties (see fig. 242).

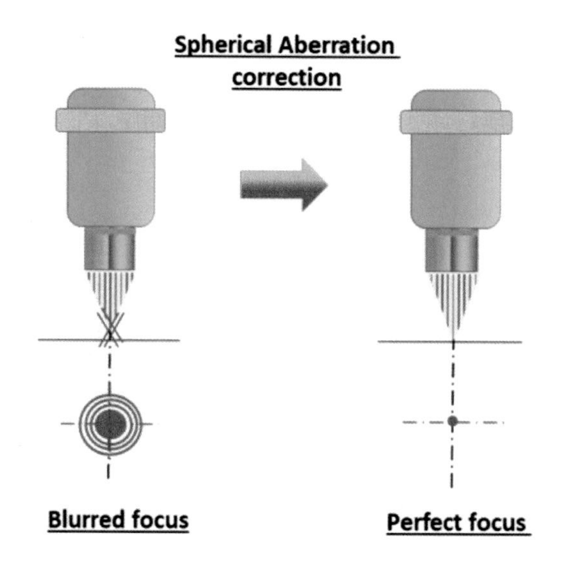

Figure 242. – Spherical Aberration correction perfecting the laser beam focus

The ability to correct the spherical aberration is called LBA (Laser beam adjuster) which is a process improvement needed for narrower streets requiring better quality and better productivity. More specific information on Aberration correction can be found on many internet sites.

Reference: The above Stealth dicing information is based on good available information from DISCO Corporation articles and literatures plus their HAMAMATSU technology partners. See also in the reference section.

17.4 – LASER MICROJET (LMJ)

One other Laser dicing system/process that was introduced to the market is the Water Jet Guided Laser systems by Synova called Laser MicroJet (LMJ). The principle of the system is a hybrid method of machining, which combines a laser with a "hair-thin" water jet that precisely guides the laser beam by means of total internal reflection in a method similar to conventional optical fibers. The water jet continually cools the cutting zone to improve cutting results and removes debris. The LMJ machine system allows precise ablation Laser process of various semiconductor materials such as silicon (Si), gallium arsenide (GaAs), silicon carbide (SiC), low-K materials and others. The system can Laser dice irregular die shape geometries. Below is a sketch explaining the Water Jet Guided Laser and one of the Synova laser systems (see fig. 242).

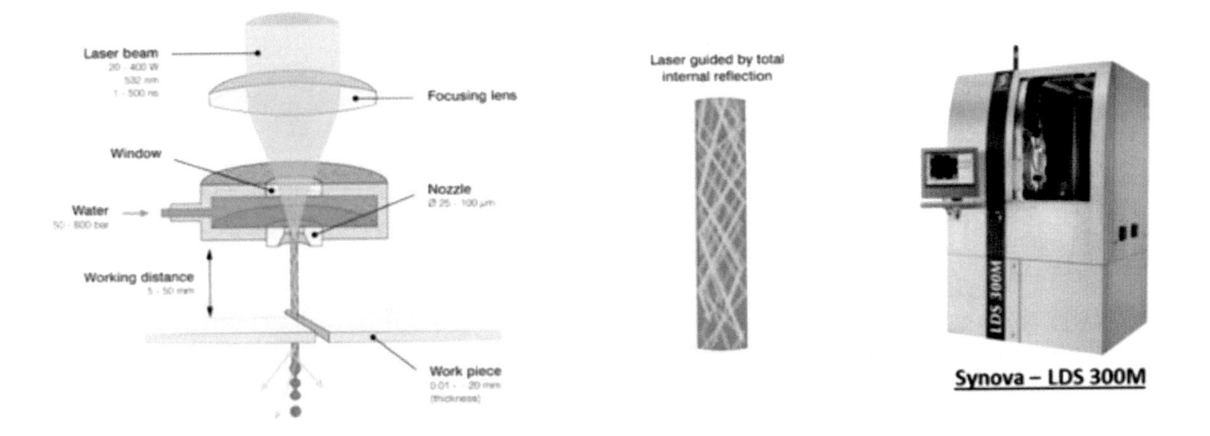

Figure 242. – Laser Microjet (LMJ) system – Reference of Synova

Below are a few sample photos using the Laser Microjet (LMJ) on Low-K and Sapphire substrates. (see fig. 243).

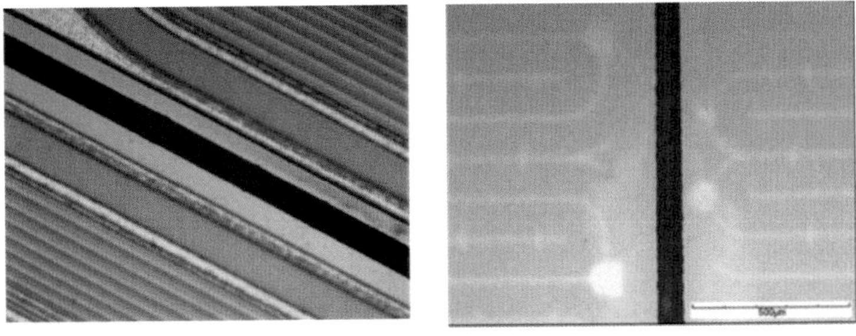

Figure 243. – Thin Low – K wafer singulation & grooving Sapphire wafers – Ref. of Synova

More information – see in the reference section.

17.5 - PLASMA ETCHING / DICING

Plasma etching is probably the newest technology process used in the semiconductor industry for singulating wafers. It is a competitor of the newest laser technologies for singulation of extra thin silicon wafers and other materials however, it was mainly aiming for the fabrication of MEMS substrates. The technology names the industry is referring to the Bosch technology process and DRIE—Deep Reactive Ion Etching. Following is a short technical review of the process principle:

Prior to the plasma etching process, the areas not to be etched on the wafer are covered by a protective mask normally photoresist. The plasma etching process generates a volatile etching effect at room temperature from the chemical reactions between the elements of the material etched and the reactive species generated by the plasma. The process is a repeating cycle of distinct steps to create anisotropic silicon etching, it consists of a three-step cycle: Film (Passivation) deposition, bottom film etching, and silicon etching.

The gasses used for the plasma process are C4F8 for deposition/creating a passivation film during the etching step to prevent sidewall etching. An SF6 + Bias is etching off the passivation film and the SF6 is etching the silicon. In the deposition process, a passivation film is deposited on the sidewalls and bottom surface of the trench/groove. In the bottom film etching step, the passivation film on the trench bottom is selectively etched. In the silicon etching step, only the silicon at the trench bottom where the passivation film has been removed, is etched. This process is repeated many times till the final depth is achieved. Fig. 244 shows the Plasma steps processes.

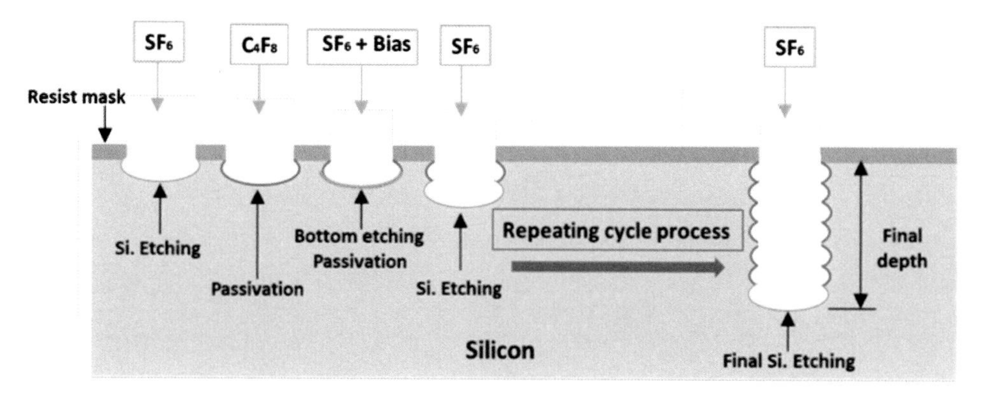

Figure 244. – Plasma etching process of silicon wafer

Per the above-described process, the plasma etching process for dicing the wafers is performed after masking the wafers with photo resist to cover the dies components leaving exposed only the street area to be plasma Etched/diced. In conventional diamond dicing and in the different laser processes each street is processed individually which is time consuming. In the plasma etching process, the whole wafer is processed at once. (see fig. 245).

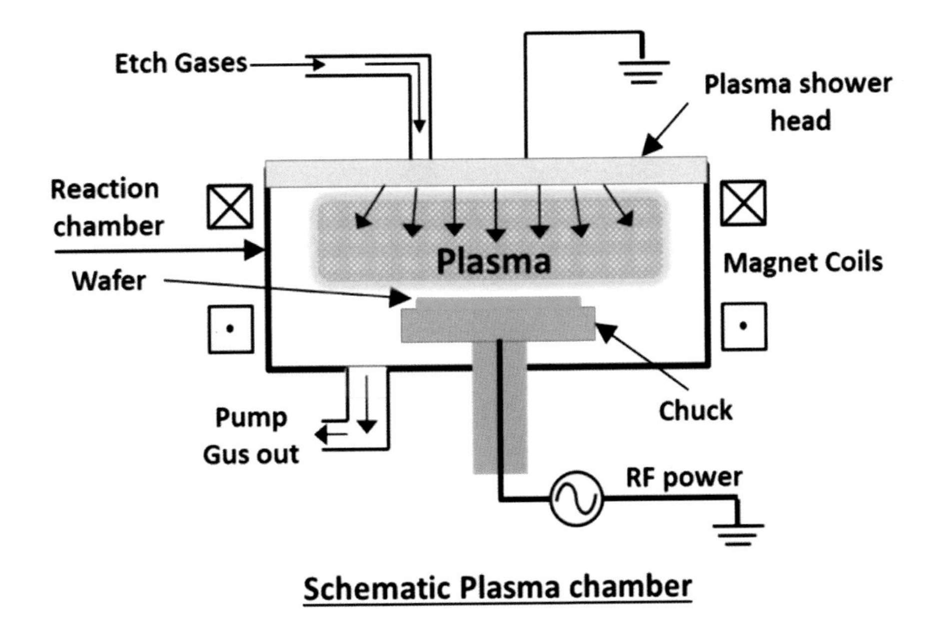

Figure 245. – Plasma Etching process

The output of plasma etching is a smooth rounded surface between the plasma step processes. This also creates a round edge on each die corner which significantly is improving the die strength (see fig. 246).

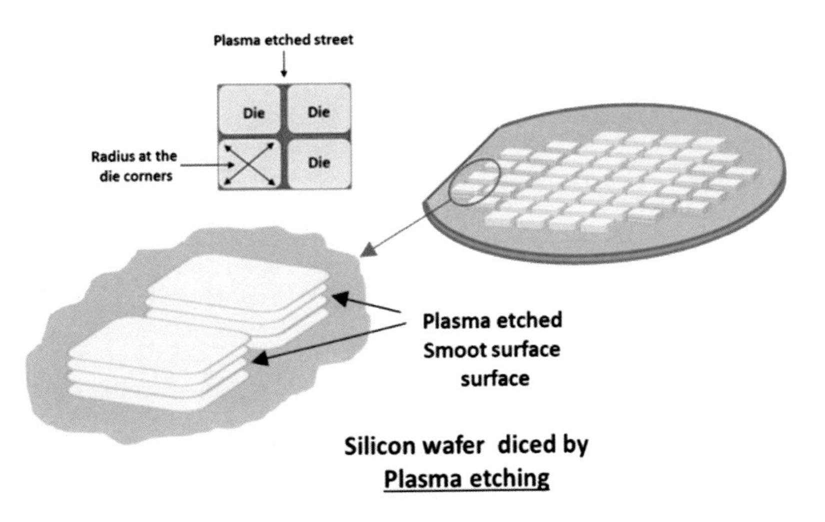

Figure 246. – Quality advantages by the Plasma etched rounded edges and corners

The plasma etching process is mainly used for MEMS fabrication and for etching/dicing thin silicon and other semiconductor wafers. There are a few options for singulation thin wafers: In most cases, a partial cut is made and then using a backside grinding process to grind the back side over the previous plasma dice kerf to separate the dies. This grinding process is called DBR (Dicing before Grinding). The other option is to start with the back-grinding process – DAG – (Dicing after Grinding) and then do the front plasma process to separate the dies. The above processes will be discussed in the Silicon application section. In general, the width of the plasma kerf is very narrow in the range of 10um and less by a depth of >0.100 mm. Following are a few advantages and disadvantages for the plasma dicing/etching process:

Advantages:

- Die density can be dramatically increased because of the much narrower plasma kerf size.
- Stronger die due to smoother edges and rounded die corners.
- All dicing streets are etched simultaneously resulting in increasingly higher throughputs.
- Compatible with solder bumps and backside metals.
- Mosaic™ and other special geometries can be processed. (Defined by the masking process) (see fig. 247).
- Dicing time of very small die sizes is not decreased like in diamond dicing. More streets to dice and lower cutting speeds are not relevant.

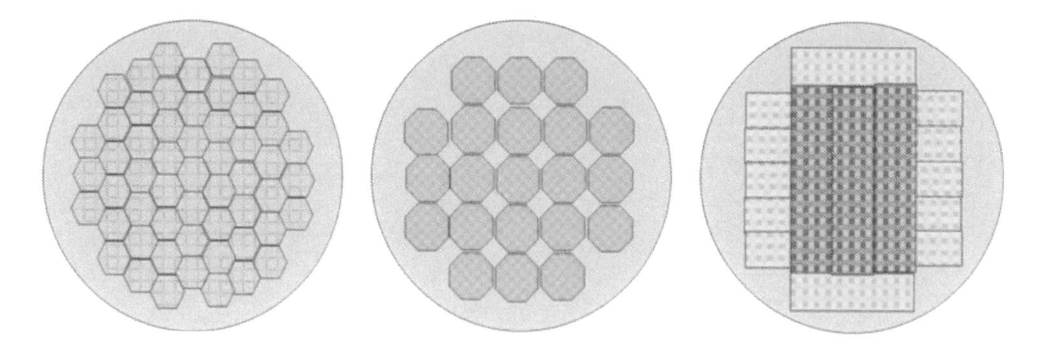

Figure 247. – Special irregular geometries made by the Plasma etching process

Disadvantages:

The main disadvantage is the need for a resist coating system, an exposure and developer system. They all are needed to perform the masking to protect the wafer active areas to leave the plasma clean area for the etching. This process is also time consuming. Much more good information on plasma etching can be found on the internet. Some guides are in the reference section.

17.6 - WATER JET PROCESS

The water jet process is used in many industrial applications and was suggested to be considered in the microelectronic industry mainly for singulation QFN & PCB type applications; however, it is not a live relevant process today. It is a very interesting process so I decided to cover it a bit so technical individuals in the microelectronic industry can add this to their general knowledge or future considerations. Abrasive Waterjet (AWJ) Is a very robust process that can cut through many materials, even heavy gauge metals. The system is made of a few components requiring a large footprint with some "messy" abrasive powders that are difficult to control in a mass production environment where extremely quality specs are required. Below is a generic sketch of the different system elements. (see fig. 248).

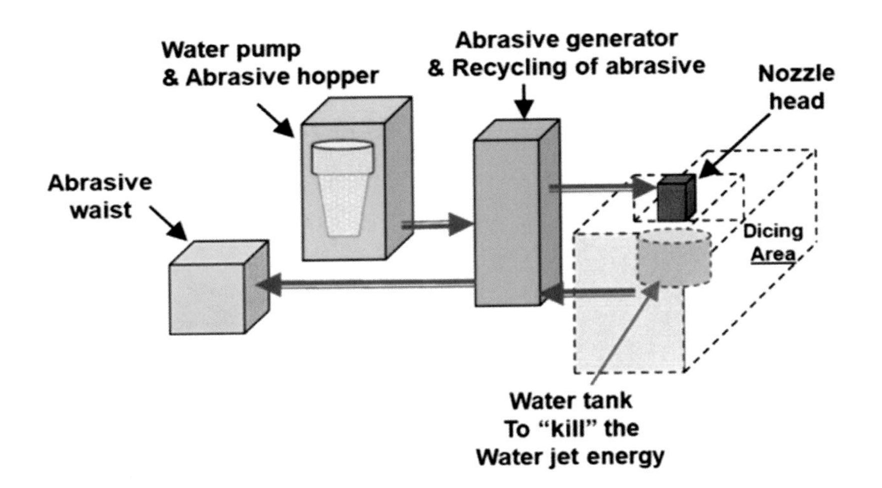

Figure 248. - Water jet process

One major problem with the Abrasive Waterjet is eliminating scratches from the abrasive particles that are spread at a very high pressure. The other problem is "killing" the water jet energy which is done using a water tank below the water jet stream. To minimize the depth of the water tank ceramic balls or similar are added to the bottom of the water tank (see fig. 249).

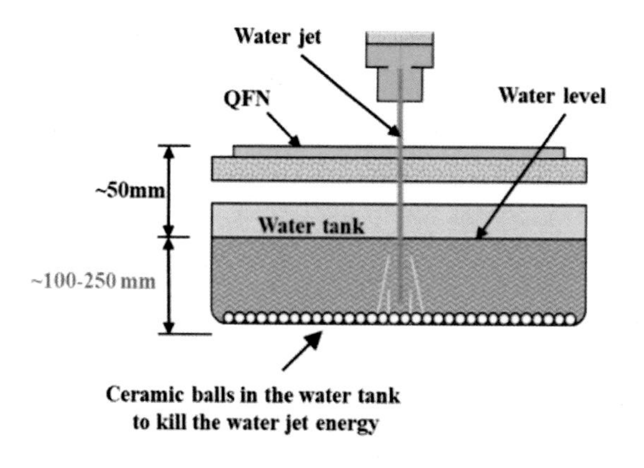

Figure 249. – Water tank with ceramic balls to kill the water jet energy

The abrasive type and size used can vary depending on the application. It can be Garnet, Sil. Carbide, and others in different mesh sizes. Other parameters controlling quality are kerf size, feed rates, waterjet pressure and the nozzle diameter. There are different methods adding the abrasive to the waterjet stream. Below is a sketch showing one design (see fig. 250).

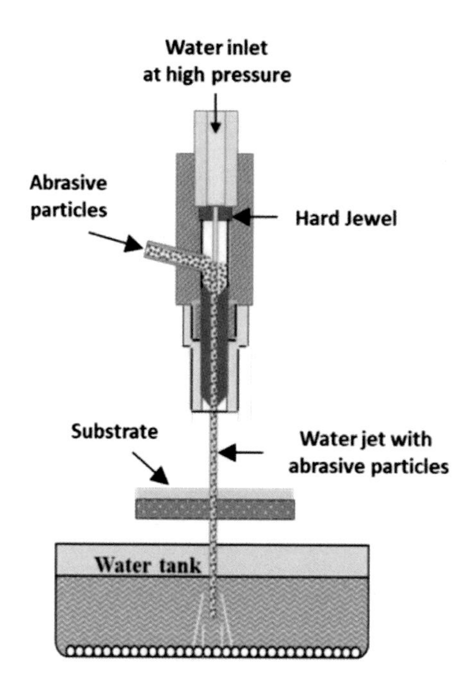

Figure 250. – Cross section of the Water Jet process

Bottom line, there were a few attempts to use the Waterjet technology for singulation microelectronic substrates, but the other technologies took over.

CHAPTER 18

DICING/SINGULATION OF MICROELECTRONIC APPLICATIONS

In the microelectronic industry, we face many applications starting from very hard and brittle substrates to soft and ductile ones. In some cases, a combination of both can be found in the same substrate. To understand the large variety of applications, only major applications will be reviewed that do reflect well the marketplace. This technical book was aimed mainly at singulation using diamond dicing blades however, new technologies in the semiconductor industry were developed requiring singulation of very difficult and demanding applications which led to new singulation technologies. Those applications will be covered as well. Following is an application display with a variety of applications, it is just to show and impress everyone on the real big mix of applications (see fig. 251).

Figure 251. – Application display covering many microelectronic applications

To be more specific regarding the different substrate's hardness. Fig. 252 showing the big hardness spread among some of the most popular microelectronic applications.

Material	Mohs hardness
Diamond	10
Sapphire	9
SiC (Silicon Carbide)	9
Alumina	9
Titanium Carbide TIC	9
High speed steel	8
Quartz	7
Silicon	6-7
PZT (Piezo Electric)	5.5-6
Glass	5.5
GaAs	5-5.5
Copper	3

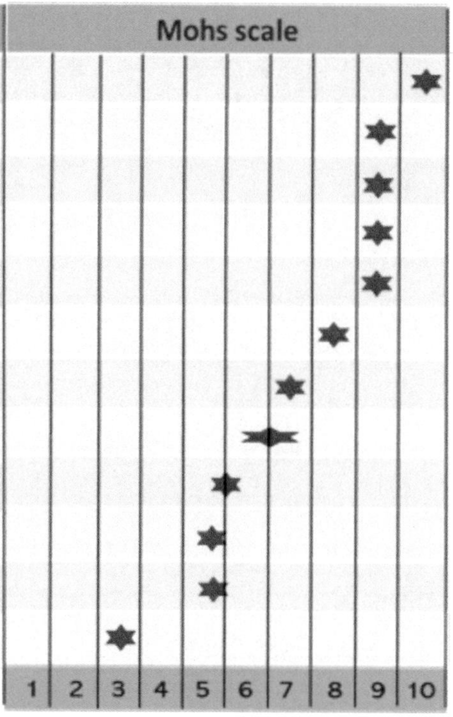

Figure 252 – Major microelectronic application materials with their hardness

Any application involves many different parameters on everything that was discussed already. It includes saw parameters, blade parameters and how both are running together and are optimized per each application spec. Fig. 253 shows in a generic mode the different parameters involved in each application.

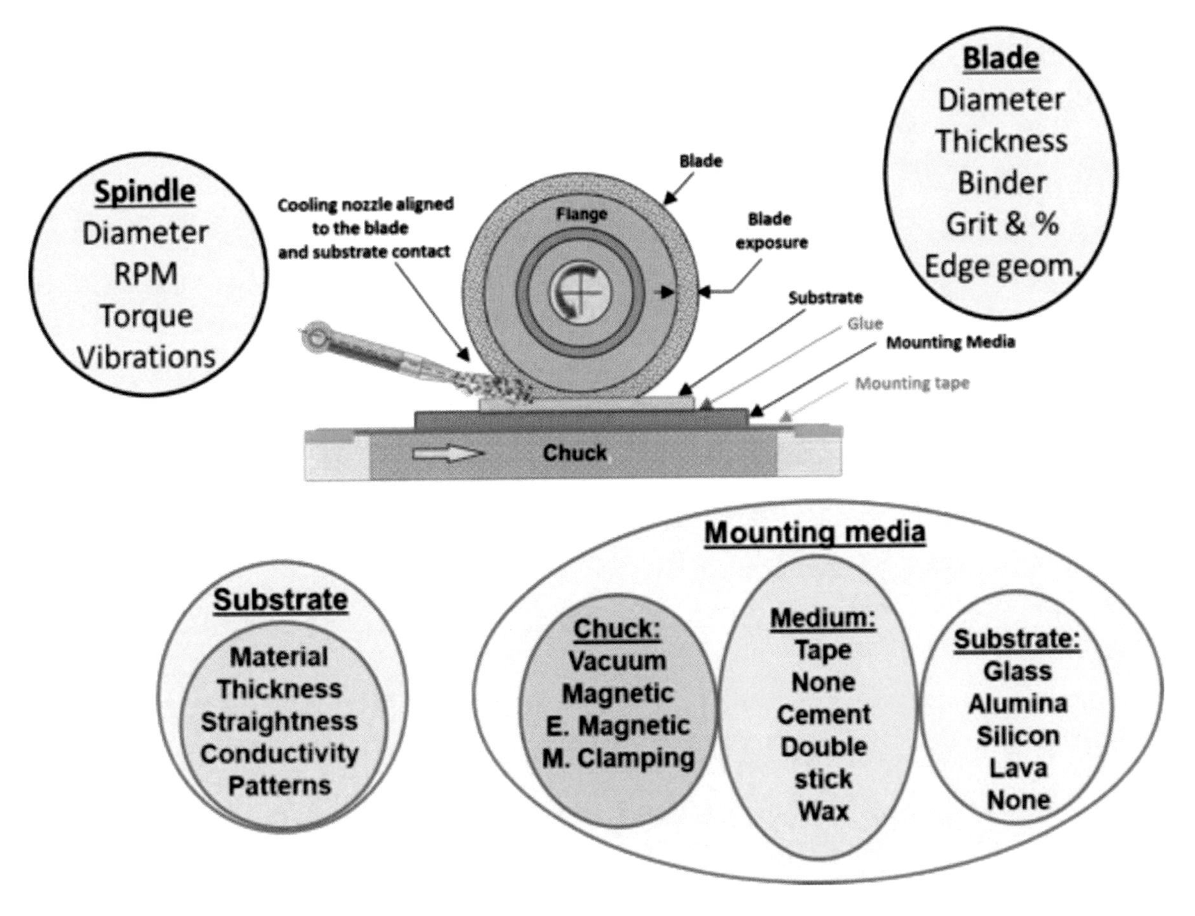

Figure 253. – Different parameters involved in any application

18.1 - GENERAL

Most parameters involved in most popular applications were discussed already at the beginning of this review, therefore no need to repeat them, however in some important or specific application areas a second short review is added.

When talking about cutting parameters, the numbers are indicated in a generic general mode and not with specific solid no.'s as every customer is optimizing their process in a different way and with different results. However, the no.'s are good starting points for new customers dealing with new applications.

One general rule in both grinding and dicing is to use a soft blade bond for hard materials and a hard blade bond for soft materials. However, in the dicing world there are exceptions, especially when using a combination of both hard and brittle material with soft and ductile materials in the same substrate.

18.2 - SILICON, SEMICONDUCTOR WAFERS

Silicon wafer
Source by Laura Ockel

Figure 254. - Semiconductor / Silicon wafer

Historically Silicon wafers is probably the no. 1 application that was diced on a bench type dicing saw. The first small silicon wafer diameters 2"–4" (50–100 mm) x .020"–.030" (0.500–0.760 mm) thick were diced only 60%–80% deep at the beginning and later cut through when tape mounting was introduced. The partial cut was at relatively high feed rate up to 6"/sec, and the cut through up to 3"/sec. The spindle speed was around 30 Krpm with minor rpm changes, higher and lower to better control vibrations and cut quality. In order to maintain cut placement accuracy, the spindle had to go through a warming up process before penetrating production wafers. This process was to maintain even and stable spindle expansion and to metallurgically maintain an accurate blade location on the - Y - axis. This process is done in many manufacturing houses till today. The blades at the very beginning were annular nickel blades 50–58 mm O.D. x 40 mm I.D. x .001"–.0014" (0.025–0.035 mm) blade thickness. The different O.D. sizes were to accommodate different blade exposures. The diamond grit was 4–6 mic. Annular blades needed a relatively long dressing procedure to perfect the blade edge runout and to expose the diamonds well. This process was done in two steps: Step 1 was dressing the blade on a Silicon Carbide Bakelite 600 mesh dressing board. Step 2 was done on a blank silicon wafer or sometimes on a production wafer starting at low feed rates. It was a tedious process requiring expensive production time. The dressing process was normally recommended by the blade vendors however every customer developed and optimized their own process. Fig. 255 shows one process that was used in the marketplace.

Dressing media – Silicon Carbide 600 mesh in a Bakelite bond
Dressing media geometry – 75 x 75 x 1mm

Dressing parameters:-	Step # 1 On Sil. Car. board	Step # 2 On Sil. Car. board	Step # 3 On Sil. Car. board	Step # 4 On blank Sil. wafer
Spindle speed –	30-35Krpm	30-35Krpm	30-35Krpm	30-40Krpm
Feed rate – inch/sec	5	.2	.5	5 cuts at .2" /sec steps up to production speed depending On cut quality
Cut depth - Inch	.002	Production depth + .002		
Cut length – Inch/mm	30 / 750	30/750	30/750	
Index –	Blade thickness x 2			

Dressing procedure of thin annular nickel blades

Figure 255. – One of the dressing processes used for annular blades

The main advantages of using annular blades were to use different O.D. sizes for different exposures and the option to downsize the blade O.D. after the blade wears down and then to use a smaller flange O.D. to achieve back the right blade exposure. In some mass production houses a rework process was developed to correct damaged O.D. blades. The blade exposure was trimmed off mechanically and a fine dressing of the O.D. was performed using Silicon Carbide dress boards to get a smooth new blade O.D. edge with well exposed diamond. This process required some skills. However, it was done in some high-volume production houses (see fig. 256).

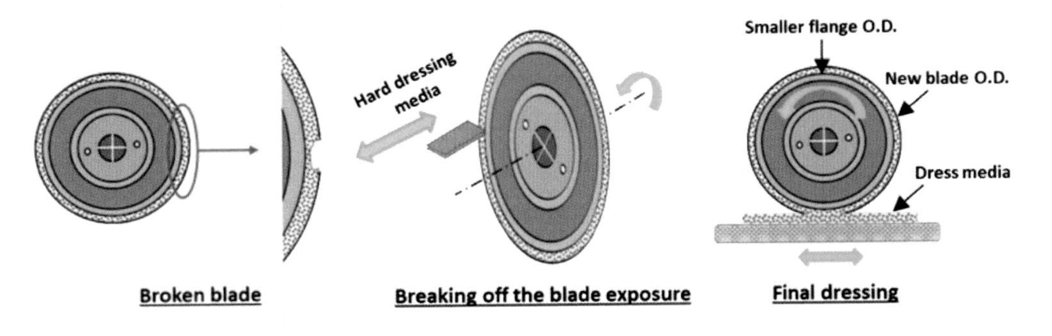

Figure 256. – Reclaiming and dressing broken nickel annular blades

The disadvantages of using annular blades were mainly the extra delicate handling process. It starts with handling the thin blade from the package, cleaning the mounting flange, mounting the blade on the flange, and mounting the flange on the spindle. Handling the flange itself was a process that required attention as small nicks and scratches on the flange edge at the blade mounting surfaces could easily deflect the blade exposure. As was discussed already, a few flange designs were used, some with the back flange as part of the wheel mount, some as a set assembly using a clamping nut and some using the snap design. All the different flange designs required special cleaning and routine QC attention. The other disadvantage was blade stiffness especially when using a large exposure. Using thin blades of .001" (0.025 mm) thick and sometimes even thinner, creates stiffness issues to perform a straight 90-degree cut. A slanted cut can also cause other quality issues like chipping on one side of the kerf, wavy cuts, and others. Quality issues require slowing down the feed rate which is a production capacity problem. Below is a recommended thin annular nickel blade exposure sketch that was established over the years (see fig. 257).

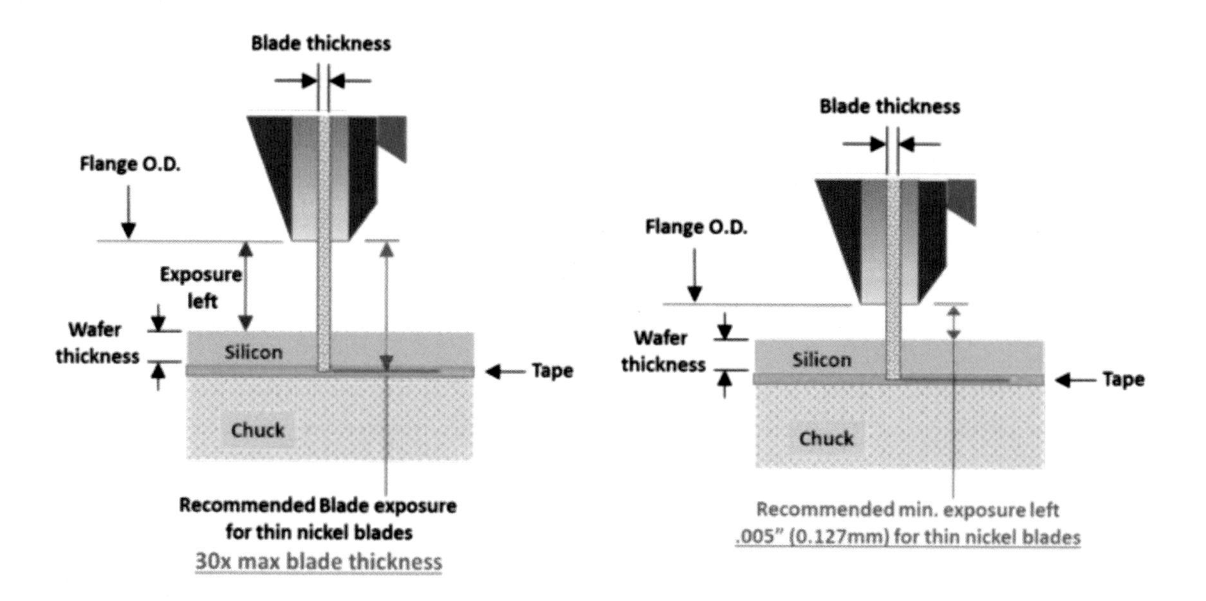

Figure 257. – Recommended blade exposure of thin annular nickel blades

Using annular blades for dicing silicon wafers was possible and was done for quite some time however it required special attention. This was leading to the development of the hub blades. A generic manufacturing process flow of the hub blades was described already in the hub blade manufacturing section. There are a few main advantages to using a hub blade for dicing silicon wafers:

Easier handling of the hub from the package to the wheel mount on the saw. Because the blade is plated on an accurate aluminum hub which is directly facing the wheel mount, the stiffness and axial runout of the blade exposure is dramatically improved. The weight of a hub blade mounted on the spindle wheel mount is much lighter compared to a flange set, vibrations are minimized = better cut quality.

Dressing hub blades were discussed already. However, at this point of the review, it is important to cover it again: Hub blades are pre-dressed in the manufacturing process so no need for the Bakelite dress board dressing process. The dressing process is simplified and is way more friendly on the saw called Override or Pre-cut. The process is automatically starting at low feed rates and gradually going up reaching production speed. Another option is also starting at low feed rates and at the same time at a shallow cut depth, gradually reaching both production feed rate and cut depth. Both options, using annular blades and hub blades can be optimized to meet cut quality specs and shorten cycle time.

Today's hub blades went through a lot of optimizations during the last 2–3 decades. There is a large selection of diamond grit sizes with different diamond % and different bond hardness. Also, thinner blades down to .0006" (0.015 mm) can be made, however for very thin wafers and narrow streets there are new technologies which were discussed already and will be covered in detail. Let's first cover today's standard silicon wafer dicing process using conventional silicon wafers and dicing processes. Most major semiconductor mass production houses are using fully automated dicing saws using either single spindles or dual spindles with automated alignments, loading and unloading tape mounted wafers, automated dicing process and D.I. cleaning and drying. Needless to say that they are all using Hub type blades.

A few important points to know regarding dicing silicon wafers using thin nickel blades:

- It is important to check and correct if needed the wheel mount axial runout and to confirm the wheel mount is parallel to the – X – movement (see figs. 101 and 102).
- On some saws a special software to fine dress the wheel mount is used to correct the axial run out to 0.001 – 0.002 mm (see fig. 258).

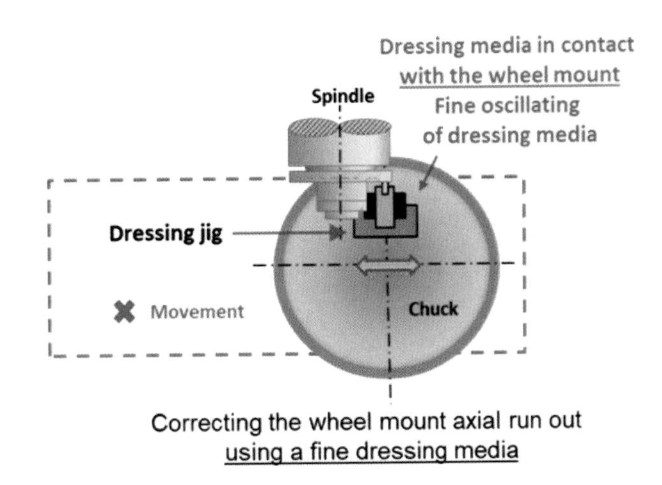

Figure 258. – Correcting by "dressing" the wheel mount runout

A wheel mount out of spec will affect the kerf width and cut quality.

On any dicing process using thin nickel blades, dicing through into tape requires special attention to the cut depth. A standard PVC tape is used for standard silicon wafers with a thickness of ~ 0.075mm (.003"). The adhesive thickness is about 0.005mm (.0002"). It is important that the blade edge will penetrate below the adhesive layer. If the Blade edge is inside the adhesive layer the blade will get overloaded resulting in wavy cuts, slanted cuts, chipping, and blade breakage (see. fig. 259).

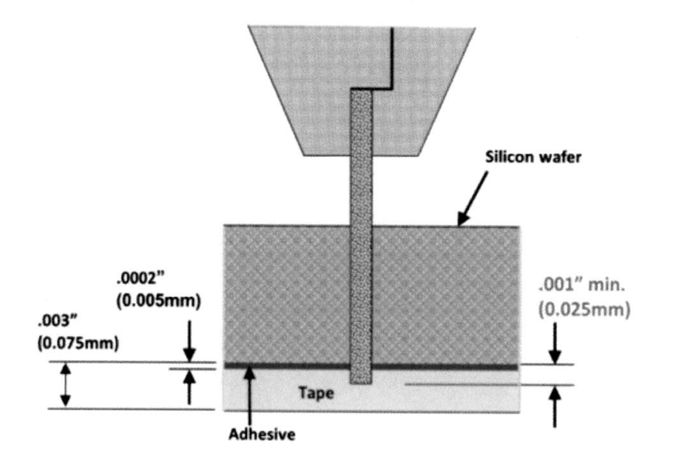

Figure 259. – Recommended Cut depth of thin nickel blades into the mounting tape

Due to continued blade wear the blade diameter is getting smaller and smaller till the blade edge is facing the adhesive layer. To overcome this problem a routine height calibration needs to be performed to maintain the right cut depth into the tape. In the new saws the software is monitoring the cut depth and makes automatic corrections if needed.

Another problem related to blade exposure is the exposure left. With thin nickel type blades running in a mass production environment, blades do wear continuously till the wafer top surface gets close to the aluminum hub. A too small exposure left may break off as those thin blades are hard and brittle and slight vibrations can cause blade breakage. It is recommended to keep min. exposure left of about 0.127 mm (.005"). This no. can vary a bit depending on the blade thickness (see fig. 260).

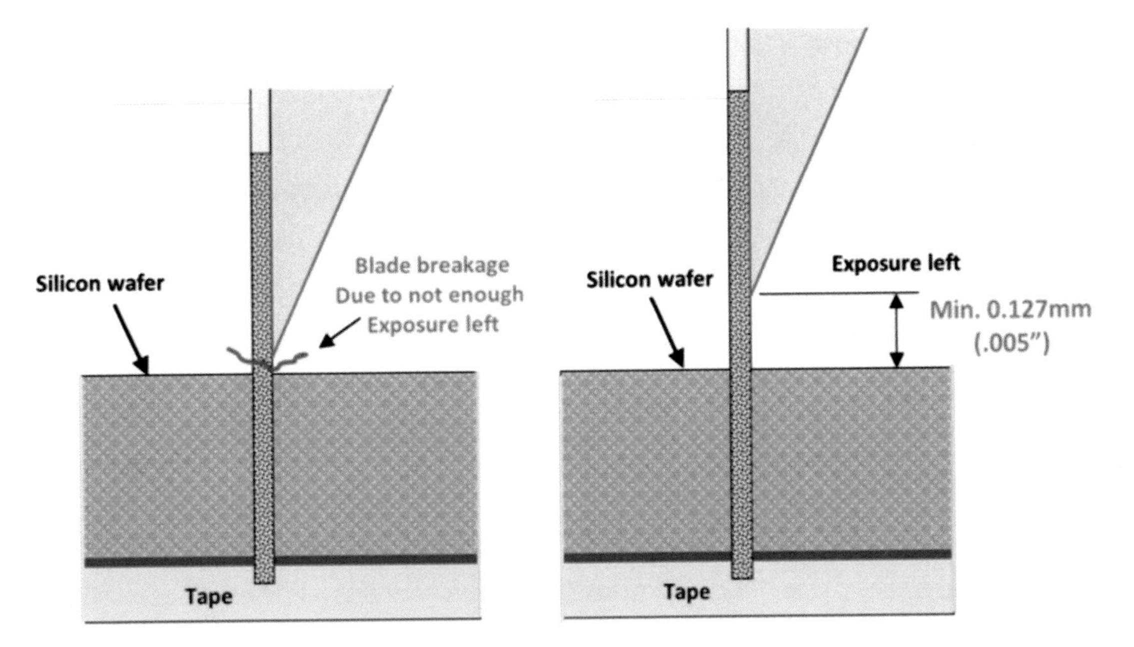

Figure 260. – Recommended blade exposure of thin nickel blades

Another parameter to remember is continued blade performance during the dicing. Customers would like the blade to perform the same quality and feed rate from start to the end of life, however this is not always the case. Depending on the application and the dicing parameters, blades are getting into overload at a certain point. This requires slowing down the dicing process mainly the feed rate and gradually come back to production speed. Some customers are slowing down the feed rate on production wafers and some on a blank wafer to meet back the quality spec. This process is modified and optimized differently between customers. If quality does not meet back the spec, the blade is replaced.

Coolant is a very important parameter in dicing silicon wafers mainly because of two main parameters, thin blades, and spindle speed. Today's air bearing spindles in most cases are running at a higher speed of 40–60 KRPM which makes the "Air knife" effect much more critical especially when using thin nickel diamond blades. As covered already, the coolant nozzle geometry, location and coolant pressure need to be optimized per application (see figs. 107–116).

As discussed already in the coolant section, the coolant in dicing silicon & GaAs wafers is D.I. water with a high resistivity of 18 MΩ-cm. This high D.I. resistivity can cause ESD (Electrical Discharge) failures. The devices that are most sensitive to ESD are generally those which include MOS – Metal Oxide Semiconductor like discrete wafers, hybrid devices, Integrated Circuits (ICs) and others. Most dicing houses are using The CO2 air bubblers system to lower the surface tension and in some cases using additives to lower the surface tension to better wash away the silicon powder debris created during the dicing. See in the coolant section a detailed review. It was found that washing the wafers after the dicing on the automated saw using the spinners at high velocity is also a key reason for getting EDS problems. This requires also to optimize the D.I. coolant and the spinner parameters. The internet is loaded with articles on this issue, see in the reference. Beside the cooling efforts to lower the chance of ESD, operators in the production lines are grounded as common practice.

Tape mounting the wafers is an important parameter to perform well during the dicing. A good bond between the wafers and the tape without any air bubbles is a must. Optimizing the curing process during the tape mounting is a key important parameter to perfect the tape mounting. See details in the mounting section.

Dicing parameters on standard Silicon wafers vary depending on wafer diameter, wafer thickness, metallization in the street's width, street width geometry, die size, blade parameters and quality spec. Today's standard major wafer diameter used in mass production houses is 12" (305 mm) but 8" (203 mm) and even 6" are still widely used. Following are typical dicing parameters on today's popular wafers diameters:

Spindle speed – 40-60krpm Feed rate – 1" – 3"/sec (25,4 – 76.2mm/sec) – depending on quality spec.

Today the 12" (305 mm) wafers are the main standard O.D. size for the large semiconductor manufacturers, however the new 450 mm (18") wafer diameter is in process of introducing into the marketplace. This process is a game changer regarding the fab requirements and will be a real singulation challenge to overcome.

The present semiconductor silicon wafers are different from wafers manufactured 20–30 years ago. Today wafers do have metallization, low-k layers in the streets and other non-friendly materials to conventional dicing that are causing blade overloading and cut quality issues. The initial solution to handle those process issues was to clean the "junk" materials by using a – V – blade or a wider blade with slightly larger diamond grit size to minimize loading. The cut was a relatively shallow cut. A second full cut into the mounting tape was performed using a thinner standard blade. This process was performing well but was cost effective due to the need to slow the feed rate on the first cut. Another higher cost was the more expensive – V – type blades. Fig. 261 shows the two-step initial dicing process.

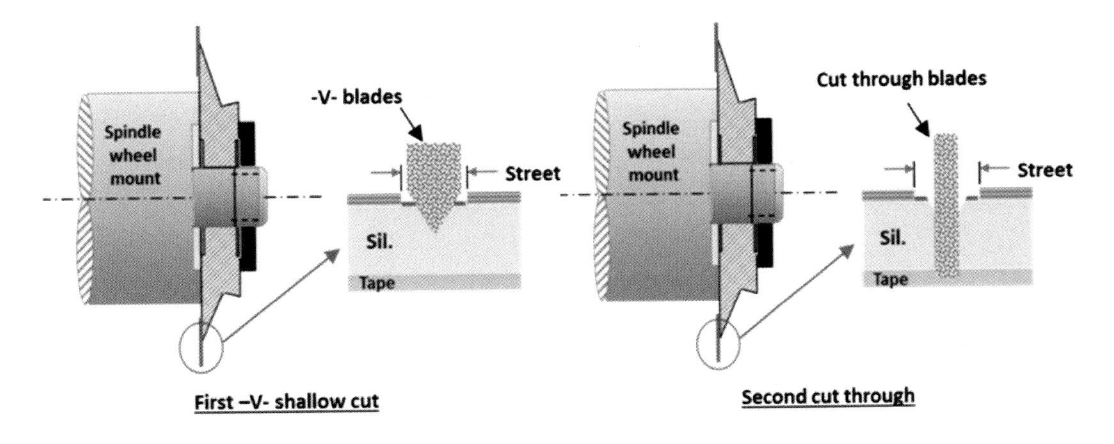

Figure 261. – Using the – V – blade first and than the cut through blade

This process was leading to introducing the dual spindle concept that could handle both cuts on the same saw and on the same wafer mounting. The disadvantage was the fact that the feed rate of both blades was not the same and this slowed the throughput as one spindle had to wait for the slower spindle to finish its dicing cycle.

18.3 - DICING LOW-K SILICON WAFERS

The next dicing development of more difficult wafers with TEG (Test elements groups) and Low-K wafers was the Ablation laser process. Low-K was developed to Increase transistor speed, reduce transistor size, and pack way more transistors onto a single die. However, the mechanical strength characteristics of the Low-K layers is low and risky of film peeling during standard diamond blade dicing. The low-K laser grooving process removes the street wiring layer, including the low-K film. There are a few options using the ablation (short pulse laser) process:

Pi (π) laser grooving (Step cut blade dicing) The laser makes two laser grooves in the dicing street, then the wafer is first diced with a shallow cut using a wide blade and finally a standard second blade is dicing through into the tape. Eliminating the test metal elements and the low-K layer in the street performs with excellent results after the diamond blades dicing (see fig. 262).

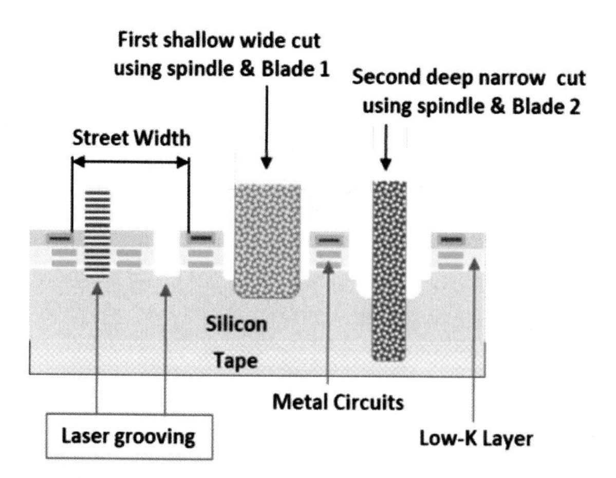

Figure 262. – 3x step dicing, laser, wide shallow cut and cut through on difficult wafers

Omega (ω) Laser grooving (One pass blade dicing) depending on the street width the laser makes one or two shallow grooves in the streets to eliminate the metal elements and the Low-K layer. Then a standard diamond blade is making a full cut through into the tape with excellent quality results (see fig. 263).

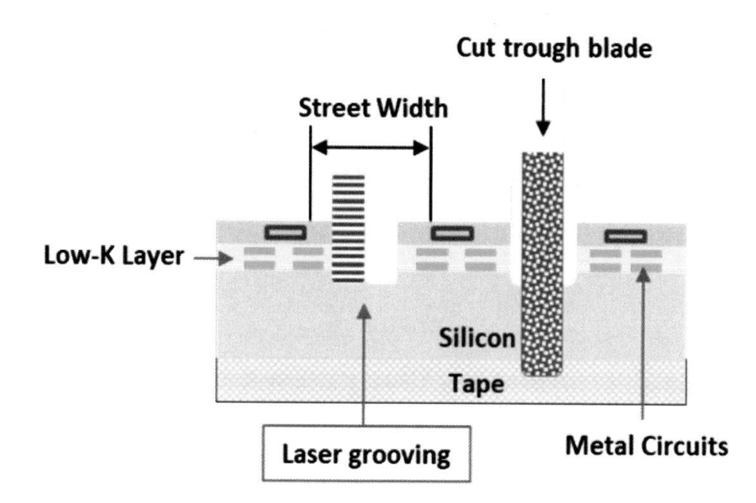

Figure 263. – 2x laser shallow grooving to dice away the low – K – and then a cut through

THINNED SILICON WAFER, LASER FULL CUT DICING.

Silicon wafers are getting thinner and thinner resulting in chipping and cracking during diamond blade dicing. DAF (Die Attach Film) mounted thin silicon wafers are increasing with tight quality demands, which is causing problems during diamond blade dicing. Diamond blade dicing thin silicon wafers (0.200 mm or thinner) are diced at low feed rate due to quality issues which is affecting the throughput. The laser process is improving dramatically the cut quality and the throughput as the laser process is around ten times faster (see fig. 264).

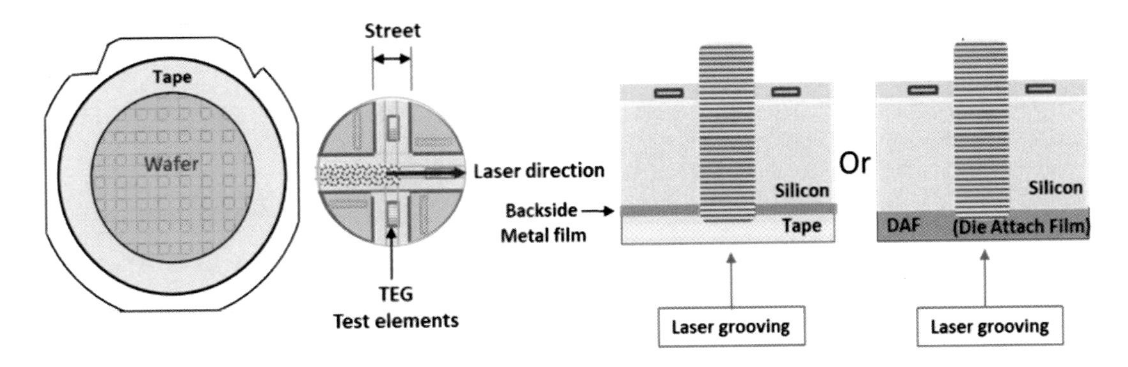

Figure 264. – Laser dicing on thinned silicon wafer. 0.200mm or thinner on Tape or DAF mounted

18.4 – DICING THIN SILICON WAFERS IN THE RANGE OF .004"(0.100MM) TO .008" (0.200MM)

Using the standard dicing process of wafers mounted on tape in the thickness range of .004" to .008" becomes a quality issue that requires slowing the cutting speed affecting throughput. For this process, the DBG (SD dicing before Grinding) was developed.

There are a few options depending on the wafer thickness and the wafer end product.

Following are a few sample options:

SDTT (Stealth dicing through tape). This process is done after BG (Back grinding) on wafers in thickness range of .006" to .008" thick and some time out of this range. Below is a generic process flow (see fig. 265).

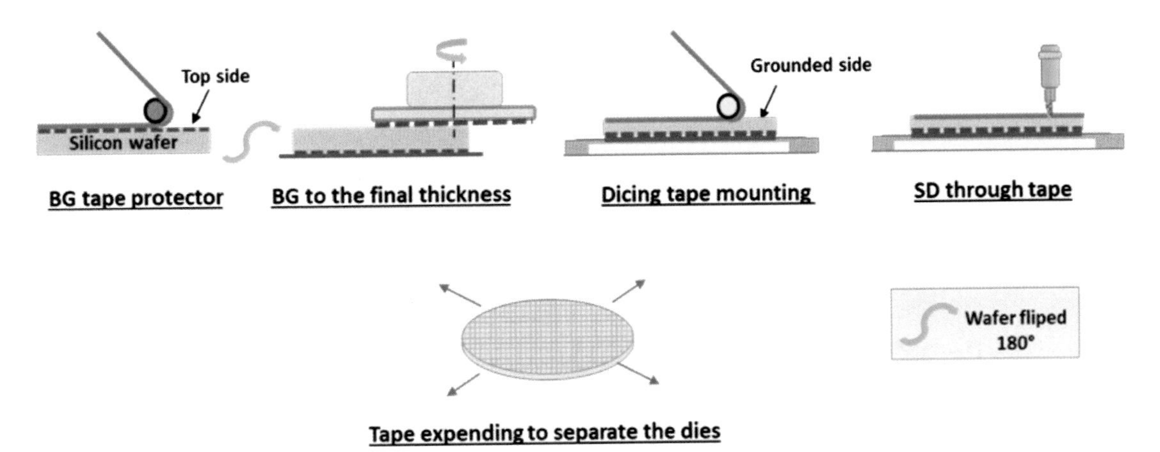

Figure 265. – Stealth dicing through tape after wafer back grinding

DBG (Dicing Before Grinding). Another process of separating dies to improve quality and to add high die strength. The wafer is first diced using the conventional diamond dicing blade to half of the wafer thickness. A protective tape is mounted on top of the diced side. The wafer is mounted on a back grinding machine with the diced side down. The back grinding is grinding the wafer below the diced area to separate the dies. The wafer is frame mounted and the protective tape is peeled off carefully on a special wafer mounter. (see fig. 266).

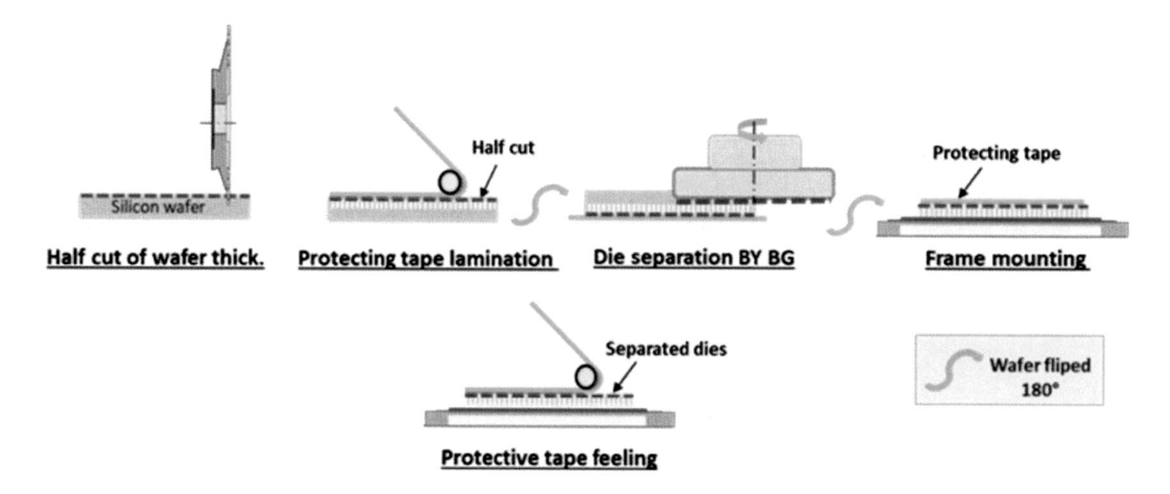

Figure 266. – Half cut, P. film on diced side, back grinding, tape feeling & die separation

18.5 - DICING THINNER WAFERS OF 0.050MM AND THINNER: (STEALTH DICING)

Flash memory and memory controllers used for smartphones and tablets are getting thinner requiring thin wafers. This becomes a manufacturing challenge to find good methods to separate dies on thin wafers 0.050 mm and thinner. SDBG (Stealth dicing before grinding) is probably the best method today to separate dies on ultra-thin wafers. This process performs with very thin kerfs improving the die strength and allowing a higher number of dies per wafer. DAF (Die attach film) is initially used for singulation dies on thin wafers. The first step is laminating a protecting film on the wafer's top side. The wafer is mounted topside down for the SD process. The SD process performs an SD layer above the wafer center and can be removed during the grinding process. After grinding, the wafer is frame mounted with the protecting film on top. The protecting film is peeled off and the framed wafer with the DAF is moved to a die separator to expand the wafer and separate the dies mounting initially on the DAF film (see fig. 267).

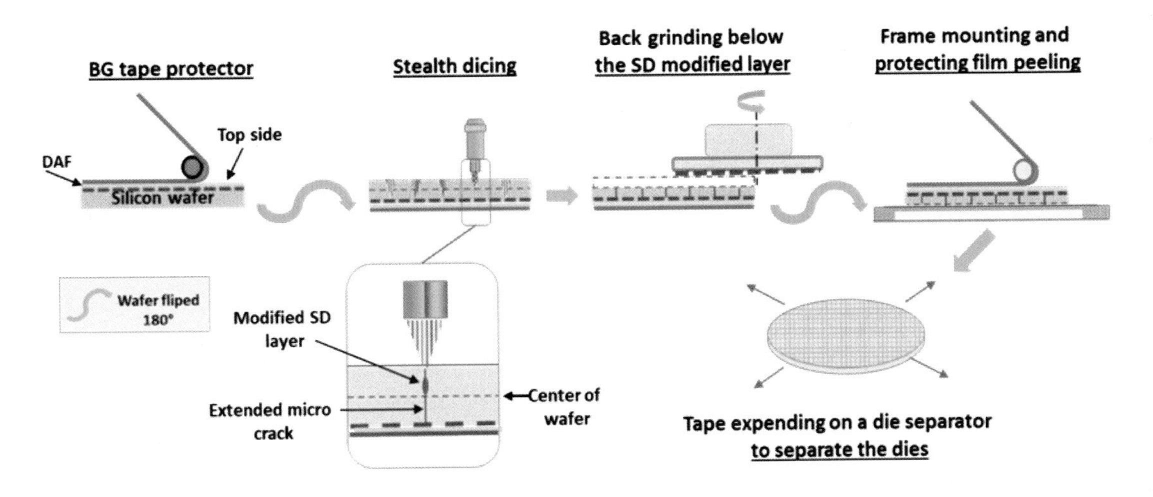

Figure 267. – Stealth dicing of 0.050mm thick wafers, BG and wafer expending to separate dies.

The above is a unique process used for very thin Sil. Wafers however it performs well also on thin Sapphire, Lithium Tantalate, Glass, GaAs, and similar materials. For thicker substrates the SD process requires in some cases to use a few SD passes depending on the substrate material and the quality spec. (see fig. 268).

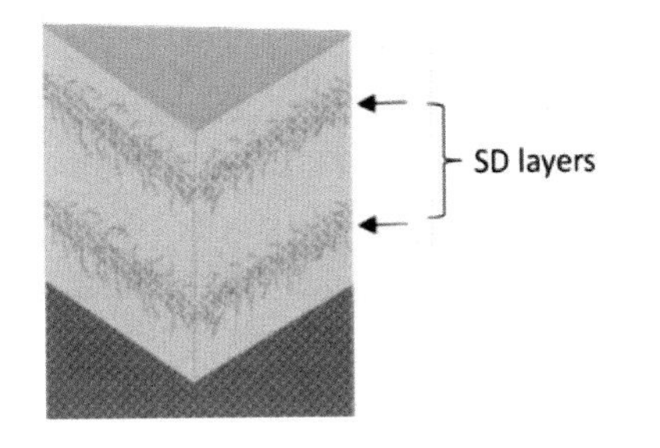

Figure 268. – SD a few pases on thicker substrates

18.6 – THIN WAFERS – DIE SINGULATION BY PLASMA ETCHING

Plasma etching is another process for singulation thin wafers. It is a complicated process but with a lot of advantages (See the Plasma Etching/Dicing review). The process is a combination of the plasma etching and the BG (Back grinding) process, a bit similar in the steps to the laser process of singulation thin wafers. The plasma etching process can be performed prior to the BG process and after the BG process. The process is different to diamond dicing and laser dicing regarding the sequence of the process. In both diamond and laser dicing the dicing blade and the Laser are traveling in every street in both X and Y directions. In the plasma process, the whole wafer is processed at the same time in one process (see fig. 269).

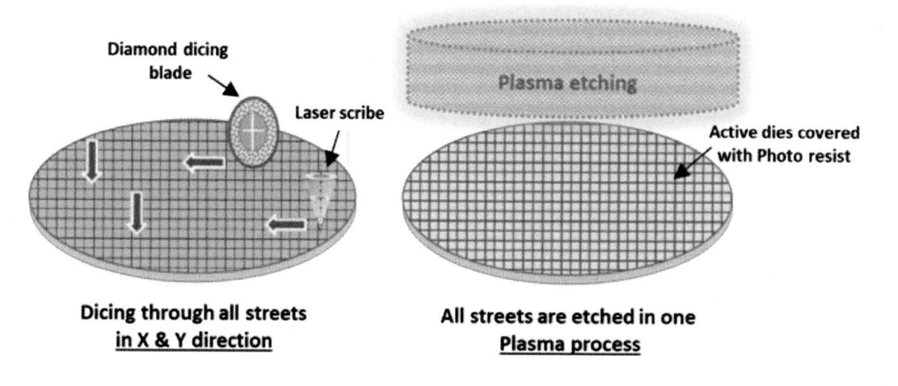

Figure 269. – Advantage of PLASMA Etch on diamond dicing and laser scribe

There is one additional required process prior to the plasma etching process which is a delicate process and time consuming. All the active die geometries need to be protected and covered with a photoresist. Only the street areas need to be exposed in order for the plasma process to etch into the wafer (see fig. 270).

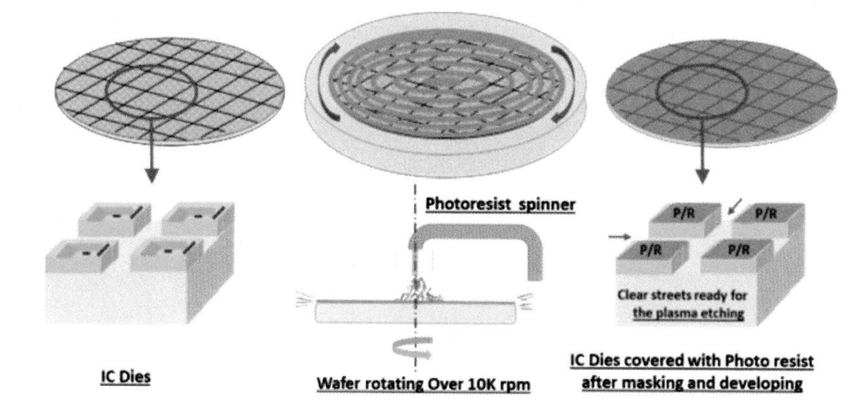

Figure 270. – Photoresist process prior to the PLASMA Etch process

Following is a review of the plasma process after film protecting the dies using the photoresist process. The areas to be plasma etched are the streets without the photoresist coating. Fig. 271 illustrates the plasma etching process.

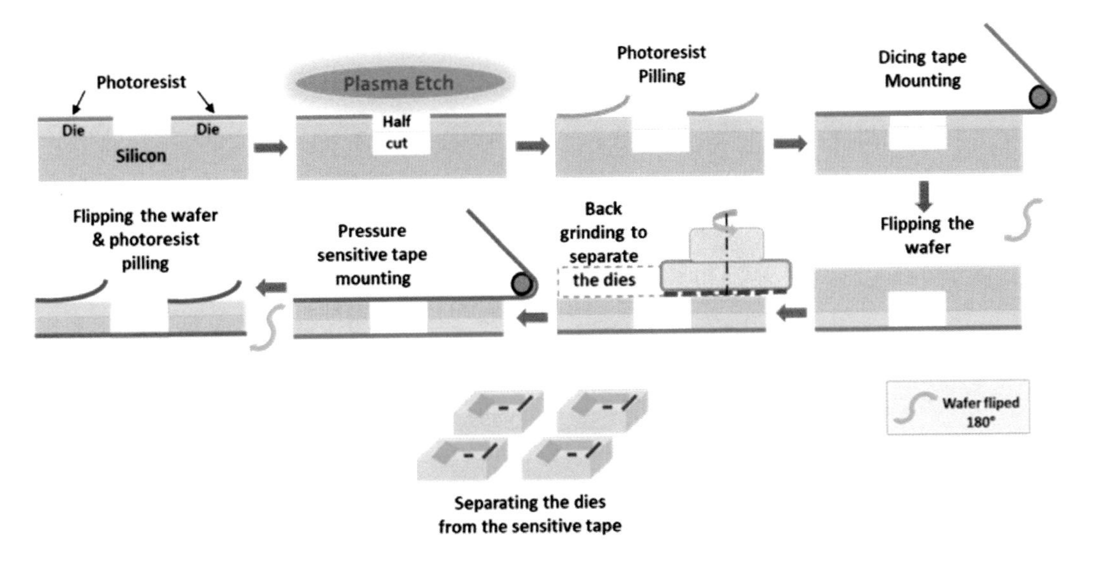

Figure 271. Plasma Etch and back grinding process to separate the dies.

Another Plasma process used for thin wafers is a stress reliever after the dicing process and the back grinding process to separate the dies. The Plasma process is rounding the die edges and smoothing the dicing surface. This process is stress relieving and adding strength to the thin dies (see fig. 272).

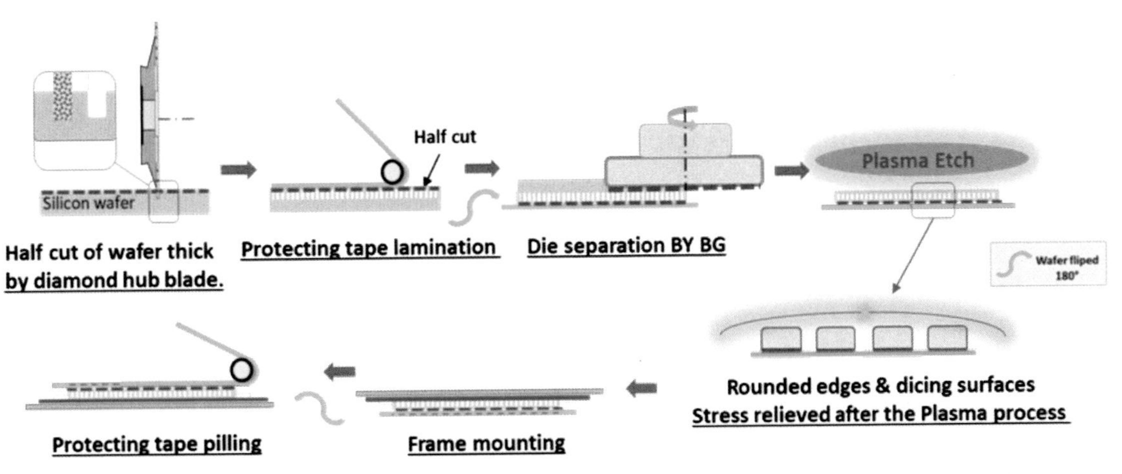

Figure 272. – Plasma etch / stress relieved after half cut and back grinding

Following are SEM showing the difference between diamond and laser dicing compared to the Plasma etch process. The round corners are the main advantage of minimizing edge stresses and improving die strength (see fig. 273).

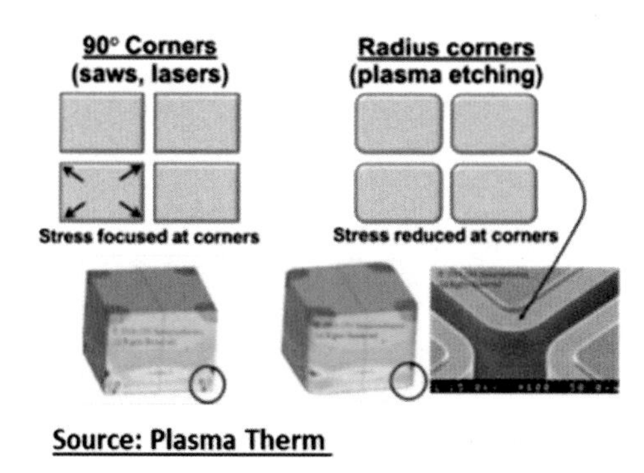

Figure 273. – Minimized die stress after the rounded edges by the Plasma Etch process

Another benefit using the plasma etch process is singulation of Mosaic™ and other special geometries. See the benefits of Plasma etch and Fig. no. 247.

18.7 – SILICON SUMMARY

A lot of silicon wafers are still diced using the conventional tape-mounted wafers using the diamond dicing hub blades. However, new demanding products requiring thin and complex wafers are making the standard singulation process very difficult and challenging. To overcome the new challenges, new technologies were developed to perform well and meet the new demands. The above-described singulation methods of silicon wafers are only examples of what the dicing companies developed and what the industry is using. The different laser singulation methods and the Plasma etching technology were adopted and are also used on other difficult hard or brittle materials that standard diamond dicing is difficult or impossible to meet today's quality specs. Some of those materials will be further covered. A lot of good information can be found via the internet for everyone who likes to explore more knowledge.

18.8 – GAAS (GALLIUM ARSENIDE)

GaAs (Gallium Arsenide)

GaAs is used in a wide range of semiconductor applications: integrated circuits at microwave frequencies, Laser diodes, Infrared light-emitting diodes, RF devices, Solar cells, and others. GaAs is a hazardous material due to its arsenic content of 51.8%wt. which requires safety consideration during the singulation process. It requires safe handling of the operators during the different singulation methods mainly personal protection and adequate equipment air ventilation and suction. Collecting and treating the waste coolant water is another requirement. Another major concern in the singulation of GaAs wafers is the brittleness of the material performing differently between channel 1 and channel

2. This phenomenon is due to the differences in crystal orientation. Standard semiconductor GaAs wafers of the 100-plane type are sensitive to quality issues due to the (111) Crystallographic orientation of 54.74° along the (110) scribe or dice plane that can result in slanted cuts and top and back side chipping (see fig. 274).

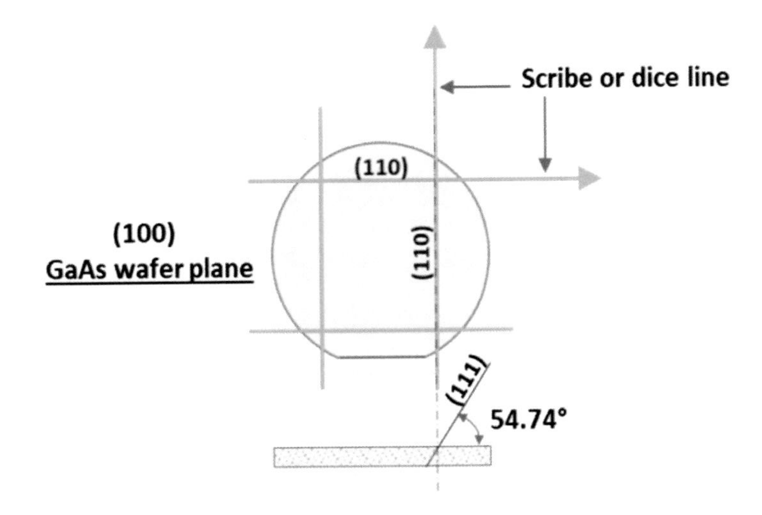

Figure 274. – GaAs 111 Crystallographic orientation along the (110) scribe / dice plane

The original method of singulation GaAs was done by diamond scribe and break, a method that is still used by some manufacturing houses. (See in the Diamond Scribe and Break process) The next singulation method introduced to the market during the early 1970s was diamond dicing using diamond annular blades and later hub blades. Due to the more delicate process, the diamond blades and dicing parameters are different from dicing silicon wafers. Following are the main process parameters used today with modifications between the different users:

Hub blade diamond grit – 1-3, 2-3 & 2-4mic.

Blade thickness – .0006" – .0012" (0.015 – 0.030mm)

Low to medium diamond %

Mounting – Blue (PVC) and UV tape.

Coolant – In most cases D.I.

Spindle speed – 20-40Krpm

Feed rate – 1-10mm/sec (Depending on cut quality)

Below is a sample photo of a diced GaAs wafer using a diamond hub blade.

(see fig. 275).

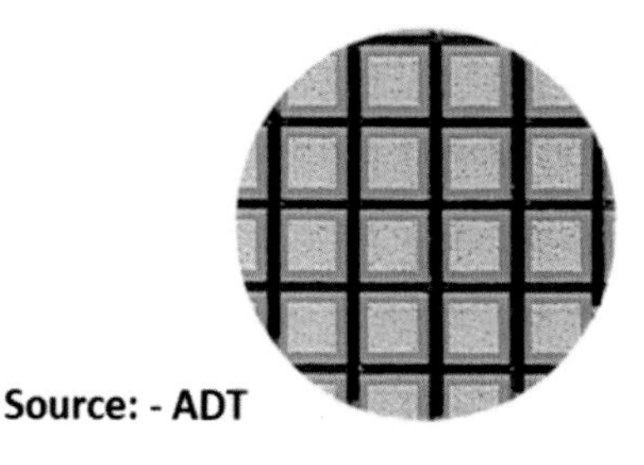

Figure .275. – Top view of a common GaAs wafer

So far the above is a review of what can be called standard singulations of GaAs wafers that was and still is used today on standard GaAs wafers. The new developed processes technology developed for the silicon market are also used on complicated GaAs wafers. Both laser scribe and break, Stealth laser after or before Back grinding and the Plasma etch process described already are used successfully on thin GaAs wafers. Other thin difficult and brittle wafers in the Semiconductor industry like SiC, Glass, Sapphire, Lita3, LiNb, Silicon on glass and others are handled similar, both using the standard diamond hub blades with the finer diamond grit, slower feed rates and the new laser and plasma etch processes on the very fine and thin wafers.

18.9 - SAPPHIRE SUBSTRATES - AL_2O_3

Figure 276. – Sapphire substrates

Sapphire is one of the hardest substrates to dice, 9 Mohs on the hardness scale almost as a diamond. Sapphire is made in a few orientations, the most popular are A-Plane and C-Plane that are relatively easy to machine and R-Plane which is more difficult to machine.

The A – Plane is used for optical devices and is resistant to high temperatures. Used for hybrid microelectronic applications.

The C – Plane is a good material base for growth of GaN for Blue and Green LED and for laser diodes.

The R – Plane a good base for deposition of silicon for high-speed IC and pressure transducer applications.

Bottom line, all Sapphire material Planes are very hard and brittle material creating real challenges when singulating microelectronic wafers in the required quality specs. The above is a general generic sapphire material information, much more detailed info can be found via the internet.

Like GaAs, the Initial singulation method was using the diamond scribe and break process which was a slow process with low yields due to the fact that sapphire does not cleave well. The next approach was to mechanically dice the wafers using diamond dicing blades mainly using resin-type binders. Due to the very hard material, a relatively large diamond is used in order to ease the penetration into the Sapphire. However, this is affecting the top and backside chipping on the brittle sapphire material and causes yield concerns. For this reason, only thicker substrates with wider streets of up to .010" (0.250 mm) are used in order to compensate for the relatively large chipping. The diamonds size used, 45mic., 53mic. and 64mic depends on the substrate thickness. Most applications are done using resin blades with coated diamond for better blade life. However, uncoated diamonds are also used to create higher wear and better cut quality. Slowing down the spindle rpm also creates a higher blade wear which improves the cut quality. The mounting media depends on the die size, relatively large dies are mounted on blue tape and very small dies on UV tape. The advantages and disadvantages of the different diamond sizes can be seen in the following sketch showing the mechanism of the resin matrix using large diamonds and smaller diamonds. (see fig. 277).

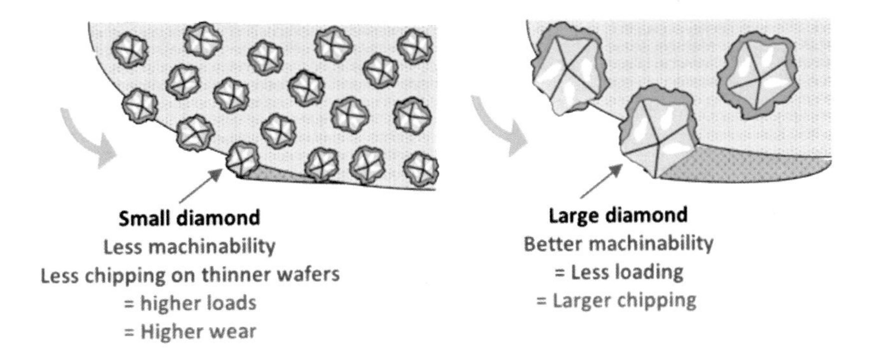

Figure 277 – Advantages and disadvantages of using smaller and larger diamonds

The Sapphire wafers are expensive, and the "real estate" cost of each die becomes an issue. This was leading using narrower streets in order to gain more dies on the wafers. In addition, the market went to thinner wafers. Both new demands become very difficult to use the standard diamond dicing process and led the way to use the diamond scribe and break process and the different Laser options.

Both Laser options can be used, the Ablation laser process and the Stealth laser process. The Ablation process used as a scribe and break process creates small debris that can contaminate the dies. To overcome contamination, a water-soluble film is applied using a spinner prior to the laser process. After the laser scribe process the water-soluble protective film is clean off using D.I. water on a spinning cleaning stage. The dies are then separated by expanding the mounting tape (see fig. 278).

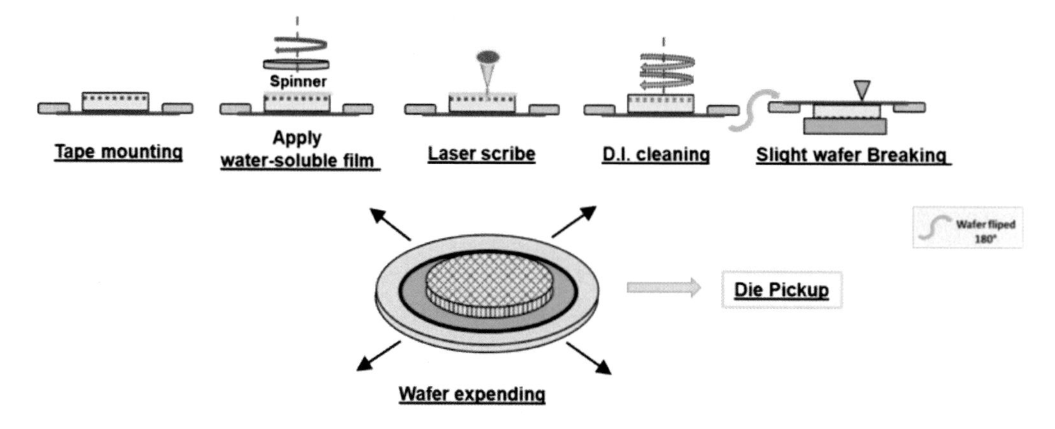

Figure 278. – Sapphire Ablation Laser scribe and the expending break process

The Stealth laser scribing is similar in steps with no need of wafer top side coating and no need of D.I. cleaning. As indicated, the laser beam is focused to transmit a wavelength to the center of the wafer. This process is performed with internal stress layers inside the wafer creating starting points for later die separation by expending the wafer and separating the dies due to the internal tensile stresses. (see fig. 279)

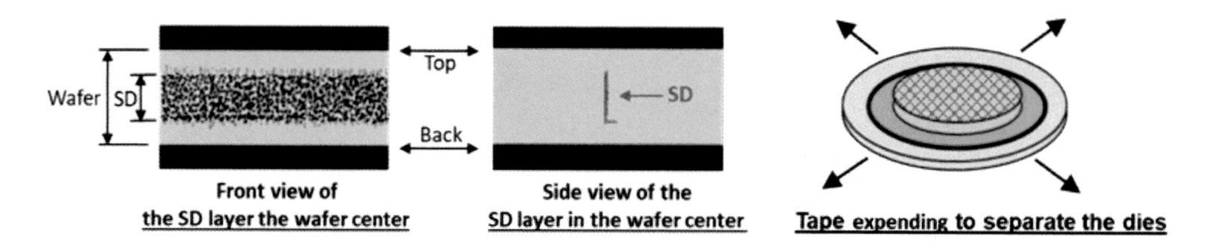

Figure 279. – Sapphire Stealth scribing and tape expending to separate the dies

Separating Sapphire thin wafers is faster than diamond dicing and diamond scribe and breaks by 5–10 times due to faster Feed rates that the Laser machines are running.

Some Sapphire LED's application is laser scribed on both sides of the wafer in order to perform with more uniform light emission. This is a great advantage for the laser process.

The above Ablation and SD laser processes are typical procedures however the processes can be modified depending on the wafer type and geometry.

Another option to separate thin Sapphire wafers is to use the Plasma etch process. However, this is less popular and is used more on MEMS devices.

18.10 - SOS - SILICON ON SAPPHIRE

The SOS wafers are R – Plane Sapphire mostly 3" (75 mm) up to 8" (200 mm) in diameter with a very thin layer of silicon deposited on top. The thickness of the silicon varies from 100 nm to 600 nm. The advantage of using the Sapphire as a base is the excellent electrical insulation. SOS wafers are closely related to SOI (Silicon on Insulator) and are used in the fabrication of high frequency devices such as RFIC's, RF Amplifiers, RF switches and durable pressure sensors. The thermal expansion coefficient of silicon is similar to R-plane sapphire making the raw material manufacturing process friendly. However, when using the diamond dicing process which was the first initial singulation process used in the marketplace it created a conflict of what blade to use. Dicing Sapphire requires a resin blade with relatively large diamonds of 45,53 or 63 mic. depending on the sapphire thickness. Such a large diamond will result with large chipping on the thin silicon layer. Some customers managed to use a 30 mic. grit but with poor yields. Additives were also used however to perform good yields the best process was to split the dicing process, first using a fine grit resinoid blade of 10 or 20mic. grit to penetrate the silicon and scribe a very shallow cut into the sapphire. Then a second process is used using a resin blade with a larger diamond grit to cut through the sapphire. To minimize the blade wear of the fine grit resin blade, a thin nickel blade can be used with 4-6mic. grit to perform 2x shallow cuts on the silicon layer. This process is more time consuming and is not the preferred way. See both options in fig. 280.

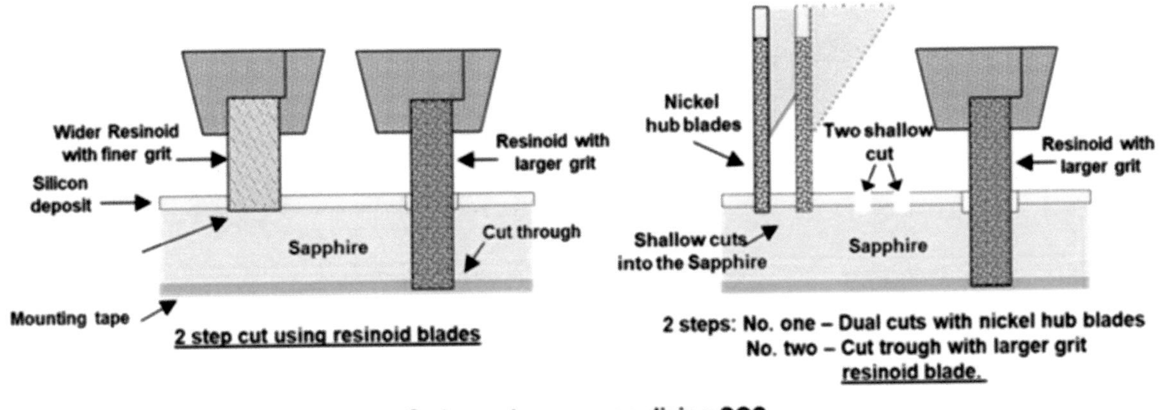

Figure 280. – 2x different two step dicing of SOS substrates

The above diamond dicing walk through is basically the initial method that was used for SOS substates and is still used in many places in the marketplace. Today the different new laser and other technologies are taking over similar to what was presented in singulation Sapphire.

18.11 - SIC WAFERS

SiC (Silicon carbide), also known as carborundum is a semiconductor material containing silicon and carbon. It is very hard and brittle, 9 in the Mohs scale almost as diamonds. SiC is used in semiconductor electronics devices that operate at high temperatures or high voltages or both. It is used in Power devices, high-frequency, and high-temperature radiation-resistant electronic devices and sensor devices. It is an ideal substrate material for manufacturing products with large ultra-high-brightness white-light, blue-ray, or laser diodes. Due to the extreme hardness and brittleness, cracks can easily be formed on the top and back side of the wafers including internal stresses that can cause micro cracks. To overcome the metallurgical properties of singulation SiC wafers, new dicing techniques were developed. Standard diamond dicing results in a very slow process and poor quality. It becomes not a practical process to use the standard diamond dicing. Following are two processes resulting in good quality and one with much better throughput. The first process is a unique process using a mechanical process with a special diamond blade mounted on a spindle with an Ultrasonic-wave oscillation mechanism. This setup results in better blade cooling, minimizing loading and less need to dress the blade due to self-re-sharpening. The cut quality is drastically improved but still with a relatively slow feed rate of around 10mm/sec depending on the wafer thickness and the cut quality spec. More info on this process can be found in the reference section. (see fig. 281).

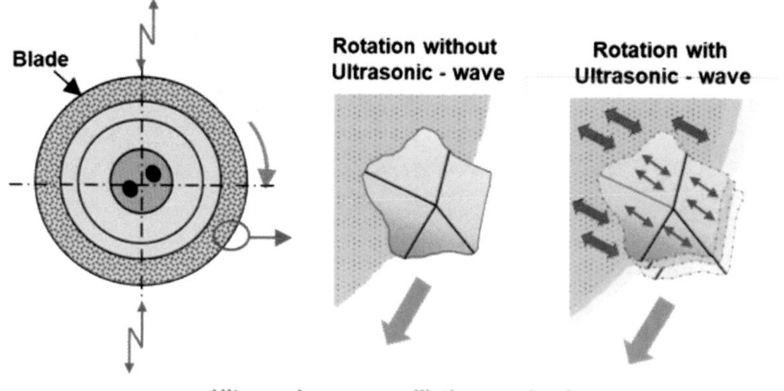

Figure 281 – Dicing SiC. substrate using a spindle with ultrasonic wave

The next SiC singulation process is the SD laser process. It is a dry and fast process without any top and back side chipping as the SD process is inside the wafer.

The no. of the SD passes depends on the wafers thickness. (see fig. 282).

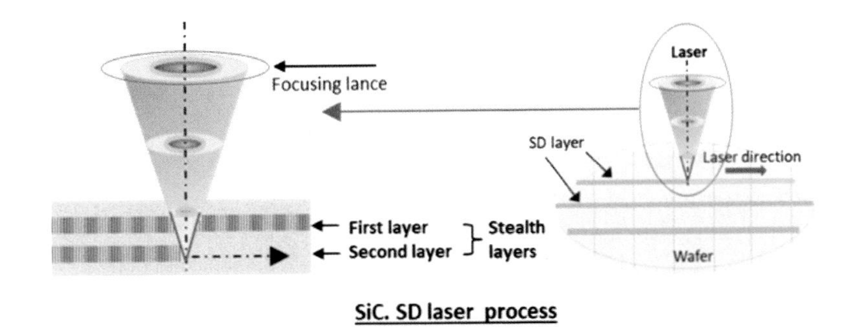

Figure 282. – Singulating SiC. substrate using the SD laser process

The feed rate of the SD process can get up to 350mm/sec per pass. Usually, to perform a good singulation process, a few passes are needed but it is still a much faster process compared to the Ultrasonic-wave oscillation process. After the SD process, the dies are separated by expanding the tape mounted wafer. (see fig. 283).

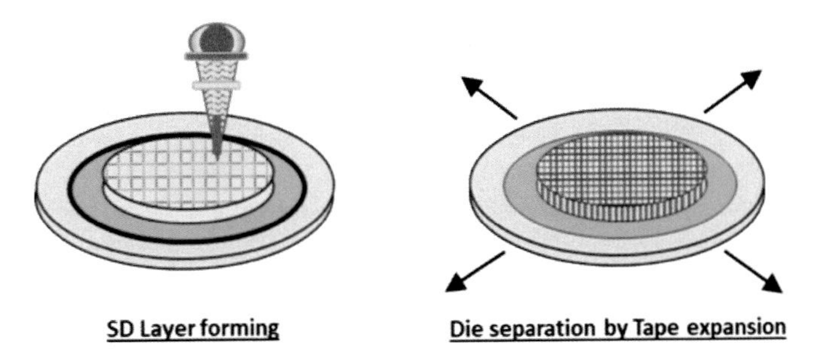

Figure 283. – Separating the dies after the SD process

The quality of the Ultrasonic Wave is a bit better however for mass production the SD laser is the preferred process. More detailed information about dicing SiC can be found in the reference section.

18.12 - MEMS (MICRO ELECTROMECHANICAL SYSTEM)

MEMS (Micro-electromechanical systems) is a process technology used to create miniature integrated devices or systems that combine mechanical and electrical components. The fabricated process is using integrated circuit (IC) batch processing techniques and can range in size from a 1 micron (mm) to millimeters. Normally the total size of a MEMS is in the range of 1^2mm.

MEMS are ultra-compact systems composed of Miniature mechanical components such as switches, sensors, actuators, and electronic circuits on a silicon wafer using the microfabrication technology of semiconductor manufacturing technology.

Today's examples of MEMS devices include airbag accelerometers for airbag sensors, microphones, projection display chips, blood pressure sensors, tire pressure sensors, optical switches, analytical components such as lab-on-chip, biosensors and many more.

There are many different miniature mechanical electrical amazing 3D geometries. Below are a few MEMS photo's showing just a few examples. (see fig. 284).

Top view of gear reduction Unit

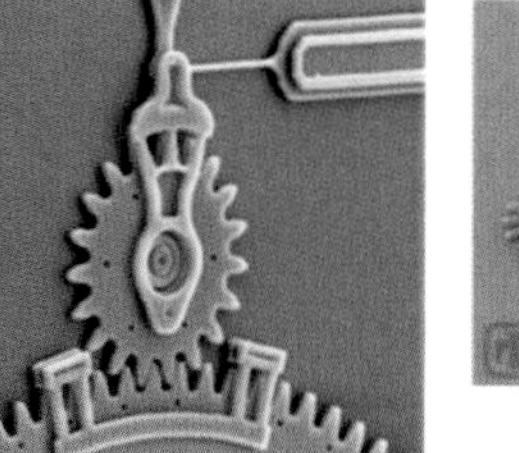

Gear Alignment Clip

Six Gear Chain

Silicon Mirror assembly close -up

"Courtesy Sandia National Laboratories, SUMMiT™ Technologies, www.sandia.gov/mstc"

Figure 284. – A few MEMS substrate geometries

As can be seen in the above photos, MEMS contain extremely delicate structures such as cantilevers, bridges, hinges, gears, membranes, sensitive high aspect-ratio topography, pressure sensitive components that will get damaged when coolant water during a dicing process is used. MEMS in many cases have moving parts that are super sensitive to contamination, small debris particles can stop any delicate moving parts.

Electrostatic Actuators are sensitive to ESD (Electrostatic discharges) and may fail when suddenly electrostatic discharges during mechanical diamond dicing and coolant flow. How to overcome those delicate MEMS issues during any singulation process was a challenge. The initial attempt was done years ago to add a protective cop to cover the MEMS die. See fig. no. 285 of a generic set-up.

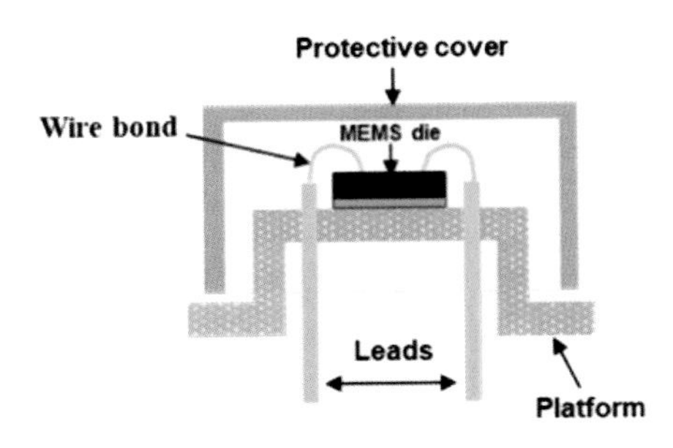

Figure 285. – MEMS with a protective cover to eliminate electrostatics

Adding the protective cover can be used only on limited MEMS products and adds a production step that is cost effective. MEMS with the protective cap can be diced using the standard diamond dicing process. Another limited option is to add a temporary protective polymer film that needs to be washed away after dicing. This process may be used only on selected applications that the delicate MEMS geometry can handle a wash process.

Scribing the wafer prior to adding the MEMS fabrication process is another option however it is a very delicate process to make sure the wafer will not get damaged cleave through during the MEMS fabrication. The scribe process can be done using diamond scribe and break process. (see fig. 286).

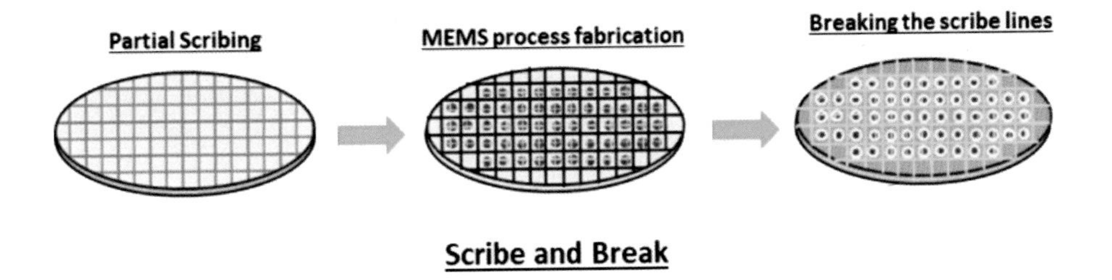

Figure 286. – 3 steps MEMS fabrication by scribing, MEMS geometry & die breaking

The last and probably the best process option is the SD (Stealth Dicing). It is a dry and clean process with no residues at the delicate MEMS areas. The MEMS dies are separated by the UV tape expansion. (see fig. 287).

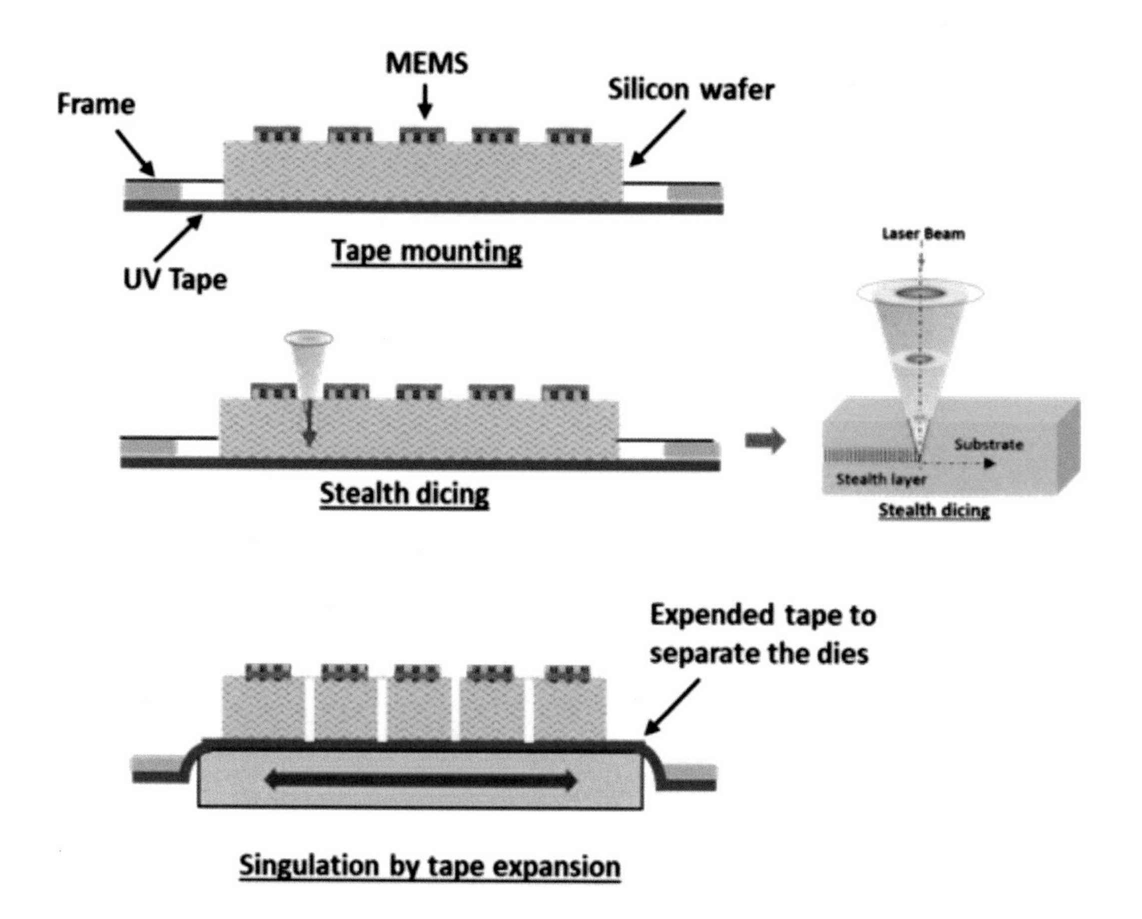

Figure 287. – MEMS fabrication, Stealth dicing and Singulating the dies by expansion

MEMS dies can be made of Silicon, Sapphire, SiC, Lita3/LiNb, Glass, Ceramic, quartz, GaAs, and other different materials. Some were discussed already. To summarize, a Cap covered MEMS can be diced using the diamond dicing process. All other open and delicate MEMS should be diced in the safest way which is the SD Stealth dicing process.

18.13 – GLASS APPLICATIONS

Glass is a relatively medium hard and brittle material which is transparent and acts as a substrate for various applications. Generic types include fused silica, borosilicate glass and alkali-free glass. Glass substrates are used in a wide variety of technical and industrial applications.

Below are only a few major generic applications:

- MEMS (Micro Electromechanical System) – Glass is used in MEMS applications as a carrier in the fabrication process of thin silicon wafers for a broad application.

- Semiconductors – Glass substrates are used as a permanent substrate that remains in the final product or as a temporary substrate in the manufacturing process of thinner materials.
- Biotechnology – Used by researchers using glass wafers to fabricate microfluidic dies. Borosilicate dies are particularly great for those applications because they are affordable and are chemically resistant.
- Integrated Circuits (IC) packaging – On a lot of IC packaging, glass wafers are being embedded within the wafer for protection. Durable packaged ICs are common in portable electronics like tablets, smart phones, and many others.
- Optical Communication's – Fiber optic communication requires a lot of glass connectors with superior surface finishes.
- Other applications using glass materials – CMOS Image Sensors (CIS), Memory and logic, Radio Frequency (RF), Power Electronics, Photonics, Technology Platform.

Some of the most common used glass substrates in a more informative description:

Below is a table with some of the most common glass substrates used in the market with their material properties. (see fig. 288).

Glass type / Properties	D263	BK7	Borofloat®33 (Brosilicate glass)	AF32ECO Borosilicate
Knoop hardness HK 0.1/20	**470**	**610**	**480**	**490**
Density (g/cm^3) @40°C/h annealed	**2.51**	**2.51**	**2.2**	**2.43**
Young's modulus E (KN/mm^2)	**73**	**82**	**64**	**75**

Figure 288. – The most common used glass substrates in the Microelectronic industry

Below are the most common glass substrates with some of their market applications.

- **D263** – Touch Panel, Lab-On-a-chip, Image Sensors, IR-Cut Filters, Optical Diagnostic, Ultra-Thin Glass applications, Sensor Covers, Fingerprint Sensors, Microfluidics.
- **BK7** – Prisms, Lenses, Substrates for Mirrors & Filter Coatings, Lenses and Dispersion Prism, Base material for Precision Optics, Demanding Optical applications, Measurement and Sensor-Technology.
- **Borofloat 33** – Fiber Optic applications, MEMS, Microfluidics, Medical & Biotechnology applications, Dielectric Filters.
- **AF32 ECO** – MEMS, Optical packaging in Semiconductor applications, (WLCSP) – Wafer-Level Chip-Scale Packaging, (TFB) – Thin Film Battery, Interposer Substrates, (UTG) – Ultra-Thin Glass, (WLO) – Wafer-Level Optics.

18.13.1 - DICING MECHANISM OF GLASS SUBSTRATES

Some of the above applications involve relatively hard and brittle materials from the glass family and some with special hard and brittle coatings like AR (Anti Reflecting) with very tight quality requirements. This involves special optimized blade technologies and dicing process optimizations. In most cases, soft resin blades with fine and special diamonds are used. For some thinner glass type substrates, special sintered blades are used with relatively fine diamonds. Optimizing resin binders are critical to perform with minimum loads and with the best-cut quality. Minimum loading in many cases means higher blade wear to expose new sharp diamonds. On the process side, optimizing the cutting parameters mainly the spindle rpm and the feed rate is needed. Standard spindle rpm used for glass is 18–30 krpm on 2" blades which is the major used size and 8–15 Krpm on the less used 4" blades. Those spindle rpms are generic no.'s and need to be optimized for each material and quality requirement. The feed rate spread is very wide, from 1–15 mm/sec and needs to be optimized per application and spec. Flange condition is very critical as a small nick on the flange edge may deflect the blade causing chipping and slanted cuts. Dynamic balancing the flange set on the spindle is also required in some cases and can improve and minimize chipping. See the dynamic balancing review described already. Using coolant additives can also be a dramatic improvement to achieve better-cut quality and improve blade life. Standard blades designed for glass can be used however in many cases fine optimizations are required. Optimizing the blade is a "fine art" and can be done only with a close involvement between the blade vendor and the actual dicing house. For resin blades different diamond sizes can be used in the range of 10–45 mic. and from different types. A major factor is using coated diamonds mainly with a nickel layer or without any coating.

The uncoated diamonds will perform with higher wear but with less loading which will perform with better-cut quality. There are also different diamonds to use between the different diamond suppliers. This may also be a factor in cut quality. As for the blade bond matrix, many variables can be used in the bond itself. The type of resin, adding additives into the bond mix to cause less loading and better bond coolant during the dicing process or to act as media to minimize the blade's wear. Needless to say, the blade manufacturing process of how to mix the powders and how to mold and cure the final matrix can all cause different blade performances.

Below is a general sketch with only a few different parameters that can affect the blade performance. (see fig. 289).

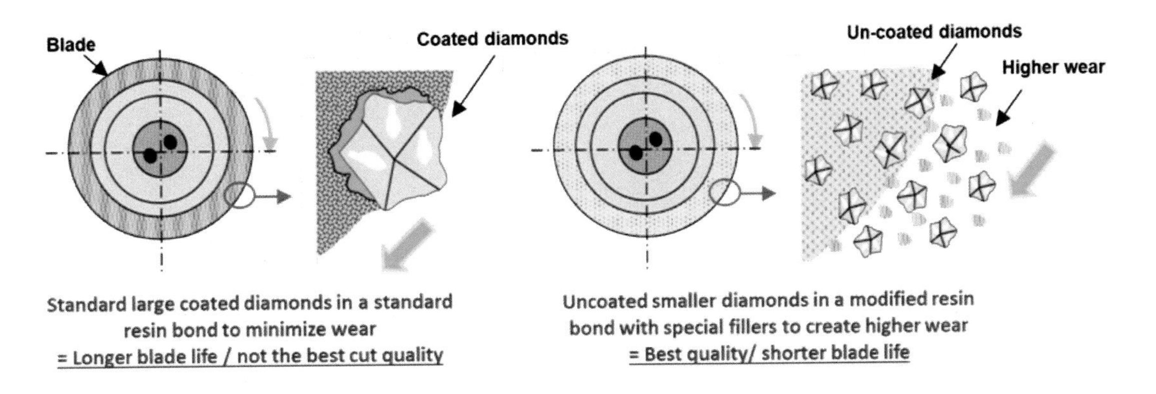

Figure 289. – Different diamonds and bond type for dicing glass substrates

In most cases, resin bond blades with the right diamonds and the best optimized bond can perform well in glass type substrates. However, when the substrates are getting much thinner with thinner streets, resin blades are not practical to use as they will wear real fast. Also, in case of harder glass substrates with tough coatings like AR, IR and similar, the higher wear of resin blades is also a problem. To solve those problems Metal sintered blades can be used mainly to overcome the problems when using very thin blades. The blade thickness using metal sintered blades can go down to 0.050 mm (.002"), however the more practical to use thin metal sintered blade thickness are in the range of 0.100–0.150 mm (.004"–.006") thick. The main problem with Metal sintered blades when dicing glass is overloading of the blades mainly because they are much harder and tend to get overloaded affecting the cut quality and in some cases tending to break. The typical diamond grit size used on the metal sintered blades for glass applications is 6–25 mic. and they are of the hard and blocky type compared to the friable diamonds used in resin blades. The main challenge of optimizing metal sintered blades is to create a higher wear by creating a more porous metal matrix that will allow diamonds to wear out and expose new sharp diamonds. This metal sintered process optimization means to have a less dense matrix in order to create more blade wear. Creating a porous metal sintered matrix requires metallurgical process changes and adding special fillers to the metal matrix. Below is a generic sketch and SEM photos demonstrating the differences between a standard metal sintered blade to a porous matrix with voids. (see fig. 290).

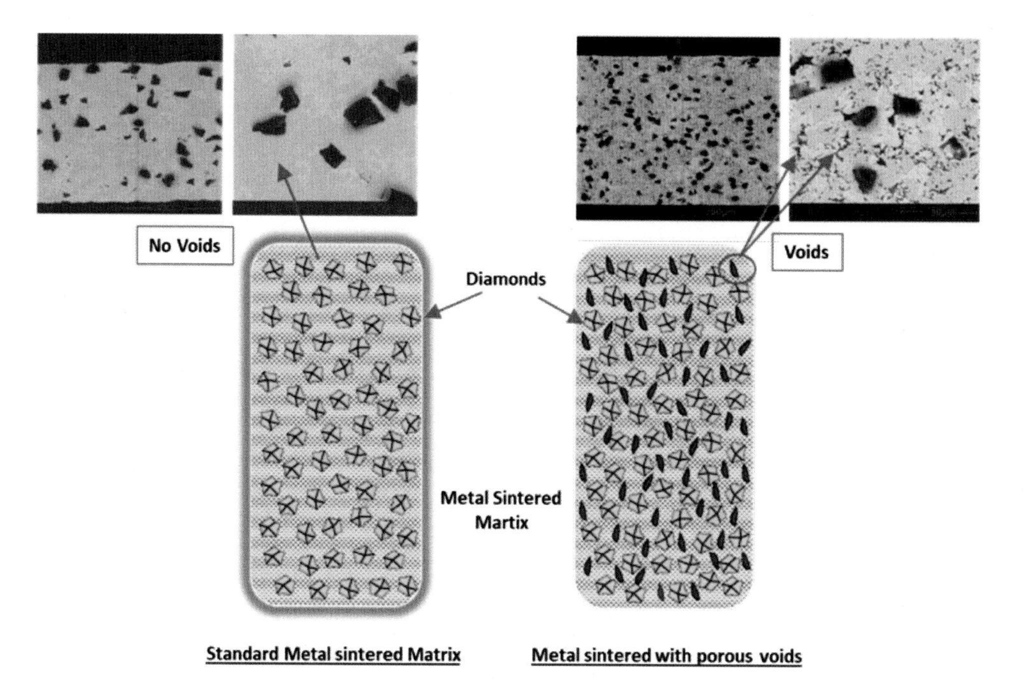

Figure 290. – SEM of metal sintered blades with a dense metal matrix and with a porous with porous metal matrix

For difficult glass applications the SD – Stealth laser is also a good option performing with good quality results and improved throughput. The stealth dicing is also used on glass out of the semiconductor industry for dicing glass panels of different products requiring round corners that conventional straight cuts cannot handle.

18.14 – SPECIAL GLASS APPLICATIONS

Following are a few examples of other glass applications requiring different dicing attentions:

The communication market has some dicing applications of Opto – Electronic Components with Glass integrated and tunable optical components using VLSI compatible technologies and other technologies such as MEMS. These technologies are used to produce platforms that are called planar light wave circuits (PLCs) and are providing dramatic size reduction of optical circuits specially for optical elements such as filters, switches, detectors, and lasers. Following are some of the main products in the marketplace: Wave guides, Optical connectors and transceivers, MEMS, DWDM (Dense wavelength division multiplexing) filters, Optical amplifiers, Tunable lasers/Laser modules, Fiber optic couplers, Optical ICs and others. The large variety of products involves a wide range of different materials with different hardness and brittleness characteristics. Among the materials are Silicon, Silica, Silica on silicon, GaAs, LiNb0, Sapphire, Quartz, InP and YVO4. The most common materials used in the communication market are silica, silicon and a combination of both. Silica glass (SiO2) contains very little metallic Impurities and can be called "pure" glass. Silica can hold high temperatures around 1000°C and is an advantage in some semiconductor applications. Some of the substrates, have special coatings like AR (Anti reflection) and IR (Infra-red) coatings that are hard and brittle causing more challenges to the dicing process. The main challenge in most cases is the surface finish required on surfaces designed to transmit light. Most specs for surface finish of the end product are in the range of 20–50nm Ra. The spec is normally achieved manually by a few lapping and polishing steps which is a time-consuming process. The requirement of the dicing process is to get the right spec or as close as possible to the desired surface finish spec in order to minimize the lapping & polishing process time. Following are a few parameters adding to a process that can meet the above surface finish requirements: Lowering the surface tension of the coolant by adding coolant additives can contribute to achieving the quality goal of the surface finish. Dynamically balancing the flange and blade assembly to 0.02 mm/sec. was found to be necessary and may be a tremendous improvement and a key factor to improve the kerf surface finish of waveguide and similar type substrates. The dynamical balancing needs to be done on the saw by measuring the vibrations and adding weights to the flange assembly. As discussed already, an optimized resin blade that wears enough to expose new sharp diamonds is a must and is probably the number one key important parameter to achieve the required specs. Below are a few examples of Opto-Electronic applications requiring superior surface finish (see fig. 291).

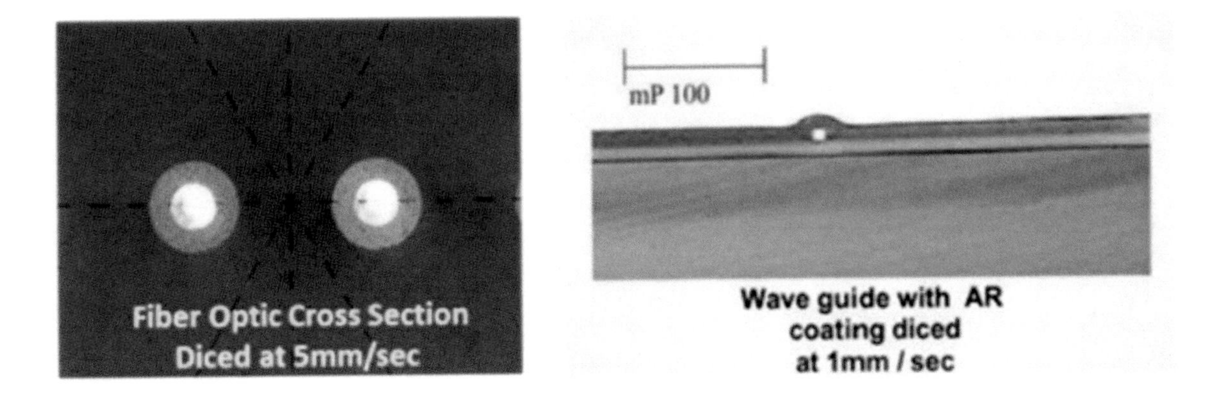

Source - ADT

Figure 291. – Optoelectronic substrates requiring superior surface finish

Some Opto-Electronic components require an 8° angular cuts that are needed to suppress back reflection in fiber optic components. Needless to say, the surface finish requires to be as described above. (see fig. 292).

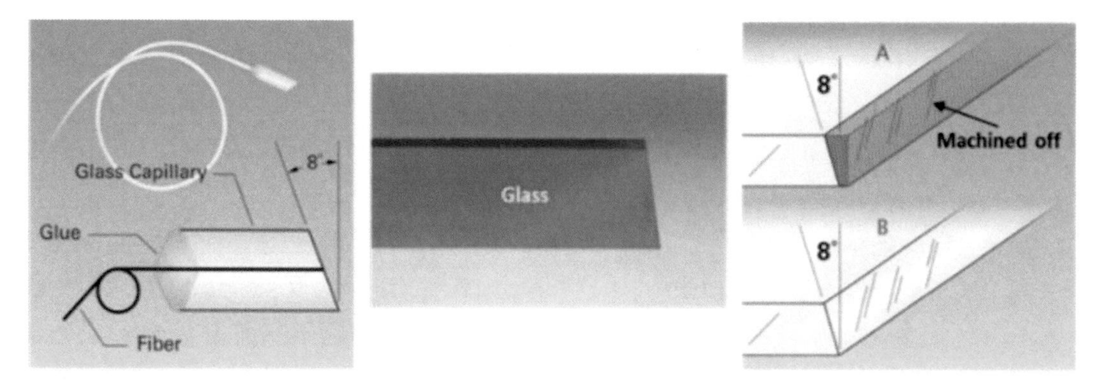

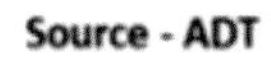

Figure 292 – Glass optoelectronic substrates diced and lapped at 8 degree angle

There are 3 options to create an angular 8° surface:

A – Grind and polish the surface after a 90-standard cut.

B – Use a jig on the saw to tilt and dice the substrate to the required 8°.

C – Use a special dicing saw with the ability to tilt the spindle to 8°.

(see fig. 293 describing the 3 options).

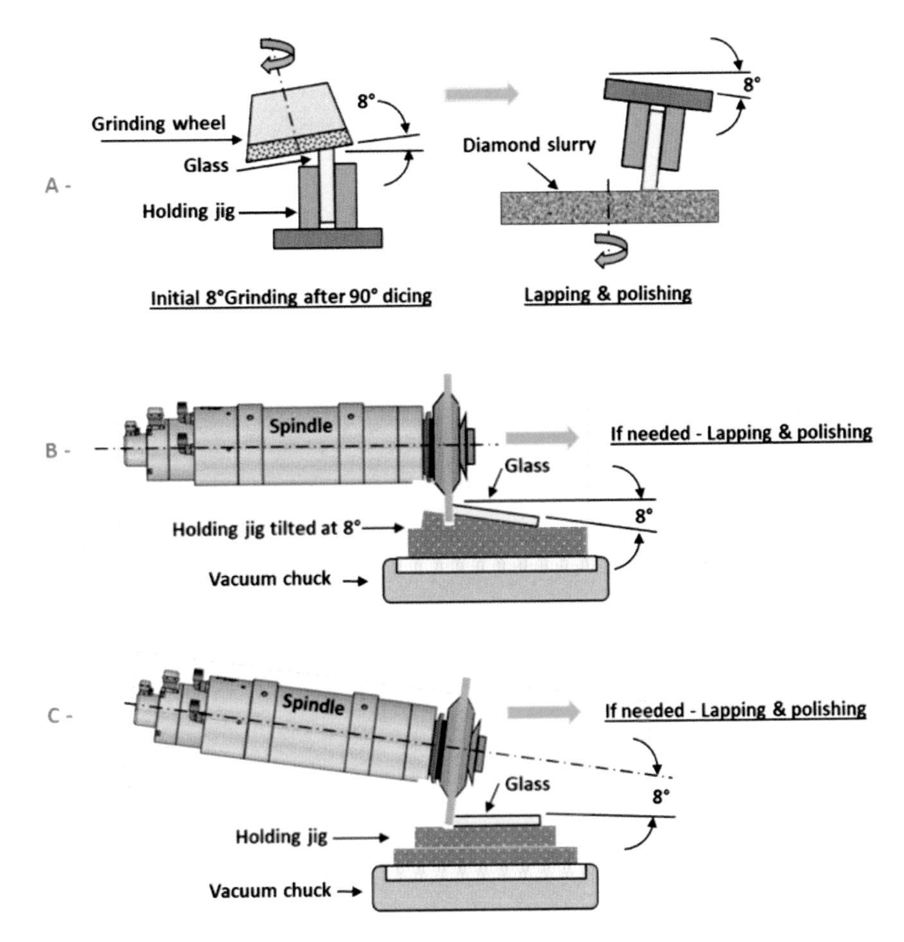

Figure 293. – 3x different methods to create the side 8 degree angle

The above dicing methods mainly using the spindle tilt and mounting the glass substrates on a tilted jig requires attention to the mounting method in order to minimize backside chipping. A protecting hard glue type coating is mounted on the back side of the substrate prior to wafer tape mounting. The idea is to have a hard backing well mounted to the glass to have a good substrate support that will minimize any substrate movement and minimize BSC. There are a few mounting methods, following is one option sketch. (see fig. 294).

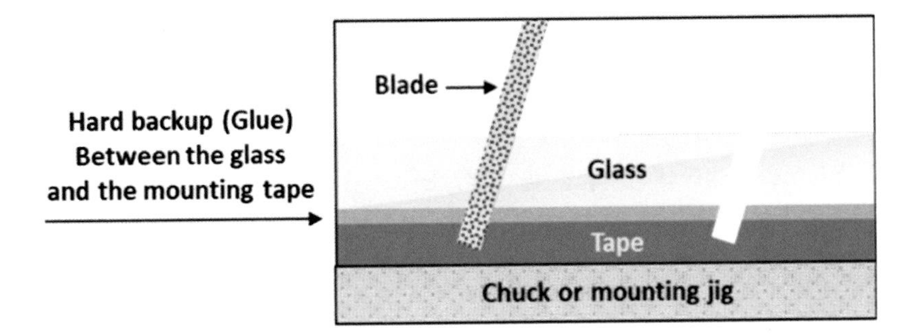

Figure 294 – Hard protective glue on the back side glass

The 8° angel is a common angle to suppress back reflection in fiber optic components, however other angles are used anywhere between 6 – 15 degrees. Major dicing saws manufacturers are offering saws with tilting spindles to the Above degrees with easy adjustments and accurate degree tolerances. See fig. 295 of a tilted spindle saw.

Figure 295. – 8 degree tilted spindle

Another glass application that is very different from the very thin substrates are the very thick glass substrates in the range of 2–20 mm thick. There are glass applications requiring cut through and applications requiring partly cut depths of a variety of cut depth. Needless to say, the cut through is more challenging. Below is a sample of a 16mm thick substrate diced almost through. (see fig. 296).

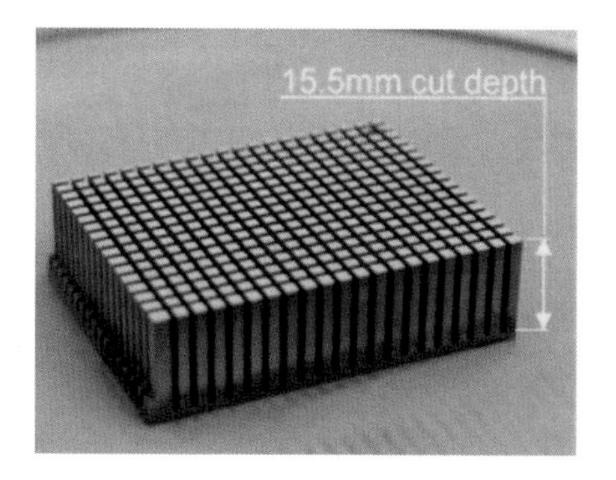

Figure 296 – 16mm thick glass diced almost through at a small index

When dicing partly through, the mounting is not an issue. However, when dicing through it is important to eliminate any movements that can cause cut quality issues and substrate and blade breakage. In most cases a wax or glue mounting is used however in some cases a thick UV is also an option depending on die size and cut depth. Blade exposure needs to be large enough and in most cases a 4" type saw is needed. Cooling a blade with a large exposure requires special cooling arrangements. See the blade coolant section with the many coolant options. What type of blade to use is critical and requires investigation and optimization. Blade thickness depends on the cut depth and is a matter of common sense and testing. The diamond grit is relatively large in the range of 45–88 micron depending on the loading and the cut quality. Spindle rpm and feed rate are also a function of loading and cut quality. Lower rpm will result in higher blade wear to expose new sharp diamonds. Changing the feed rate is done together with optimizing the spindle rpm and a function of optimizing the cut quality and the best throughput. The type of blade in most cases is a resin blade in a thickness range of .010"–.030" (0.250 mm–0.760 mm) depending on cut perpendicularity and blade loading. A few resin type blades can be used, A standard shape, a resin serrated, a unique SPG – Special side groove resin blade and a steel core resin blade. The standard blade geometry may perform well if the loading is not high and the quality spec of cut perpendicularity is open however, if the spec is difficult to achieve; other blade geometries can be used. One good option is to use a resin serrated blade. Serrated blades do minimize the loads However, they wear faster but perform with less loading and do improve cut quality mainly cut perpendicularity. Below is a thick optic glass prism diced using a serrated resin blade. (see fig. 297).

Figure 297. – Edge of a Thick glass Optic Prizm

Resin serrated blades are widely used however, they lose the blade edge geometry. In order to solve this problem and make sure the device profile has a straight edge geometry, a special thick mounting substrate is glued to the production substrate. This is needed for the blade to penetrate deep into the base mounting substrate resulting with a perfect profile on the production glass substrate (see fig. 298).

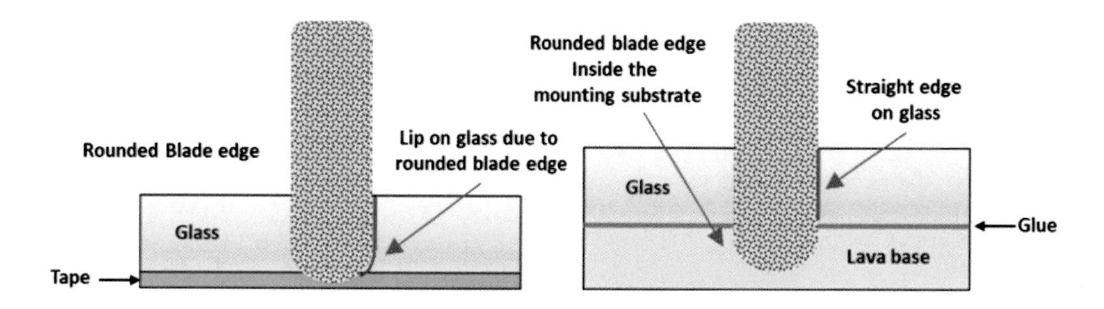

Figure 298. – Tape mounting compare to thick base mounting affecting the cut perpendicularity

Below is a thick optic prism mounted on a thick lava base substrate diced using a serrated resin blade. (see fig. 299).

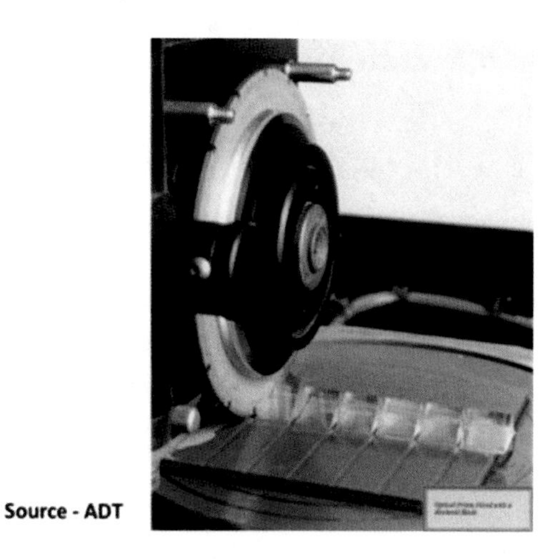

Figure 299 – Thick Optic Prizm mounted on Lava base and diced using a resin serrated blade.

Special grooved blades have shallow grooves on both sides of the blade. The side groves are designed to minimize the side friction and to cool better the blade during dicing. This blade geometry proved to perform better not only on thick glass but also on thick ceramic and similar substrates. The SPG blade can be used with standard flange sets and with high cooling flanges. The high cooling flange set is designed to flow the water coolant from the center of the flange to the edge of the SPG blades via the grooves on the blade sides. See in the flange section, the high cooling flange design. Fig. 300 shows the SPG blade geometry.

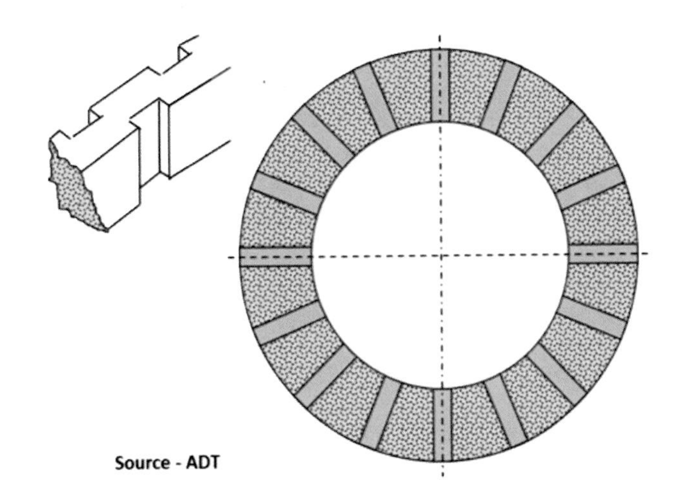

Figure 300. – SPG – Special grooved blade geometry

The last option dicing real thick substrates and maybe the best solution is using a steel-core resin blade. The Steel-core blade is a very stiff blade performing very well on cut perpendicularity. (see fig. 301).

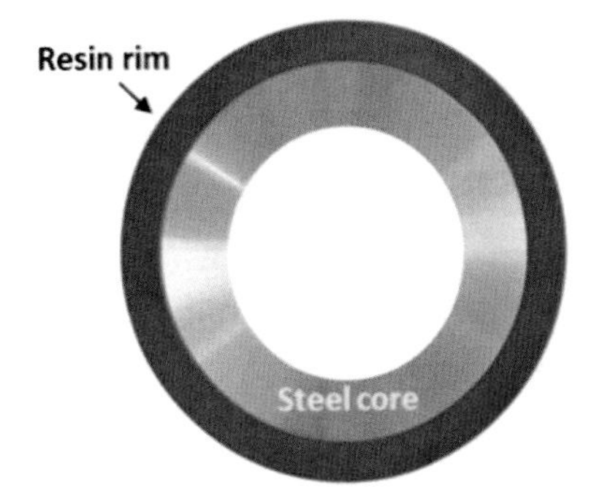

Figure 301. – Steel Core Resin blade

Glass-type substrates have become a major product in the semiconductor industry mainly due to its excellent chemical, mechanical, electrical, and optical properties. Glass is playing an increasingly important role in the semiconductor market, adopting various functions in integrated circuit (IC) semiconductor devices. A lot of good information on the many different glass-type products can be found via the internet. The major companies are Shott, Corning and others, some exploring new and special products for the semiconductor and microelectronic industries.

18.15 - SILICON ON GLASS (SOG)

Another application related to glass is silicon on glass (SOG) used in a few applications mainly In Pressure Sensors, MEMS, and others. As Silicon on Glass is a bit similar to Silicon on Sapphire, which was reviewed already, the SOG will be discussed in a short mode. There is one main difference between dicing glass and sapphire as glass is not as hard and brittle as sapphire. This makes the dicing of silicon on glass easier. Pressure sensor is easier to dice as it is possible to use a resin blade with relatively small size diamonds. As described in the silicon on sapphire application, silicon needs a small diamond in the range of 6 mic. The glass needs a larger diamond size of 30–45 mic. However, it can be diced at slower feed rates using a 15–20 mic. with low diamond concentration. This will cause more chipping on the silicon however it can be a compromise if the spec allows it. It all depends on the total thickness of the sensors. In the past pressure sensors were quite thick in the range of 4 mm (.157") thick with different die sizes. A 4mm thick substrate requires a relatively thick blade which can overload and create chipping and perpendicularity problems. This issue was resolved for some users by splitting the dicing into 2 steps. Step one was dicing the silicon using a nickel blade with 4-6mic. grit and step no. 2 using a resin blade with 30–45 mic. grit. Today's most silicon on glass pressure sensors are much thinner and are easier to dice. The MEMS Pressure sensors are in most cases more sensitive to diamond blade dicing and are singulated by Stealth dicing or Plasma etching similar to what was reviewed in the MEMS section. Below is a photo showing the thicker pressure sensors. (see fig. 302).

Figure 302. – Pressure sensor substrates

18.16 - QUARTZ

Dicing quartz is close to dicing glass. However, it is harder than glass, with 7 Mohs in the hardness scale. Quarts is brittle and tend to cause chipping during dicing. It is used widely in frequency control applications due to its temperature stability and low cost. Quartz provides the best performance in narrow band SAW filter, SAW resonator, SAW oscillator, clocks, and data recovery unit. SAW filters (Surface Acoustic Wave filters) is one major application using Quartz. Below is a generic sketch of a SAW filter. (see fig. 304).

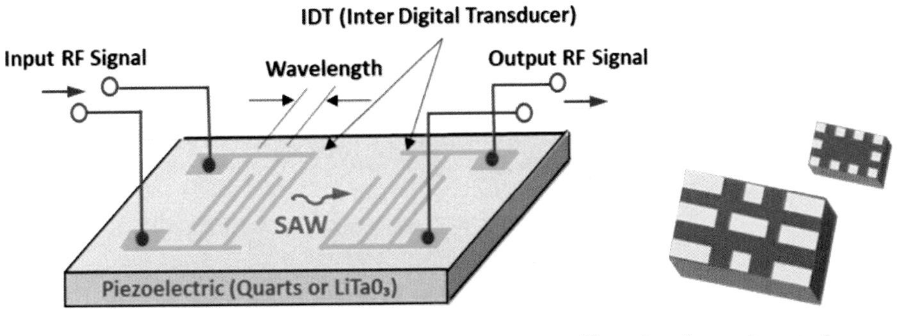

Figure 303. – Generic SAW filter substrate

SAW filters are widely used in cell phone and base station applications for filtering Surface Acoustic Wave filter. SAW filter is a filter whereby the electrical input signal is converted to an acoustic wave by so-called inter-digital transducers (IDTs) on a piezoelectric substrate such as quartz. There are many different SAW type geometries that can be diced both with 2" dicing saws and with 4" dicing saws. It all depends on the device thickness and quality spec. 2" blades will perform with better cut quality and 4" blades with improved blade life. Due to quality demands in the last years the industry is aiming more for 2" dicing saws. Resin is the main type of blade used in a high production mode. There are new resin matrices developed to minimize loading and to perform with better cut quality. Blade thicknesses are in the range of .006"–.010" (0.150–0.250 mm) with diamond grit sizes of 20–45 mic. Spindle speed vary depending on cut quality and spindle vibrations: 2" spindle 20–35 Krpm, 4" spindles 8–15 Krpm. Feed rates is also a matter of quality and is in the range of 2–10 mm/sec on both spindles 2" & 4".

Mounting depends mainly on the substrate thickness. Thicker substrates can be mounted on glass or other base using wax or glue. This requires much more handling for mounting and dismounting after the dicing. Thinner substrates are normally diced using tape mounting. Blue tape is preferred due the lower tape cost however on very small dies UV tape is used.

Below is a 2 mm thick Quartz wax mounted on glass diced on a 4" saw using a resin blade .008" thick x 35 mic. grit. Attention to the same quality on both the Quartz and the glass base. (see fig. 304).

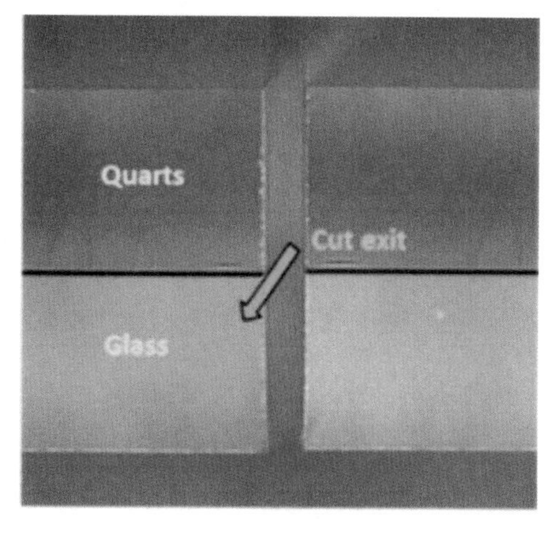

Source - ADT

Figure 304 – Dicing Quarts on glass base

18.17 - LED (LIGHT EMITTING DIODE) SUBSTRATES

LEDs are gradually changing the world of light technology, forcing manufacturers to change their manufacturing procedures and continuously improve to meet new market demands. Most "white" LEDs in production today use a 450 nm–470 nm blue GaN (gallium nitride) LED covered by a yellowish phosphor coating usually made of cerium-doped yttrium aluminum garnet (YAG: Ce) crystals that have been powdered and bound in a type of viscous adhesive. This special phosphorous coating is a crucial material in the manufacture of white LEDs and is the reason for the "magic" white LED light. Below are a few cross sections of typical LED technologies. (see fig. 305).

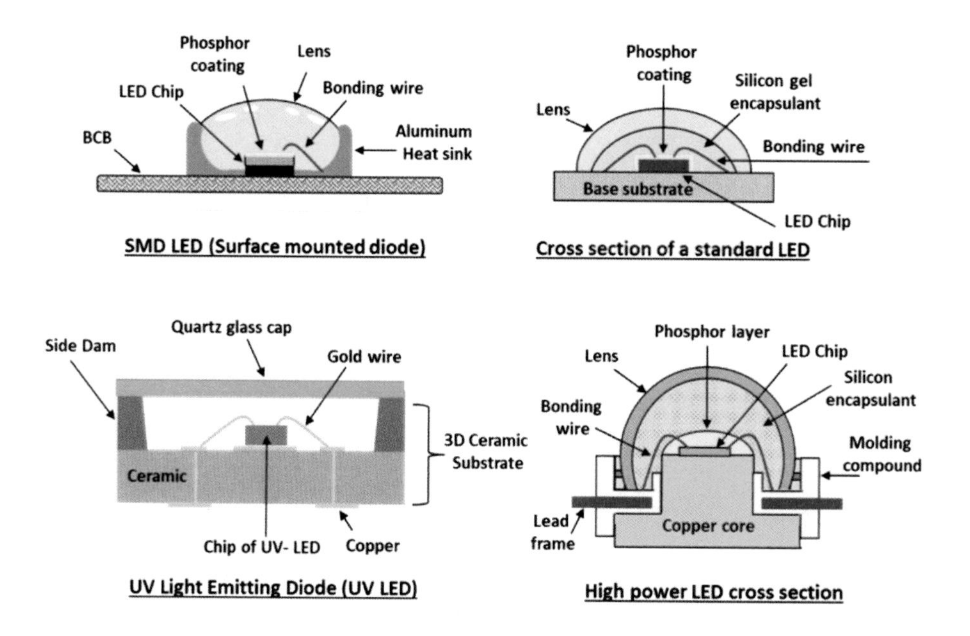

Figure 305. – Cross sections of typical LED's technologies

The above cross section sketches are not part of separating the LEDs using diamond dicing, they are the end product after the manufacturing process. Almost all LEDs are manufactured in an array geometry substrate and need to be singulated on dicing saws. In general, there are three types of LED base substrates: PCB type, EMC (Epoxy molding compound similar to QFN) and Ceramic type.

PCB. It sounds like an easy one but does have some issues. Dicing through just PCB is relatively easy and can be diced with a few blade matrices. The main issue in some substrates dicing PCB is the need to dice through copper & gold plated vias. The copper is soft and creates smearing and burrs that can cause rejects. There are a few options to solve this problem, part is on the LED manufacturers side to optimize the copper properties and to minimize the thickness of the copper plating.

Optimizing the copper plating process is not an option in many cases, so the solution needs to be on the dicing side. The best blades found to perform well are thin nickel blades with diamond grits of 10, 13 and 17 mic. In addition, the nickel hardness and diamond concentration are other options to evaluate and optimize. PCB can be diced using 2" and 4" spindle and the dicing parameters need to be optimized on the saw. Spindle rpm on 2" spindle can run between 25–30 Krpm and on 4" spindle 10–18 Krpm. Feed rates are relatively high and should be optimized per the application spec. Customers are using feed rates up to 6"/sec (150 mm/sec). Almost all BCB substrates are diced using tape for mounting. The type of tape depends on the LED/device size. Any LED size of 4x4 mm and smaller is normally diced using UV tapes. Another parameter that can help is using a cooling additive to lower the surface tension of the coolant in order to better cool the blade and the substrate. The coolant can also be chilled using a chiller simultaneously with the coolant additive. Fig. 306 shows a clean cut with min. burrs in the via and a problematic via with a lot of burrs.

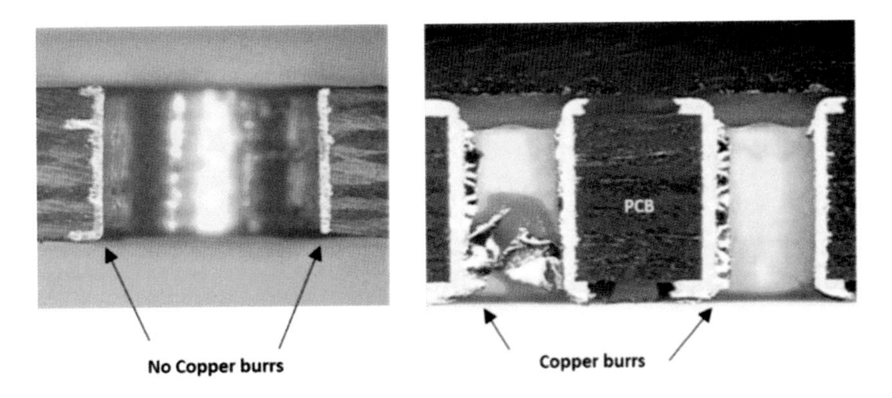

Figure 306 – Cross section of vias with copper burrs and without copper burrs

18.18 - EMC (EPOXY MOLDING COMPOUND) SUBSTRATES

The EMC substrates are in a way similar to dicing QFN substrates, which will be discussed later in detail. Most EMC substrates are made on a lead frame configuration. (see fig. 307).

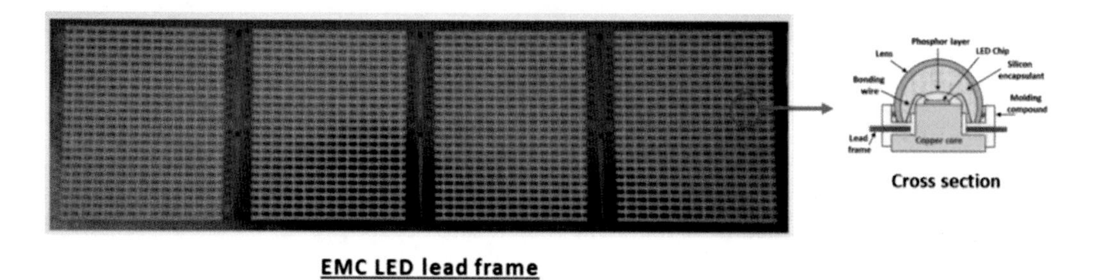

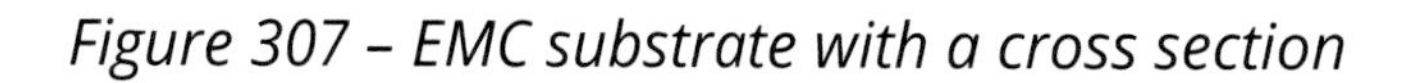

Figure 307 – EMC substrate with a cross section

EMC LED Lead frame, sometime called "Pre-mold LED lead frame," applies to high-power LED package providing a combination of copper and high-performance thermosetting plastics or in different warding, an Etching copper frame + EMC process.

This design has the advantages of high heat resistance, High-density, Anti UV, High current, small volume and is suitable for large scale production. The main dicing problems with this type of substrate are mold chipping and copper burrs. (see fig. 308).

Figure 308. – Copper burrs on a street cross section of an EMC led substrate

Dicing the EMC LED lead frames can be done with nickel blades and metal sintered blades x 30–63 mic. grit. Both with 2", 3", and 4" diameters. Blade thickness depends on the street width and the substrate thickness. As mentioned, a similar product to the EMC LED Lead frame is the QFN, which will be discussed in more details.

18.19 - CERAMIC LED LEAD FRAME

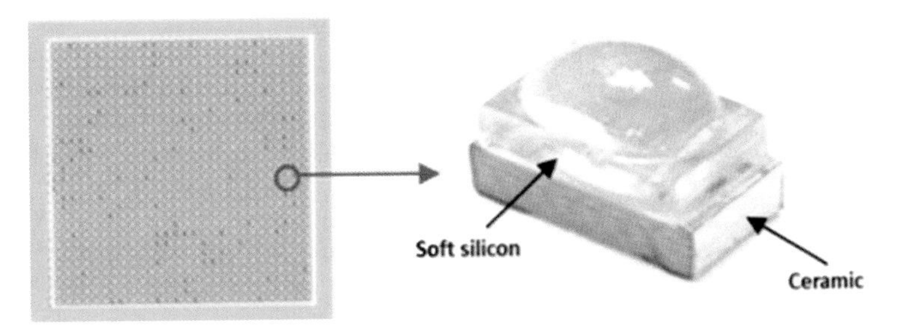

Figure 309. – Ceramic LED substrate

The Ceramic LED lead frame base is normally made with Alumina 96% and 99.9%.

However, the industry is aiming for softer ceramics to ease the dicing process. The substrate design can consist of a few different materials creating quality problems. One problematic material is a soft silicon layer that tends to smear outside of the ceramic LED sides.

A good way on the manufacturing side is to eliminate the soft silicon in the street area or to use a harder silicon that will minimize blade loading and smearing. Another potential problem is copper smearing. (see fig. 310).

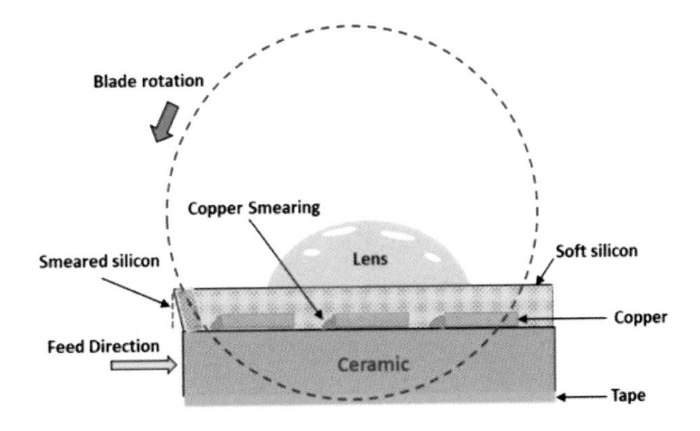

Figure 310. – Smeared soft silicon at the exit of each cut

Silicone-based housing materials have a relatively high thermal stability. This broader temperature stability range makes silicone a suitable candidate for LED housings. Blade exposure can also be an issue, in some cases a large blade exposure is needed in order to eliminate any contact between the flange edge and the lens. (see fig. 311)

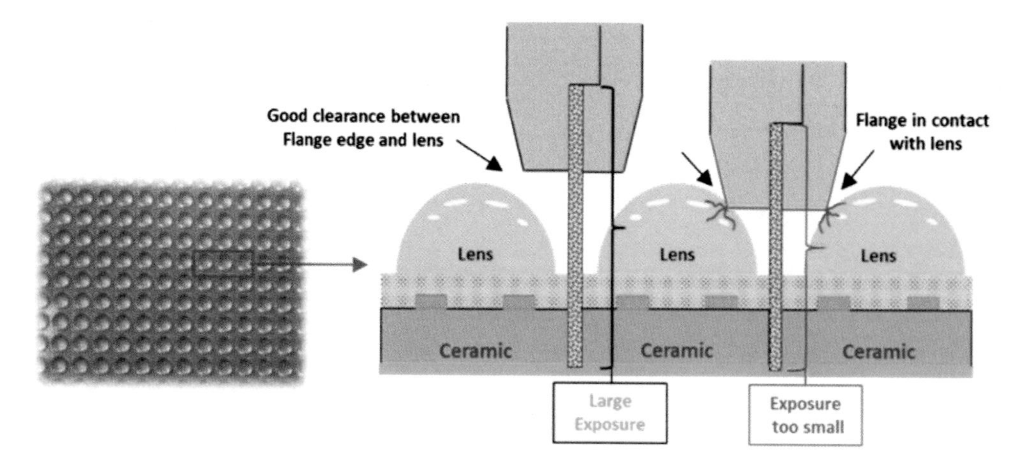

Figure 311. – The important of blade exposure to eliminate lens damage.

Resinoid blades in 2", 3", and 4" diameters are used for dicing ceramic LED substrates. Blade thickness can range between .005"–.010" x diamond grits between 45–88 mic. grit depending on the substrate thickness and the street width. A few resin matrices can be used depending on the blade loading and quality specs. The resin matrices can be from very soft to hard and can be optimized depending on the spindle loads and quality requirements. It is always a compromise between quality and through-put.

Spindle speeds are also flexible and need to be optimized depending on the dicing spec. The spindle RPM on 2" and 3" O.D. blades range from 25–35 Krpm and 10–15 Krpm on 4" blades. Feed rates normally used are 2–25 mm/sec depending on production quality and the through-put goal.

The LED technology is an amazing market improvement of many different products like the housing light, the car industry, military products, and many others. A lot of good information can be found on the internet. See in the reference section.

18.20 - HTCC AND LTCC - HIGH AND LOW TEMPERATURE COFIRED MULTILAYER CERAMICS

Source - Adamant Namiki Precision Jewel Co - Japan.

Figure 312. – HTCC and LTCC substrates

HTCC is a ceramic based on alumina (Al2O3) and aluminum nitride (AlN). LTCC is a ceramic with glass mixed into alumina and is also called glass ceramics. The differences between HTCC and LTCC:

LTCC is the cost-effective and upgraded version of HTCC with firing temperatures in the range of 700°C –960°C. Likewise, HTCC is a conventional ceramic tape casting similar to the processing of LTCC substrates but with higher sintering temperatures of 1600° C.

The lower fabrication temperature of LTCC is achieved by the additive manufacturing process of low melting point glasses, ceramics, or a combination of both. The lower sintering temperatures of LTCC is an advantage as it allows to add in the manufacturing process lower resistive metals like silver, gold, platinum, palladium, and others as conductors. LTCC (Low Temperature Co-Fired Ceramics) is a multi-layer glass ceramic substrate which is Co-fired with low resistance metal conductors at low firing temperature (less than 1000°). It is sometimes referred to as "Glass Ceramics" because its composition consists of glass and alumina. In simple wording the manufacturing process is basically making multilayer circuits and then laminating them together. Material composition of LTCC is 50% Al_2O_3 and 50% glass. HTCC is 92% Al_2O_3.

HOW IS LTCC MADE?

LTCC is prefabricated green foils. Depending on the given design, holes are punched at given positions using laser or mechanical punching and then filled with a metal paste. These so-called vias form the vertical contact in the multilayer structure. The metal pastes are then screen printed onto the green foil to form the planar electrical conductor paths. After the metal structures have been applied, the green foils are pressed together in a specified sequence and sintered between 850 °C–900 °C. Sensitive structures or surface components can also be applied or fitted. Subsequently, the LTCC substrates are separated/singulated by diamond dicing and tested for their functionality.

Figs. 313 and 314 show two generic manufacturing flows of the LTCC.

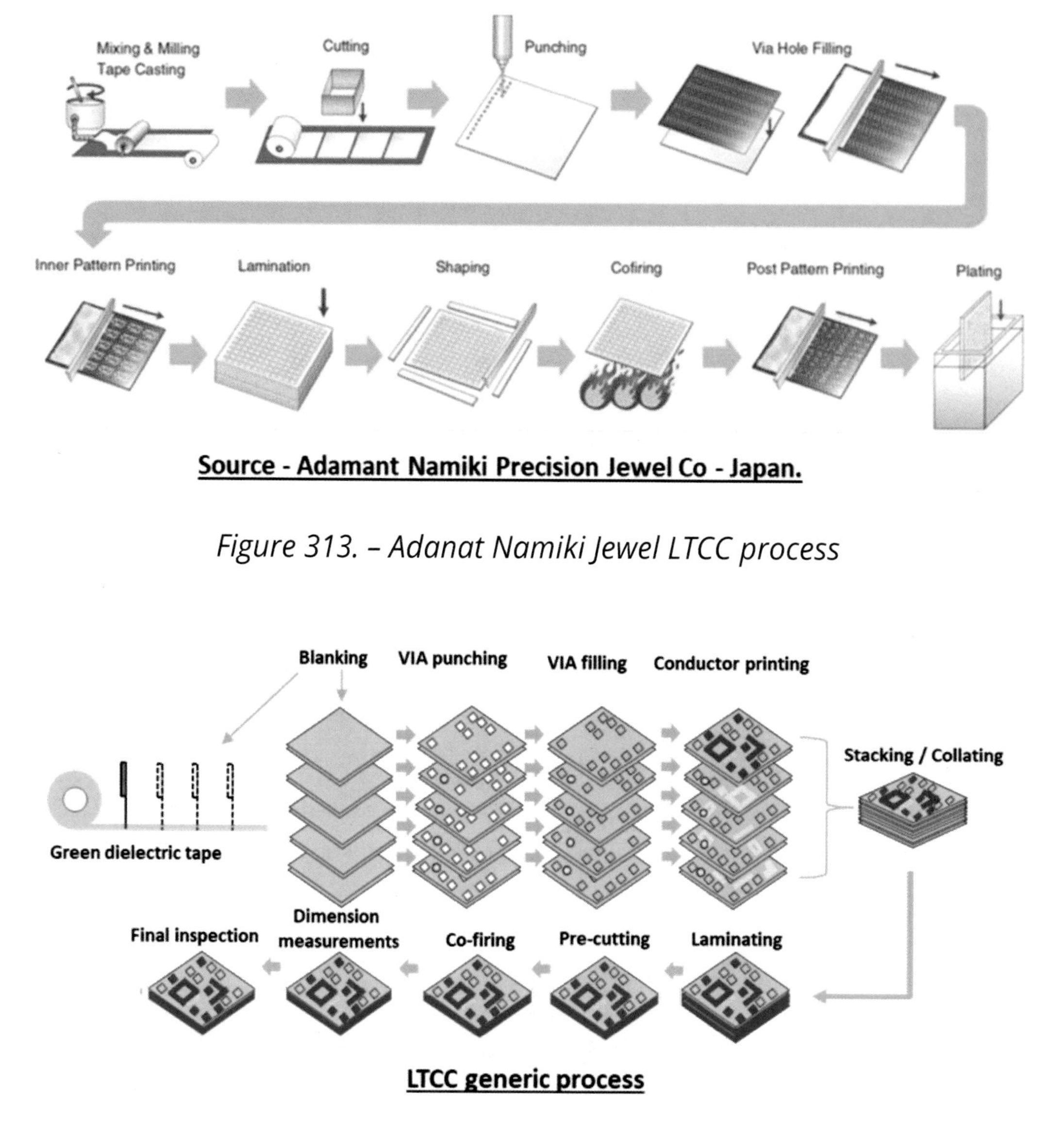

Figure 313. – Adanat Namiki Jewel LTCC process

Figure 314. – Another generic LTCC manufacturing process

The above generic process flow varies depending on the end product design. It can be with more or less specific end product parameters.

Main LTCC application:

MEMS Sensor Packages

RF SiP / FEM Substrate

Frequency Device Package

Medical Equipment

Industrial Equipment

LED Chip Carrier

Below is a typical LTCC end product showing many different elements designs on one multilayer device. (see fig. 315).

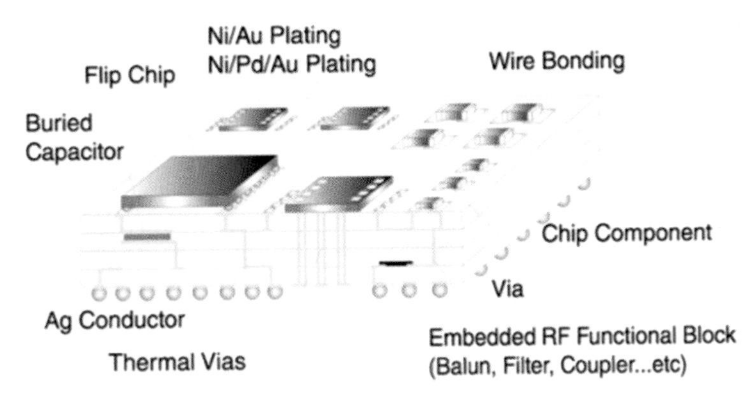

Source - Adamant Namiki Precision Jewel Co - Japan.

Figure 315. – LTCC with the many different design elements

HTCC – High Temperature Co-Fired Ceramic is a popular material choice for hermetic packaging due to its desirable electrical properties, high mechanical strength and good thermal conductivity. It is harder compared to LTCC, in the range of 1207 (Knoop) kg/mm^2. Firing sintering temperature – 1600°C.

Main HTCC application:

- High Frequency/High Temperature Electronics
- High Reliability Industrial/Commercial
- Information
- Medical
- Military
- Optoelectronics
- Telecommunications

Singulating both the LTCC and HTCC is a relatively common diamond dicing process. Mounting in most cases is done with UV tape, mainly to be able to dice deep enough in order to minimize backside chipping. LTCC is more sensitive to chipping and microcracks due to the material combination of alumina and glass, which makes it more brittle. Resin blades are the best matrix to use with a diamond grit of 15 to 30 mic. depending on substrate thickness. In many cases, more wearing resin matrices are needed to be optimized in order to minimize loading. Using uncoated diamonds for better cut quality will perform with higher wear, losing diamonds faster and exposing new sharp diamonds. (see fig. 316).

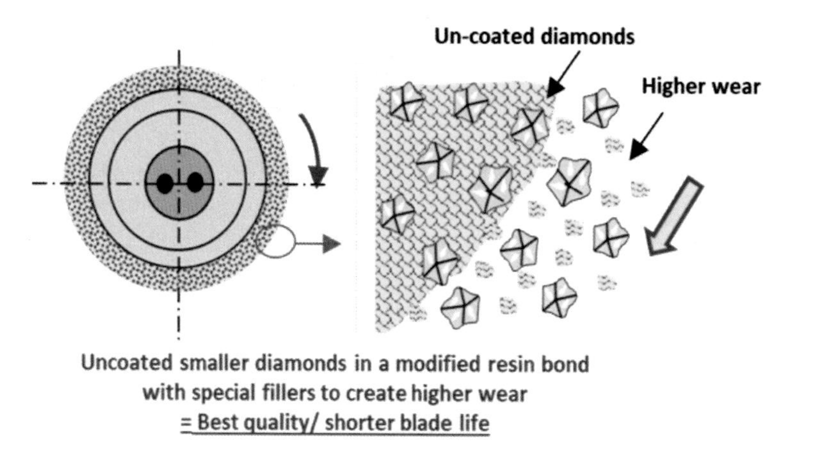

Figure 316. – Modified resin bond and uncoated diamonds to create higher blade wear

HTCC is more similar to dicing hard alumina and is a bit easier achieving the good cut quality. More will be discussed in the dicing hard alumina section.

Both LTCC and HTCC can be diced on 2", 3", and 4" spindles. It all depends on the substrate thickness and quality requirements. 2" blades vibrate less than 4" blades and will perform with better cut quality but with higher blade wear. Spindle speed: 2" & 3" O.D. blades – 25–35 Krpm, 10–15 Krpm on 4" O.D. blades. Feed rate depends on cut quality and can range from 2–30 mm/sec.

Blade thickness depends on the substrate thickness and the street width and can range between .006"–0.20" (0.150–0.500 mm).

18.21 - ALUMINA SUBSTRATES

Figure 317. – Different Hard Alumina substrates used in the Microelectronic industry

Alumina – Al2O3 ceramic substrates is one of the most popular substrates in the microelectronic industry. It has excellent heat resistance, high mechanical strength, abrasion resistance, and small dielectric loss. The name alumina is very generic, as there are a lot of different types and applications using alumina as a base material for the above-mentioned reasons. There is a large spread of ceramic purities in the different alumina products, starting at 85% up to 99.9%. Let's concentrate on the most popular alumina's, 96% and 99.6%.

ALUMINA 96%:

This is a standard for thick film substrates which is widely used in the manufacturing of Hybrid microelectronic circuits. It is a low-cost substrate, with excellent electrical insulating properties, mechanical strength, good thermal conductivities, chemical durability, and dimension stability. Fig. 292 is an example of a Hybrid Circuits with a PCB on a ceramic substrate.

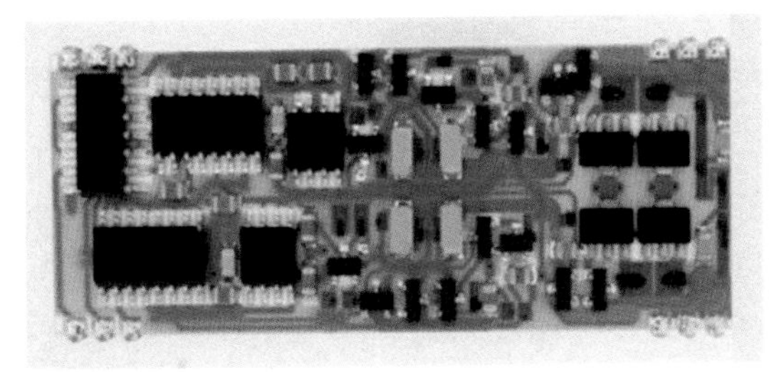

Figure 318. – Hybrid Circuits with a PCB on a ceramic substrate.

ALUMINA 99.6%:

A smooth and low porosity material, 99.6% Alumina is used mainly for thin film devices. The high purity and smaller grain size of the 99.6% Alumina allow the material to be smoother with less surface defects and to obtain a surface roughness of less than 1u-in. 99.6% Alumina offers high mechanical strength, low thermal conductivity, excellent electrical insulation, good dielectric properties and good corrosion and wear resistance. So, for more dense and smooth base substrates the 99.6% Alumina is the way to go. The hardness of Alumina 99.6% is 14.1 (1440) - KNOOP (kg/mm2) using a 1000gm weight or ROCKWELL 45 N – 83 Compare to 96% Alumina - 11.5 (1175) - KNOOP (kg/mm^2) using a 1000gm weight or ROCKWELL 45 N – 78. The much harder 99.6% Alumina makes the dicing much more challenging.

DICING ALUMINA 96%:

For many years—probably since the late 60s, 96% alumina has been diced on many applications. At the beginning, substrates were in the range of 1"x 1" and gradually went up to 5"x 5" and many in between. The thickness ranges from 0.10 mm up to 6 mm and in some cases close to 10 mm. The different thickness and substrate sizes affect the mounting method and the dicing blade and cutting parameters. Let's discuss the majority applications which are dicing 0.20 mm - 2 mm (.008"-.080") thick . Resinoid blades are almost the only blade to use. They can be of 2"-3" O.D. or 4" which in many cases is an advantage due to the less wear, performing with better blade life. Blade thickness depends on the street width and substrate thickness. Blades thickness can range between .004"-.012" (0.100 – 0.300 mm). The diamond grit on standard substrates range between 45 - 63mic. However, it can go up to 88mic. and down to 30mic. all depends on the quality spec requirements and the throughput requirements. There are other parameters that can be optimized in order to perfect the cut quality and will be discussed in the 99.6% alumina. The majority of the substrates are tape mounted using blue tapes and on the smaller die sizes using UV tape. However, on some difficult substrates, a much more solid mounting is needed. See in the Alumina 99.6%. Dicing parameters depend on the blade diameter and quality goal. On 2"-3" O.D. blade spindle speed ranges from 25-35 Krpm and 10-16 Krpm on 4" O.D. blades. Feed rate is a matter of processes optimization and quality spec. 1-20 mm/sec needs to be optimized for quality and throughput.

DICING ALUMINA 99.6%:

Dicing 99.6% Alumina is more challenging. The harder and more dense Alumina requires better blade optimization and better dicing parameter optimization. The substrate mounting also needs special attention especially when using thicker blades.

No need to repeat the dicing parameters and the blade parameters as they are close to dicing 96% Alumina and need anyhow to be optimized. As for the mounting, it requires some understanding. The resin blades used are designed to wear enough to minimize loading in order to perform well on cut quality. The blades have in some cases different fillers and lower diamond concentration to perform with higher wear and freer cut. This can cause too much blade side wear and cause the blade edge to get rounded. Having a rounded edge on a shallow cut depth creates a lip at the bottom side of the substrate which eventually causes backside chipping. This is very critical when dicing on tape and using a wider blade. (see fig. 319)

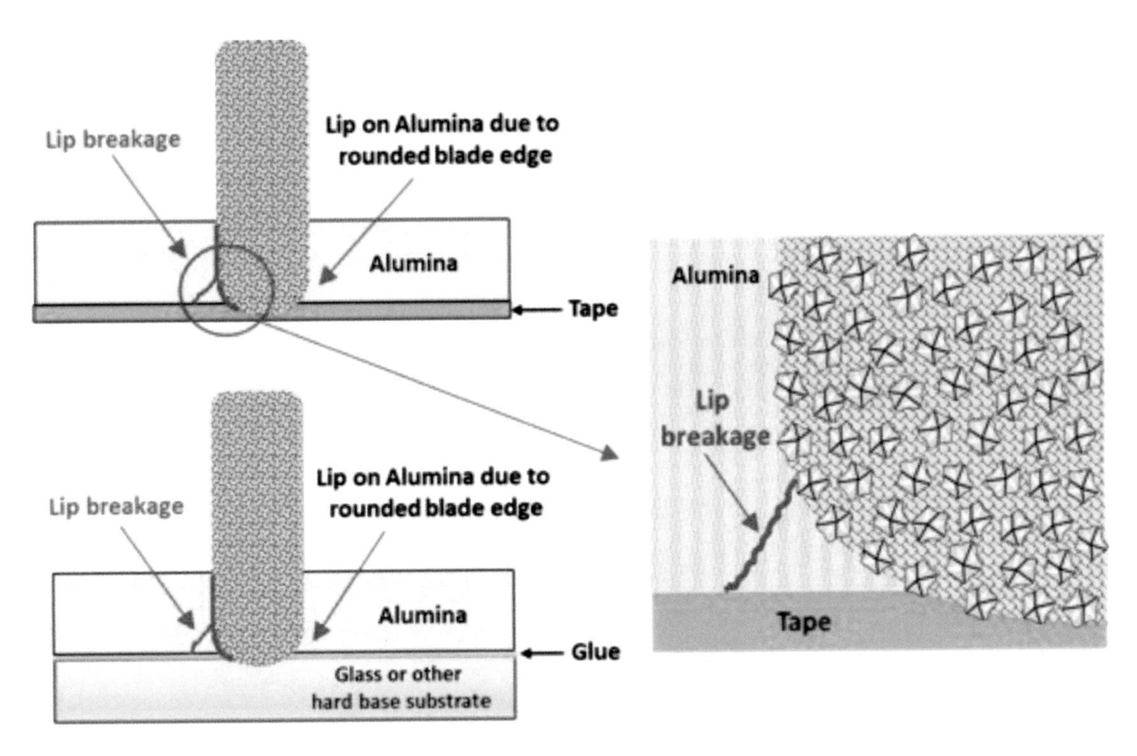

Figure 319. – Lip effect when dicing Alumina substrates on tape

To overcome this specific issue is to use a thick UV tape or use a wax or similar mounting to a hard base for mounting in order to cut deep enough so the rounded blade edge is way below the bottom side of the substrate. (see fig 320).

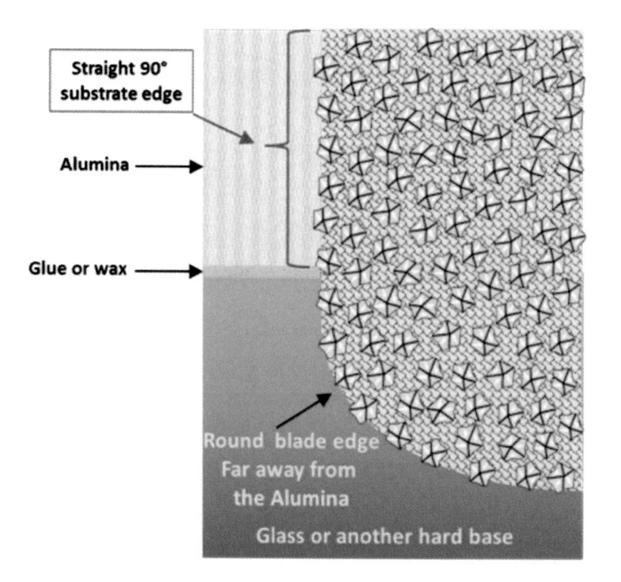

Figure 320. – Alumina substrate waxed on hard base to cut deeper and eliminate lip effect

Another important step that needs to be repeated is making periodic height sense on the saw to make sure the rounded blade edge remains well below the Alumina. A very strong mounting is also important to minimize vibrations that may cause die movements that can cause top-side chipping. This is even more critical when dicing very small dies in the range of less than 3 mm x 3 mm. (see fig. 321).

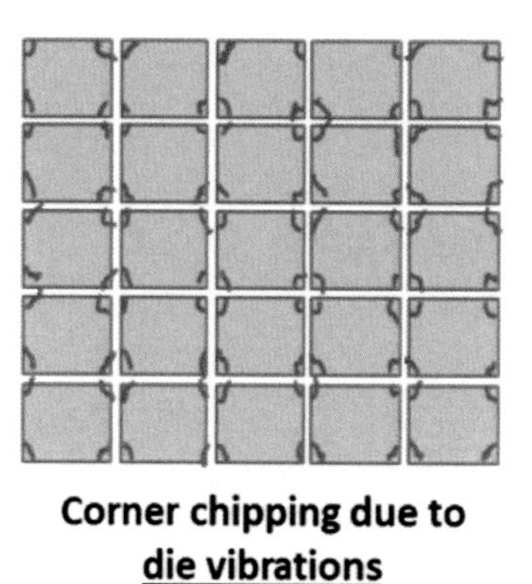

Figure 321. – Chipping at the cross section of the dicing streets due to substrate vibrations

The above-described issues are similar to other substrate material and the solutions can be adopted to those substrates.

18.22 - ALUMINUM NITRIDE [ALN]

(Gray in color)

Figure 322. – AIN – Aluminum Nitride substrates

Aluminum Nitride, AIN Ceramic formula is a newer material in the technical ceramic's family. While its discovery occurred over 100 years ago, it has been developed into a commercially viable product with controlled and reproducible properties. It features a combination of very high thermal conductivity, 5 times better than Al2O3 and has excellent electrical insulation properties. AIN hardness is 1170 Knoop (kg/mm^2 / 100g) better mechanical strength than standard Al2O3. This makes AIN great for use in power and microelectronics applications. AIN ceramic products are mainly used in high-density hybrid circuits, microwave power devices, semiconductor power devices, power electronic devices, optoelectronic components, semiconductor refrigeration and other products as high-performance substrate materials and packaging materials.

Dicing and mounting properties of AIN are similar to ordinary Alumina but needs to be optimized per the required specs.

18.23 - DICING THICK ALUMINA SUBSTRATES

Another Alumina that needs special attention is dicing very thick substrates regardless of the Alumina type. Dicing very thick Alumina substrates of 6–15 mm thick needs a different dicing approach. The loading during the dicing of very thick substrates is high and requires high power spindles usually in a 4″ configuration. The 4″ spindle/blades require not only to deal with the load but also to deal with the large exposure needed that cannot be available in 2″ and 3″ O.D. blades. The blades that are used are in the range of 4″–5″ outside diameters. The type of blades used for dicing the thick Alumina substrates are relatively thick Resinoid blades and in some cases Vitrified blades that are very hard to minimize blade wear, however for better quality more wearing resinoid blades are preferred. Blade thickness range used are .012″–.030″ depending on the substrate thickness. The diamond grit size on those blades is larger in the range of 63–105mic. Using larger diamond grits may not be enough and requires matrix and diamond concentration optimization. In addition, special blade edge geometries are used to minimize loading. Following are a few options that are used in the market: Resin serrated blade, SPG (Special grooved blades), resin and Vitrified steel core blades that can also be with serrations.

The steel core blades are probably the best choice mainly because of the good stiffness that performs with good cut straightness, a problem that occurs when dicing thick substrates. Fig. 323 shows the different blade geometries:

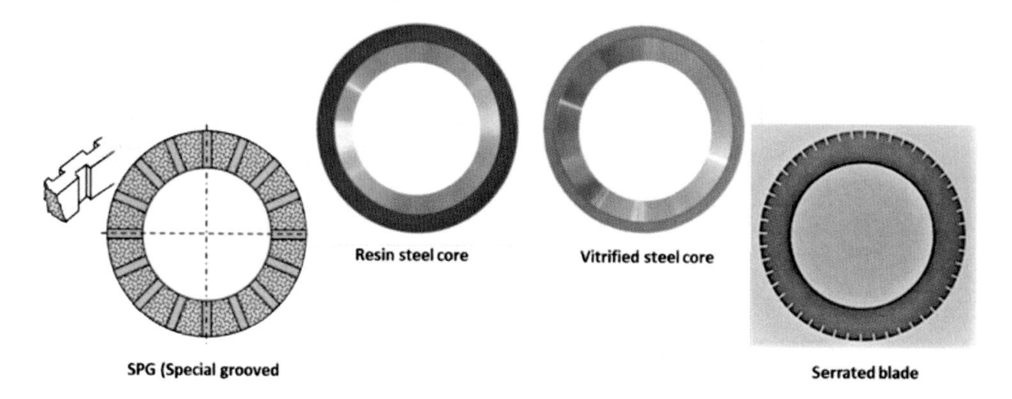

Figure 323. – Typical blades used for dicing thick Alumina substrates

Coolant is another option to help dicing through the thicker alumina substrates. Coolant additives lower the surface tension of the coolant, helping the coolant penetrate the deep kerf cooling the blade and the substrate. Using the 4.6" SPG blades with the high cooling flange is another great help when dicing thick Alumina substrates. Fig. 324 shows the high cooling flange set-up and Fig. 325 shows a generic diced thick substrate.

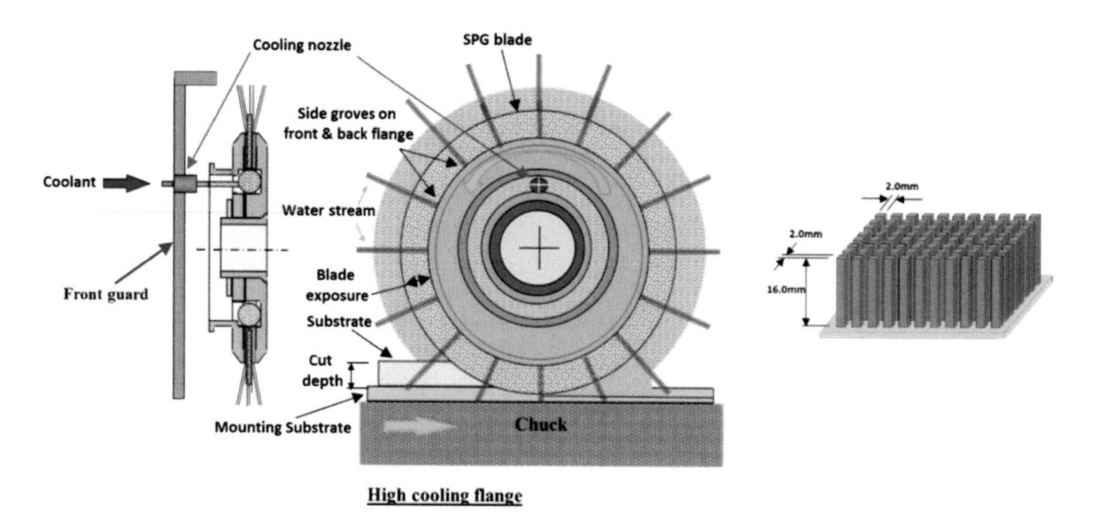

Figure 324. & Figure 325. – HCF flange used for dicing very thick substrates

18.24 - MLCC (MULTILAYER CERAMIC CAPACITOR)

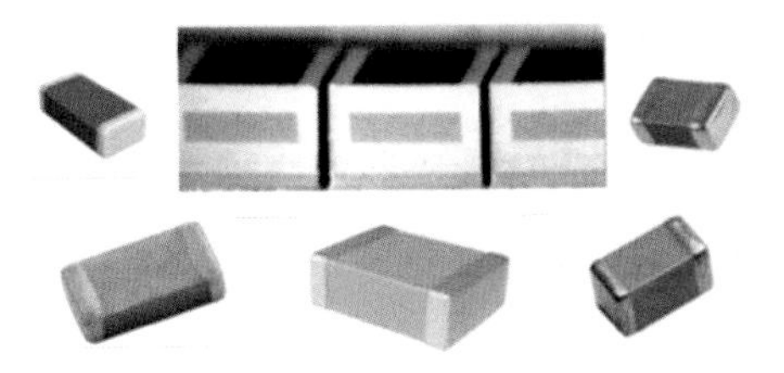

Figure 326. – Typical MLCC – Multilayer Ceramic capacitors

MLCC (Multilayer Ceramic Capacitor) is an SMD (Surface Mounted Device) type capacitor that is used in wide ranges of capacitance. MLCC is a better capacitor than other capacitors due to the better frequency characteristics, higher reliability, higher withstanding voltage and more. Multilayer ceramic capacitors consist of many closely spaced parallel electrodes between thin layers of a high-capacitance dielectric ceramic material. Fig. 327 shows a generic Multilayer Ceramic Capacitor.

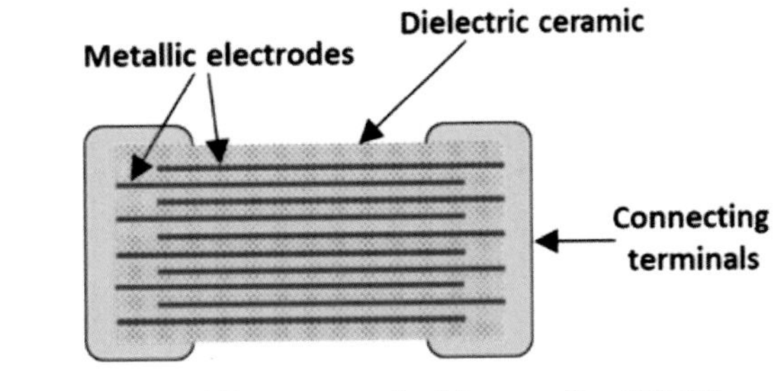

Figure 327. – Typical cross section of MLCC substrate

The initial process is preparing thin green ceramic sheets, screen printing, stacking many layers and laminating. The next step before firing and termination of the MLCC sides, is separating the individual capacitors. In the past most capacitors were separated by Guillotine which was a problematic process because of quality issues and the thickness limitation of 1–5 mm max. The main quality issues of the Guillotine process were losing the side geometry, getting a top build-up of the green ceramic. (see fig. 328).

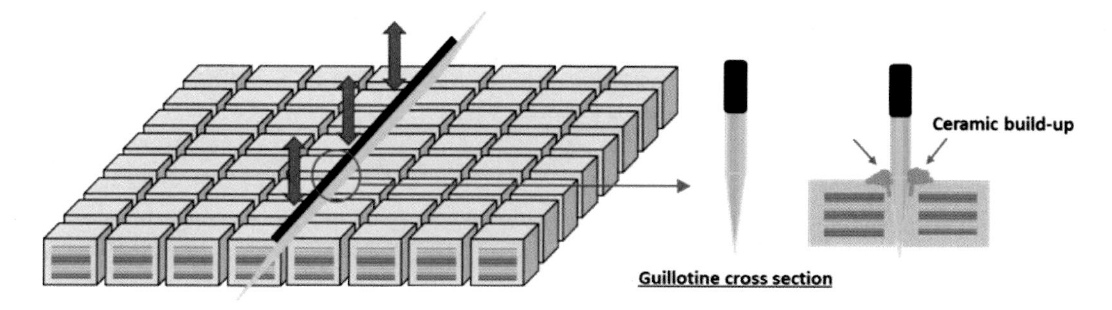

Figure 328. – The Guillotine separation process

The newer separating process is using the dicing process. A few dicing blades can be used, nickel serrated annular blades, steel-core nickel blades and Tungsten carbide saw blades. The diamond grit used with nickel serrated blades and steel-core blades are 30–70 mic. and with thickness of .005" to .012" (0.127–0.300 mm). Most blades are in the range of 4"–5" O.D. The Tungsten Carbide blades are only used with green and dry green ceramic that coolant cannot be used as the coolant will dissolve the ceramic bond. The Tungsten carbide saw blades do perform with good cut quality but with a very short life as the sharp edges are lost very rapidly. The main problem with the nickel serrated blades and the steel core nickel blades is a powder build up on the blade sides which are causing wider kerfs and metal smearing of the metal layers in the MLCC causing electrical shortage. This problem requires frequent dressing to clean off the powder build-up. Another problem is mounting the MLCC green substrates. Tape mounting is an option however it is challenging to remove the diced devices after dicing. Another interesting mounting process is mounting the green ceramic on an ASH FREE paper mounted on a solid metal base. The metal plate with the diced MLCC is put in an oven for the next firing / sintering process. The MLCC diced parts are reaching the right hardness and the ASH FREE paper is burning completely while separating the mounted MLCC parts. The metal plates are loaded with a lot of parts and can be in large sizes of about 15 x 25 cm. Fig. 303 shows a generic MLCC dicing process.

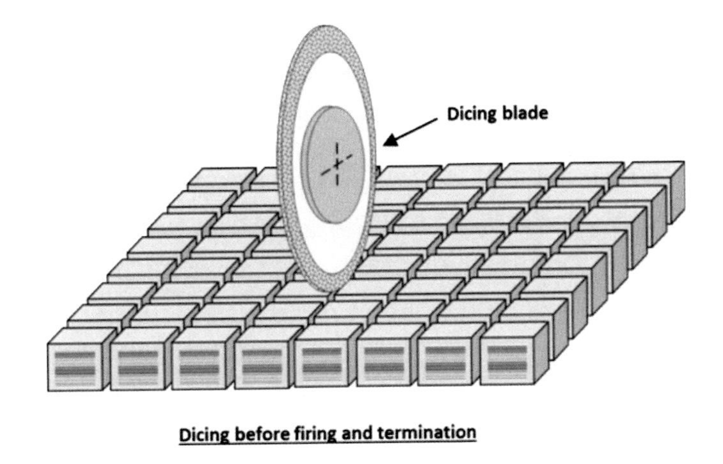

Figure 329 – Initial Green ceramic dicing before firing

The MLCC substrates are very soft at this stage so a high feed rate of 50–250 mm/sec (2" x 10"/sec) can be used. The spindle speed used on the 4" systems range between 10–18 Krpm. The next process after firing / sintering and separating the devices is adding the terminations on both sides of the MLCC devices. The process includes a few steps of Deeping and Plating the MLCC sides. (see fig. 330).

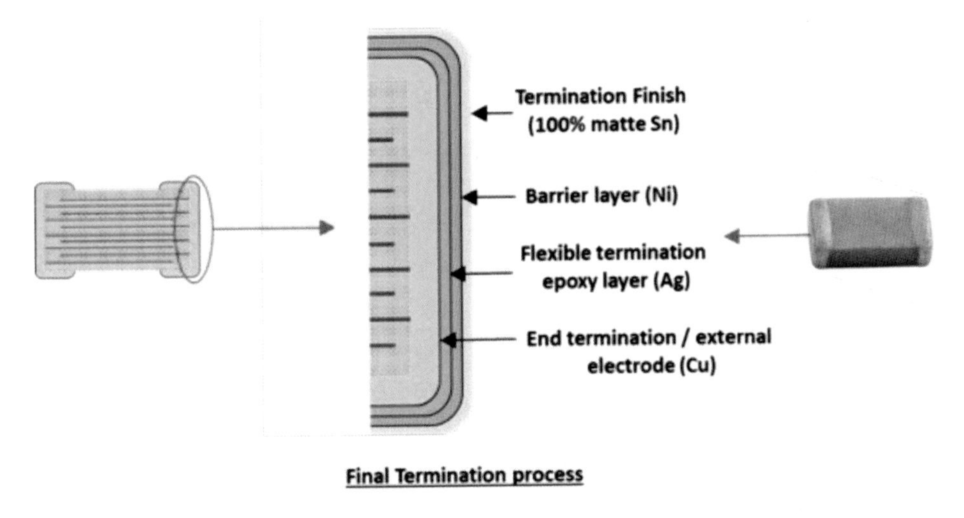

Figure 330. – Cross section of the different materials covering an MLCC

18.25 - TANTALUM CAPACITORS

Tantalum capacitors are another interesting technology. Among many market applications, some Tantalum capacitors are uniquely used in medical products, military, and space related products. Some of the special tantalum capacitors require special dicing attention. The Tantalum in most cases is a powdered sintered product mixed with organic materials. The dicing involves in many cases multiple dicing steps. On some substrates, initial wide grooves are diced in the Tantalum material and then filled with different epoxy type material. A second cut is diced into the epoxy fillings. See a generic sketch in Fig. 331. Tantalum applications are relatively brittle and tend to wear the blade causing it to lose the blade geometry. The blades used for dicing Tantalum capacitors are mainly resin type with relatively large diamonds of 53–105 mic. The blades in most cases are 4.6" O.D. x .010"-.030' thick. The blade resin matrix used in many applications is with higher diamond concentration and special matrix additives to maintain the edge geometry. In a few Tantalums dicing applications the SPG (Special grooved) blades and the special high cooling flange sets are used in order to minimize loading. (see fig. 332). Coolant additives and low temperatures coolant are also used. There is a lot of good information that can be found via the internet covering the principles of manufacturing and electrical related usage of the tantalum capacitors.

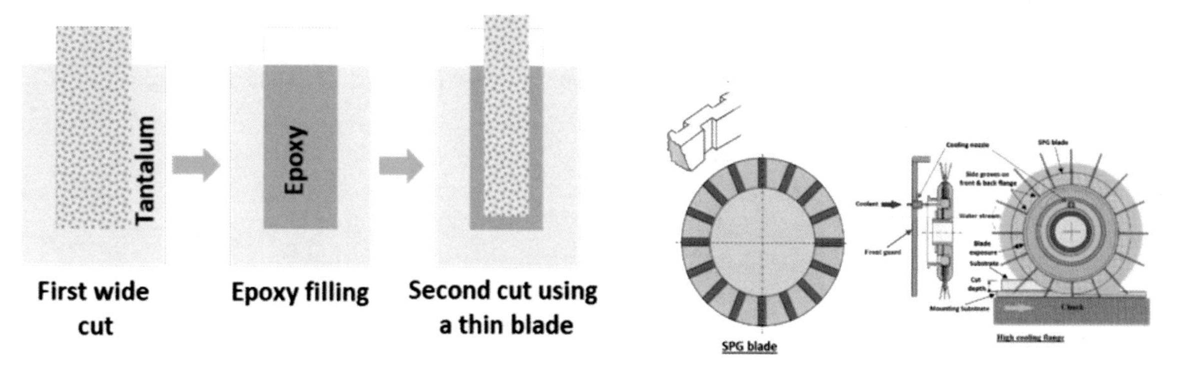

Figure 331. & Figure 332. – One popular Tantalum capacitor dicing process

18.26 – PZT – ULTRASOUND / TRANSDUCERS / SENSORS

Ultrasound Transducers / Sensors is a widely used product in many different applications. This review will concentrate only on medical Ultrasound Transducers in medical instruments which are used in our day-by-day life. Today, every organ has a specific analyzing medical transducer that involves a specific medical purpose.

HOW DOES AN ELECTRICAL SIGNAL CHANGE INTO AN IMAGE IN AN ULTRASOUND SYSTEM?

Ultrasound waves are emitted rapidly from the transducer. The sound waves travel through tissues and fluids, some of the sound waves are reflected back to the transducer. Analyzing the back reflected sound waves, the ultrasound machine creates an image via a computer of the tissues. The principle of ultrasound imaging is simple: sound waves are sent into the tissue and the reflected waves are used to create an image of the tissue. (see fig. 333).

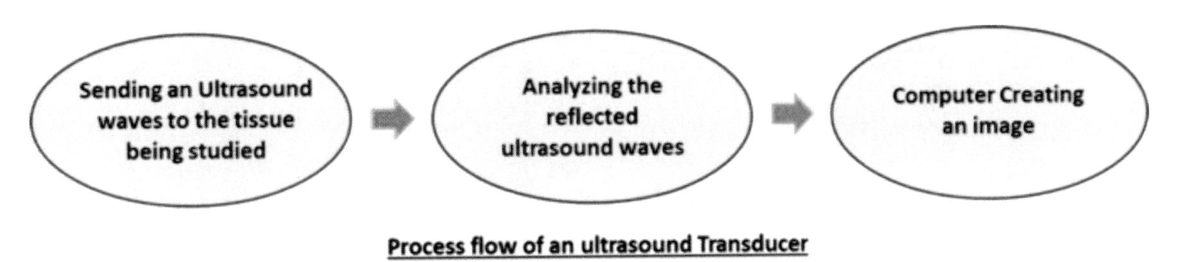

Figure 333. – Principle of the ultrasound imaging process flow

Piezoelectric Crystal is the main technological "gimmick" in the Ultrasound Transducers. Fig. 334 shows a generic Ultrasound Transducer.

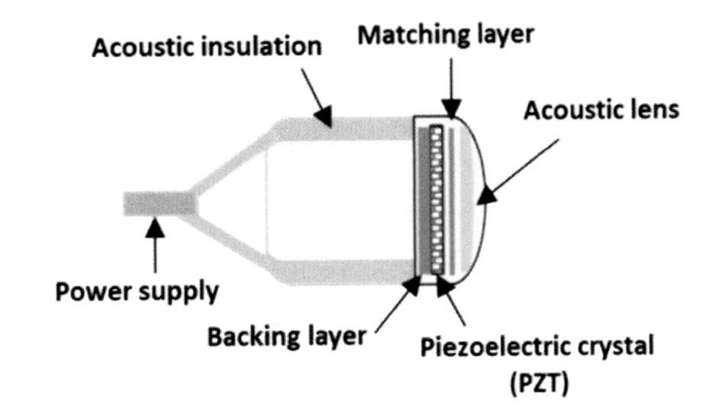

Figure 334. – Generic Ultrasound Transducer cross section

18.26.1 – THE PZT DICING TECHNOLOGY:

This review will cover only the dicing process of the PZT.

PZT (Piezoelectric) is a material that can produce electric energy upon application of mechanical stress. PZT is composed of the two chemical elements lead and zirconium combined with the chemical compound titanate. PZT is formed under extremely high temperatures. The PZT substrates vary between the many different medical applications which require dicing parameters to be optimized. The geometry of the different PZT is basically of thin substrates in most cases however, the dicing is a real challenge.

There are some easy wide cuts around the edges of the substrates, but the main "heart" of the ultrasound transducers consists of a large no. of very thin cuts in a very small pitch / index to create the ultrasound effect. The cut depth can vary on the same substrate with different cut depths. Cut perpendicularity spec on the deep cuts is in the range of 0.005–0.008

mm x kerf variation from top to bottom of 0.003–0.005 mm, both parameters are tough to achieve with thin blades in the range of 0.030–0.040 mm. (see fig. 335).

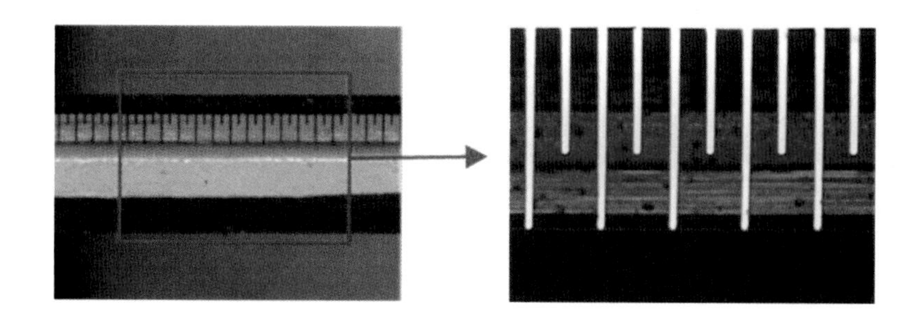

Fig. 335. – Images of a typical PZT diced substrate

The blades used on most Medical PZT applications are annular nickel blades in the range of 50–58 mm O.D. x 0.015mm - 0.075mm (.0006"-.003") thick x diamond grits of 4–6 mic., 6–8 mic., 8–10 mic., 8–12 mic. Those thin blades require a relatively large exposure that are not available with hub blades. The flanges have special set screw holes designed to be dynamically balanced on the saw, normally in the range of 0.1–0.15 mm/sec.. The dynamically balancing is very important to minimize vibrations and meet tight kerf dimensions and cut straightness to perform with no breakage between the thin cuts and the small pitch between the cuts. The nickel blade matrix, diamond size and diamond concentration are closely controlled to meet the above requirements.

NICKEL BINDER:

Because the exposure ratio to the blade thickness in a lot of PZT applications is over the standard and in addition min blade vibrations are needed to perform with a very narrow and straight cut the nickel matrix of the thin blades need to be with improved micro grains of the nickel binder. This is achieved by optimizing the plating process using special power supplies with special optimized plating parameters. Fig. 336 demonstrates fine and coarse nickel micro grains.

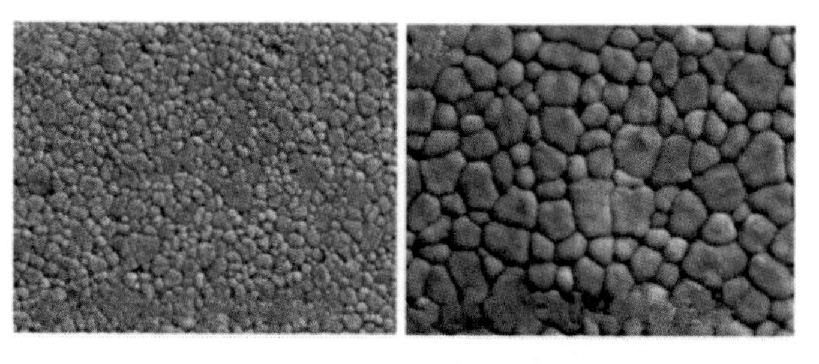

Fine and coarse micro grain of nickel plating

Figure 336. – Fine and coarse nickel bond micro grains

DIAMOND GRIT SIZE AND DIAMOND CONCENTRATION:

Because of the tough and demanding dicing geometry, minimum loading during the process is required. The diamond grit size and diamond concentration are major factors in how the blade performs with minimum loading, kerf width, and cut straightness. The diamond grit size is slightly larger than the diamonds used for dicing silicon wafers. 6–8 mic. and 10–12 mic. are the most common grit sizes used in this process. Diamond concentration is usually lower than standard but needs to be optimize per PZT application.

DRESSING THE NICKEL BLADES:

Specially designed silicon carbide dressing boards with finer mesh sizes are used in the PZT dicing process. The mesh size used for silicon wafers is 600mesh however on PZT dicing applications 1000, 2000, 3000 and even 4000 mesh are used. 1000 and 2000 mesh are the most popular. The blades are rotating at 20–30 Krpm for both the dressing and dicing. The idea is to expose the diamonds and "grind" the blade edge to have a perfect runout to the spindle, Normally the cut depth is in the range of 0.5 mm–0.6 mm x small index a bit wider than the blade thickness. The dressing is performed by an index of a few mm to 10 mm. 20 -40 cuts are made, part at 10 mm/sec and part at 20 mm/sec. The feed rate after the dressing is much slower down to 1–8 mm/sec depending on the cut depth and how sensitive are the thin walls between the cuts. The above parameters change between manufacturers depending on the application. Normally the no. of cuts per one PZT substrate are high and can reach 200–350 cuts on one substrate. This requires online dressing during dicing the PZT. Depending on the cut depth and the quality every 50–100 cuts a single or multiple cuts are made in a side silicon carbide 1000 or 2000 mesh dressing media. This is to expose new fresh diamonds and to wear out dole diamonds. After the dressing process a height sensing of the blade is done before continuing to dice the PZT substrate. Fig. 337 demonstrates a PZT substrate 12 mm wide x 75 mm long after dicing.

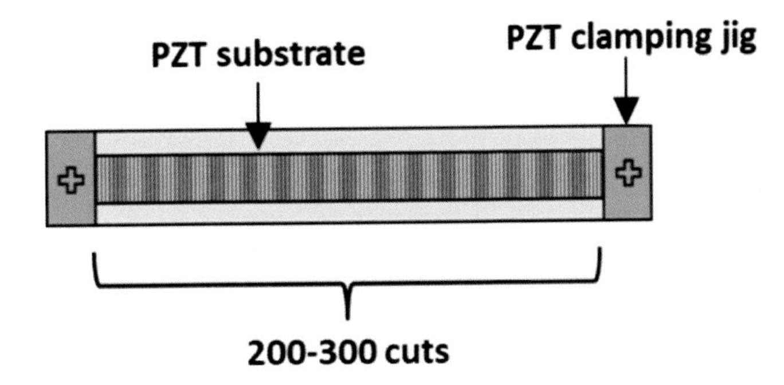

Figure 337. – Top view of a typical PZT substrate

Other parameters that help to meet cut quality and blade life is using coolant additives and chilled coolant. A few coolants additive vendors are available, and they are normally mixed with DI water at a ratio of about 600:1. The coolant mix is chilled down to about 10C. The coolant flow is normally 0.7–1.0 L/min and is carefully monitored to minimize blade deflecting and blade loading. Blade loading is monitored by the spindle amps and varies depending on the saw/spindle vendor. The main point is continuing monitoring all the dicing and saw parameters in order to maintain good yields. There are more specific product and dicing parameters to explore however, confidential information of the different companies dealing with this unique process need to be protected.

18.26.2 - SC [SINGLE CRYSTAL] PZT

Single crystal PZT is a relatively new material used in the medical Ultrasound industry.

The major advantages of single crystals (PMN-PT and PZN-PT) over conventional PZT 5H are their high piezoelectric and electro-mechanical coupling constants, which are attractive for high performance transducers. There are a few different types of Single Crystal PZT, however for this short review only a generic type will be discussed as SC.

The – PMN / PT SC is manufactured by growing a boule (Ingot) size geometry which is growing along the <001> direction. (see fig. 338).

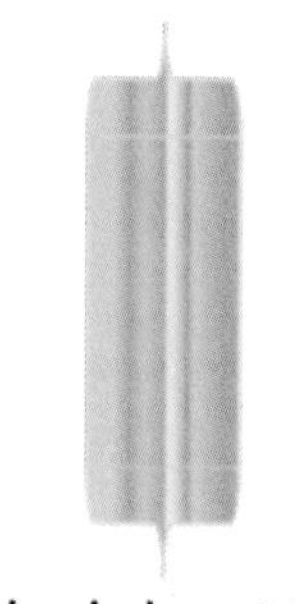

Growing single crystal boule

Figure 338. – A grown single crystal PZT Ingot

The next steps are common production steps similar to manufacturing silicon Ingots. It should be noticed that not all of the Single Crystal boule is with the right electrical and mechanical characteristics to perform well in an Ultrasound / Transducer (see fig. 339))

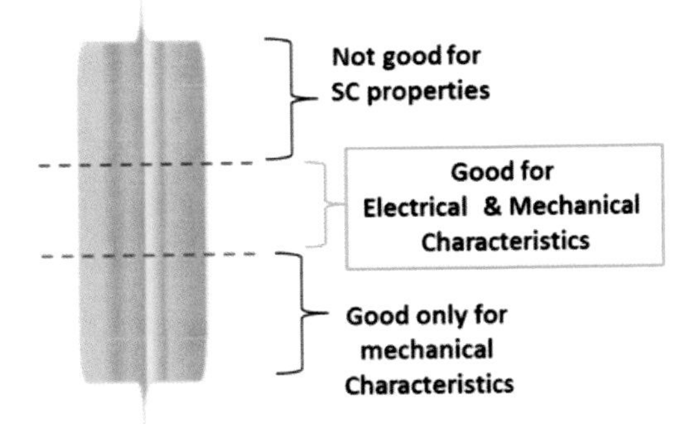

Figure 339. – The center area of the good electrical & mechanical characteristics to be used

As can be seen, only part of the SC boule can be used for good production.

The above info may not be 100% accurate as new improved developments are constantly made to improve yields and cost. The next step is slicing from the good part of the boule slices to be over size from the final thickness. This process is done using an ID slicing machine (see fig. 340)

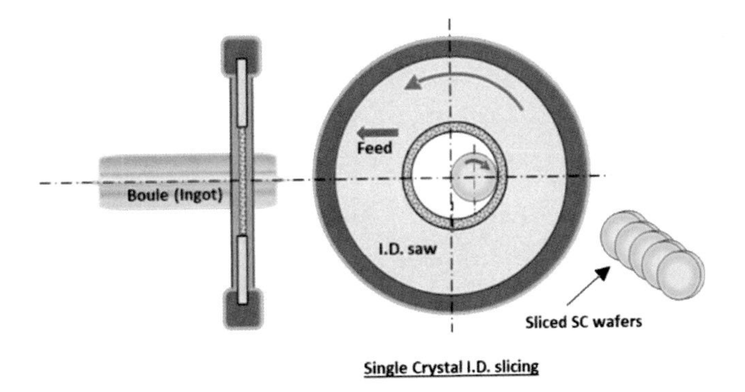

Figure 340 – SC Ingot sliced using an I.D. saw

The wafers are inspected and marked for micro cracks. The good areas are diced outside the micro cracks area to the final outside dimension. The damaged areas are scraped. The diced wafers are then ground and lapped to the final thickness and surface finish. (see fig. 341).

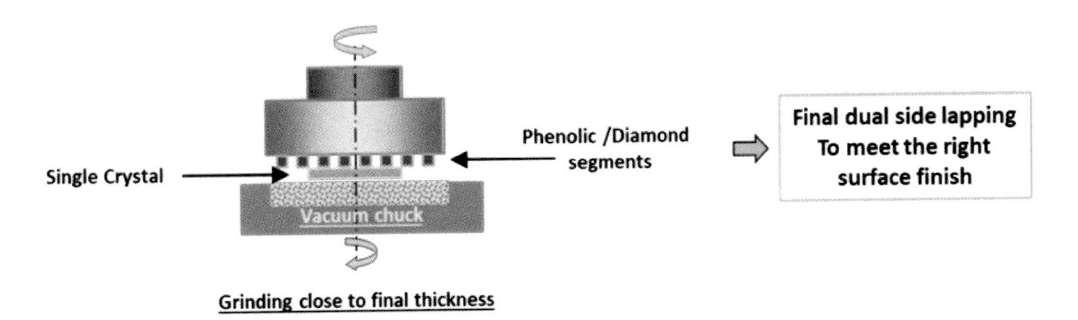

Figure 341. – Grinding and final lapping process to meet the final thickness

In order to release internal stresses after the grinding and lapping the SC parts are annealed in an oven. (see fig. 342).

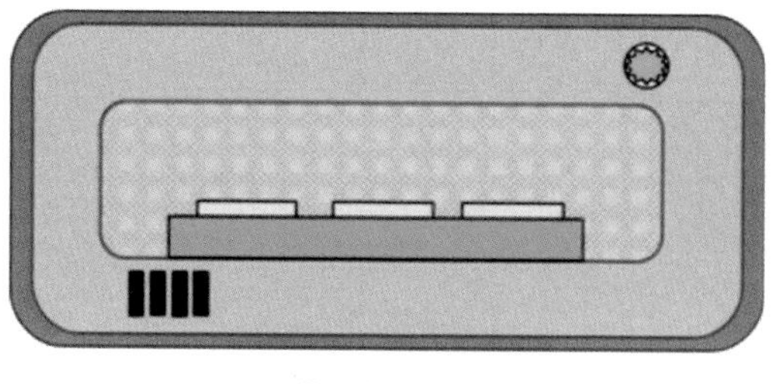

Figure 342. – Annealing to reduce stresses after the grinding and lapping process

The last process before testing and shipping the SC PZT parts to the Ultrasound / Transducers manufacturer, the parts are going through a sputtering process of Nickel/ chrome/gold per the different manufacturer's orders. The above walk through the SC process is only a generic view and may be different between vendors, it is just to demonstrate the difference between SC and the standard PZT material. The SC material is way more brittle than the standard PZT and the manufacturing process is much more expensive. However, the SC material is a breakthrough for some ultrasound transducers. In the manufacturing lines of the ultrasound transducers, the end product geometries are similar to the standard PZT Transducers. There are many cuts made, some with way over 300 cuts on each device, with a small pitch between the cuts, and in some cases with multiple cut depths. The cut quality is way more difficult to maintain, which makes the dicing process much slower by reducing the feed rates in the range of 1–3 mm/sec. The spindle speed is reduced to around 10 Krpm mainly to minimize vibrations. The rest of the process is similar using coolant additives and making sure the flange/blade is well in the dynamic balance spec. The blades are nickel blades, both annular and hub blades with fine diamond grits of 3–6 mic., 6–8 mic. and close to 10 mic. Fig. 343 shows a generic topside quality of a diced SC PZT.

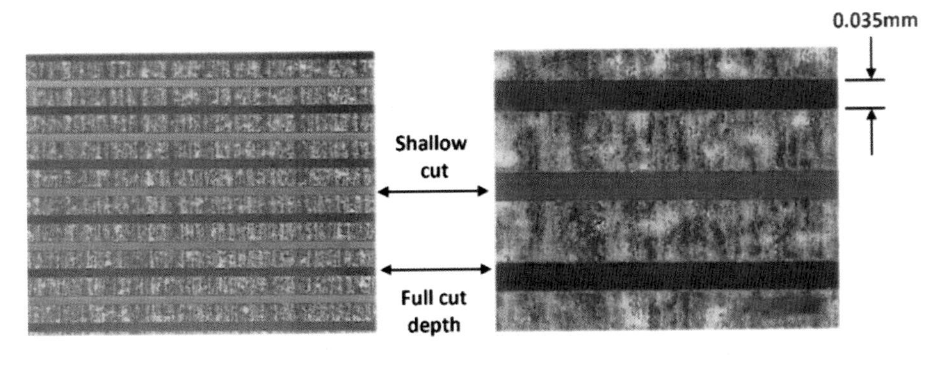

Figure 343. – Topside view of a typical SC diced substrate

The same remarks mentioned on the standard PZT are the same with the SC PZT. There are more specific product and dicing parameters to explore however, confidential information of the different companies dealing with this unique process need to be protected.

18.27 - INK JET

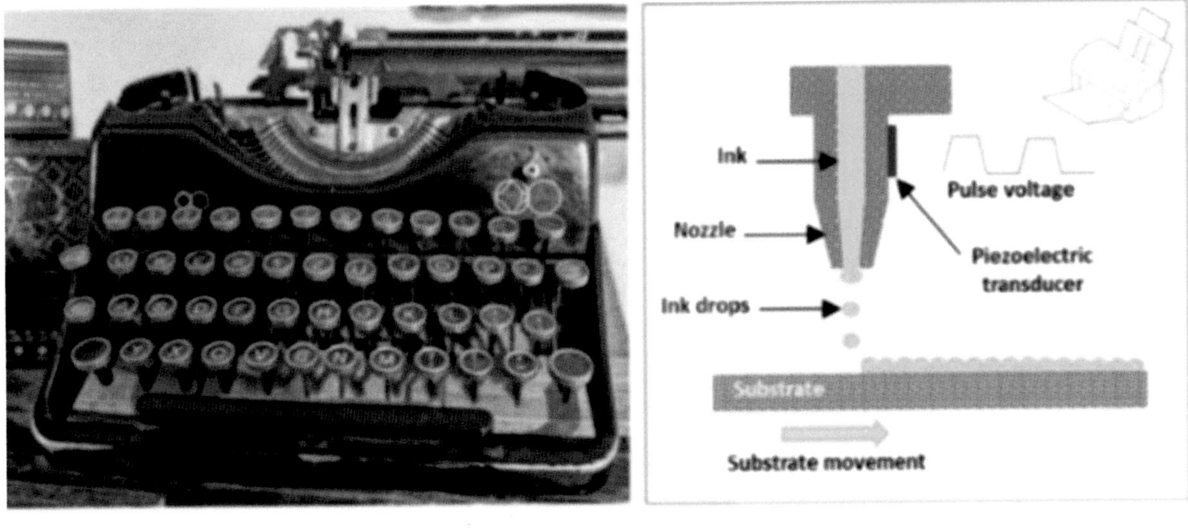

Figure 344. – Old and New printers

This application will be less covered as it has similar characteristics to dicing PZT and silicon semiconductor applications. The Ink Jet technology is done today by many printer manufacturers. It is amazing how low the prices of the printers are. In simple wording – "get the printer cheap" but buy the ink from us. The Ink jet technology is amazing with some miniature parts made of PZT, silicon or both depending on the manufacturer design. The Inkjet printers are classified based on their modes of ink ejection, continuous mode, or Drop-on-Demand (DOD) mode. A lot of technical information can be found via the internet, see in the reference section. The cut depth is very different from dicing PZT, the cuts are much shallower but the cut quality and cut location are very important. 2" O.D. nickel annular and hub blades are used with thickness of .0008" up to .005" with small diamond grits 2–4 mic., 3–6mic., 4–8mic. up to 10mic. grit. Some of the parts to dice are tape mounted and some are mounted using mechanical jigs. On some designs thicker resin blades are used however it is on the lower side. The silicon application side is handled similar to semiconductor wafers. One important side of the application is cut quality at the Ink channel areas where the ink needs to flow out as smooth as possible. Dynamic balancing is used to minimize blade vibrations. Coolant additive is also used in some applications to lower the surface tension of the coolant and to wash better the dicing residue away. All the above are to maintain the best cut quality which is the main important parameter.

18.28 - BGA - BALL GRID ARRAY

Figure 345. – BGA type substrates

A Ball Grid Array Integrated Circuit is a surface mount device (SMD) component that possesses no leads. This SMD package employs an array of metal spheres that are made of solder called the solder balls for connections to the PCB (Printed Circuit Board). These solder balls are affixed to a laminated substrate at the bottom of the package. There are many different types of BGA design. Following are two examples: Fig. 346 demonstrates a BGA die/chip that is mounted on the substrate and connected to the substrate by wire bonding technology.

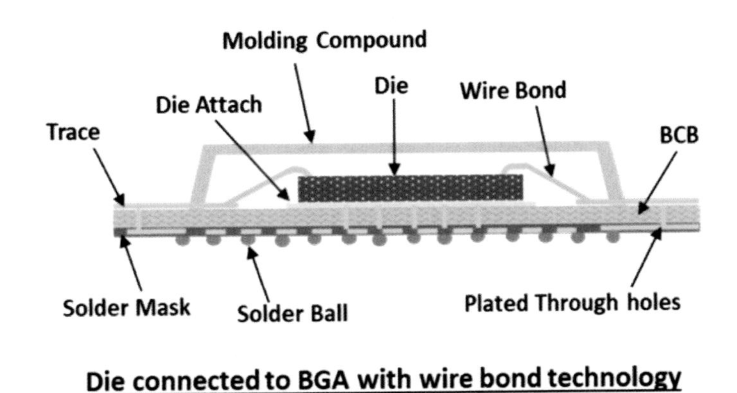

Figure 346. – Cross section of a die/chip mounted to a substrate by wire bonding

Another BGA displays how the BGA die is connected to the substrate using flip-chip technology. (see fig. 347).

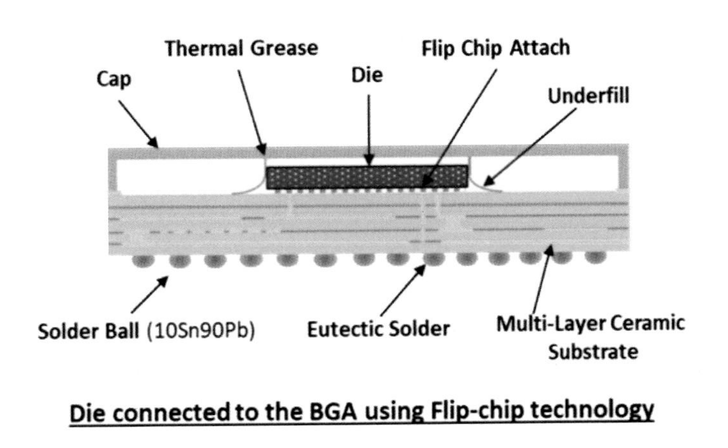

Figure 347. – Die connected to a BGA substrate using the Flip – chip technology

See in the reference more good sites explaining in detail the BGA technology and usage. BGA substrates are diced today in a mass production array geometry for many products. The following applications are using the BGA process: notebook computers, Personal Digital Assistants (PDAs), high density disk drives, camcorders, digital cameras and maybe the largest application is cellular phones.

The dicing process is done on fully automated Tapeless saws using metal sintered blades. See the tapeless mounting section. This dicing process has a pick and place process with on-line QC. Devices that are out of spec are rejected and put into a scrap station. The dicing process involves online dressing by starting using slow feed rates on the production substrates and gradually ramping up to production speed that is quite high up to 200 mm/sec depending on the device size. Blade diameters being used are 2″ and

3" (50 mm–75 mm) with Metal sintered bond using diamond grits of 30-63mic. depending on the substrate thickness and cut quality spec. Blade thickness is in the range of 0.008"-0.014". Following are the cutting parameters: Spindle speed - 2" O.D. - 30–40 Krpm, 3" O.D. - 20–30 Krpm, Feed rate - on both 2" and 3" O.D. - 3"- 8" sec (75 mm - 200 mm/sec). Quality specs vary between the different product and customers, following are some generic quality parameters:

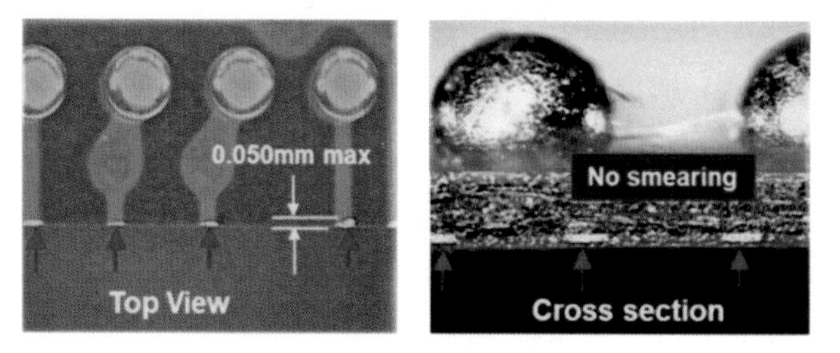

Figure 348. - No electrical shortcuts between the copper leads

Edge chipping - < 0.050mm

Copper Burrs - No electrical shortcut. In some cases, a specific no. is indicated < 0.050mm. (see fig. 348).

Package size: - Nominal +/ - 0.050mm

Package symmetry: - <0.100mm. (see fig. 349).

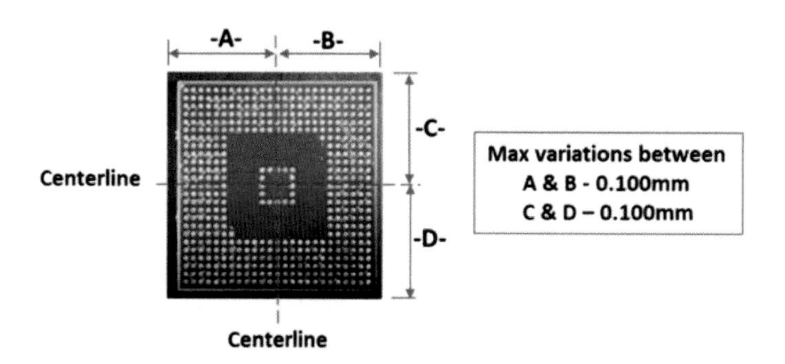

Figure 349. - Package symmetry geometry spec

Dicing BGA has a few major issues to be discussed. The first one is the device side profile which is affecting the cut perpendicularity and the device size. The initial development of dicing BGA was done using nickel blades. Nickel blades can easily dice through plastic type, PCB, FR4 and similar substrates however, due to their hard matrix and low radial wear it affects the side wear. The side wear is gradually affecting the kerf profile width, or in another word slanted side kerf walls. This phenomenon is causing the device to lose its dimension. (see fig. 350).

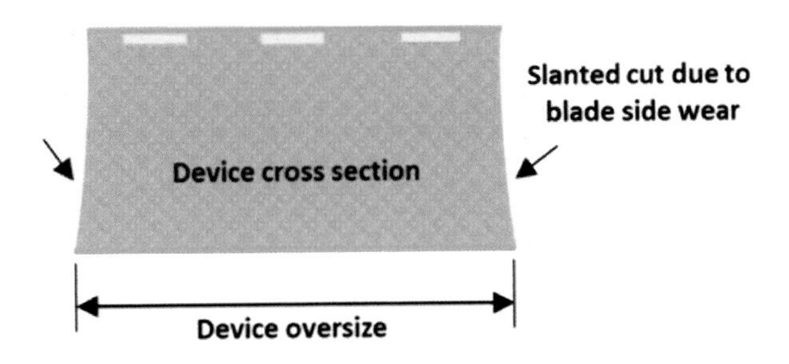

Figure 350. – Blade side wear causing slanted cuts and package out of spec

The only way to overcome this issue is to create more radial wear in order to maintain min. side wear. The idea is to optimize both edge and side wear in order to minimize total wear and maintain a good blade life. This was difficult to achieve with a standard nickel blade. The best blade to perform with good cut quality and good blade life is a metal sintered blade with diamond grit of 30 to 53 mic. grit. Fig. 351 demonstrates the right wear behavior compared to an extra edge wear causing too much side wear.

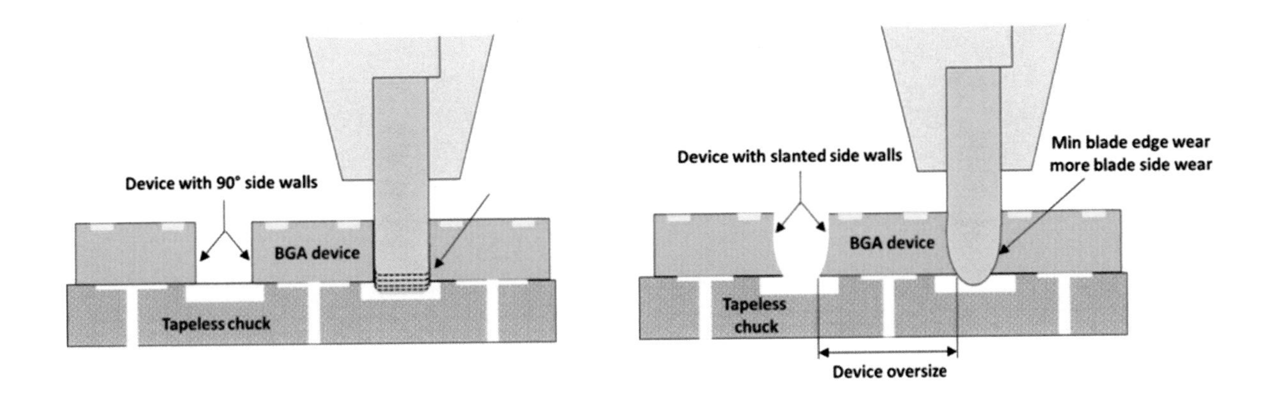

Figure 351. – Min. blade edge wear compare to high edge wear causing out of spec BGA device geometry

Metal sintered blades are the majority blades used to dice BGA on high volume production tapeless fully automated saws however, on some applications nickel serrated blades are also used as they have a higher edge wear compare to non-serrated blades and can maintain straight cuts with a good blade life. The nickel blade can be Hub serrated or annular serrated with diamond grits of around 30–40 mic. The market is pushing for more UPH, which is a tough demand to maintain good quality at feed rates of up to 200 mm/sec. One of the main tasks is to make metal sintered blades with diamonds well exposed on the blade edge but with min. diamond exposure on the blade sides. Too much diamond exposed on the blade sides creates too much side wear as the diamonds are wearing out of the blade matrix and this leads to having oversized devices and a premature blade replacement.

Fig. 352 shows a metal sintered blade with diamond exposed on the blade surface and another metal sintered blade with diamonds not exposed on the blade surface.

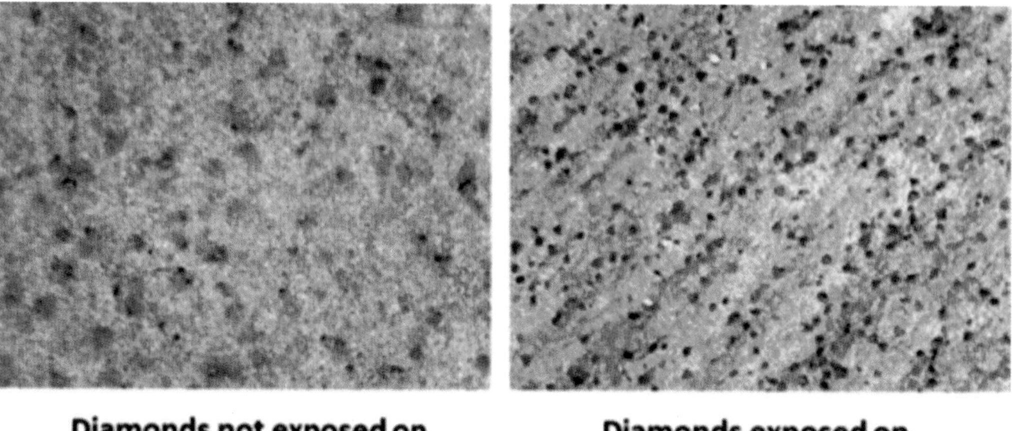

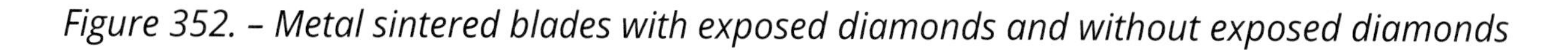
Figure 352. – Metal sintered blades with exposed diamonds and without exposed diamonds

Another way to cause more radial edge wear is to optimize the metal sintered blades matrix by adding fillers and optimizing diamond concentration. This optimization is delicate not to cause too much edge wear that will affect the blade life. The above-mentioned potential issues and the market pushing to dice faster and faster is a real pressure on all blade manufacturers to have a continuous development program in order to maintain their market share. Continuous efforts that the blade manufacturers are dealing with, is to measure the kerf profile of any new blade optimization in order to maintain the customers/market specs. Fig. 353 demonstrates a kerf profile inspection.

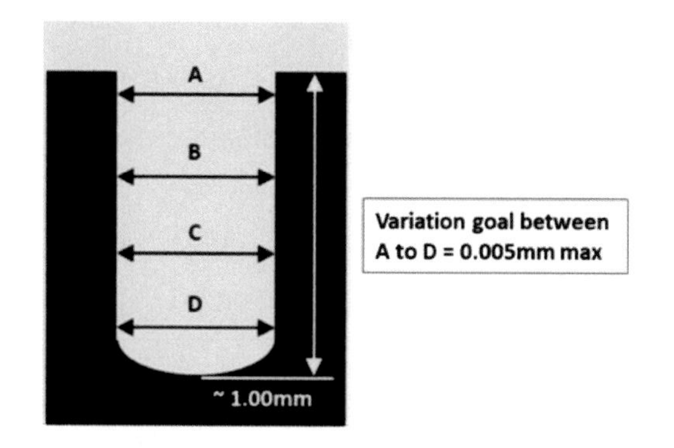

Figure 353. – Blade cut profile inspection

The second problem is protrusions during the dicing.

This problem is related to blade loading during dicing and to poor device clamping on the vacuum tapeless mounting during dicing. Protrusion in simple wording is a cut that does not completely cut through the substrate, leaving a small substrate material at the end of the device. (see fig. 354).

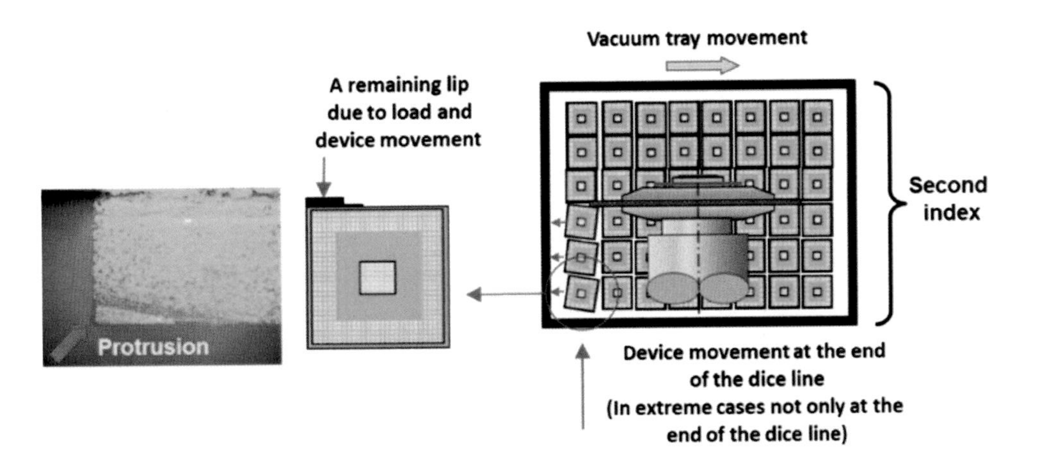

Figure 354. – Device protrusion due to devise movement during dicing

STEPS TO MINIMIZE PROTRUSION DURING DICING:

- Improve drastically the vacuum on the tapeless chuck – Probably no. 1
- Improve the flatness of the BGA substrate for better vacuum clamping.
- Slow down the feed rate – if possible.
- Adjust the cooling nozzle & coolant pressure. A too high pressure may cause device movements during dicing.
- Use a more wearing (Softer) blade to minimize loading.
- Re-design the index dicing sequence on index no. 2 – usually the shortest index (see fig. 355).

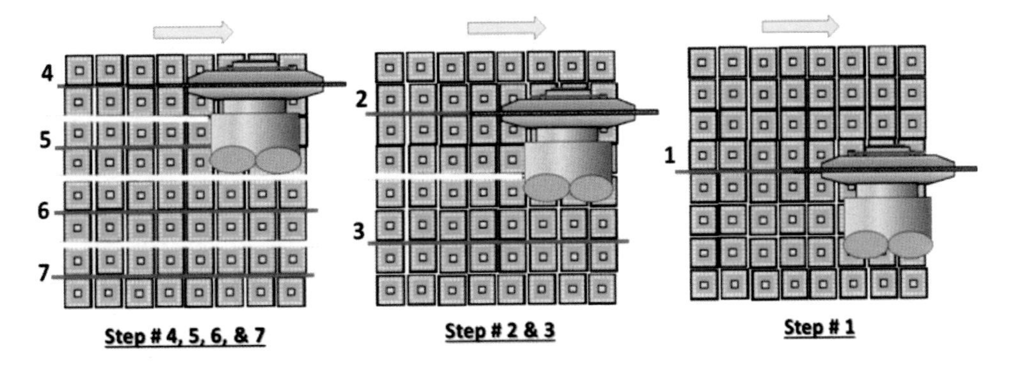

Figure 355. – Suggested cut mode to eliminate the protrusion problem

The idea is to maintain during the dicing process a larger clamping area for better clamping. BGA is one of the largest applications in the marketplace, probably no. 2 after semiconductor silicon wafers

18.29 - CBGA (CERAMIC BALL GRID ARRAY)

Ceramic Ball Grid Array substrate is a square-shaped or rectangular ceramic package that uses solder balls for external electrical connection instead of leads. The solder balls are arranged in a grid or array at the bottom of the ceramic package body. (see fig. 356).

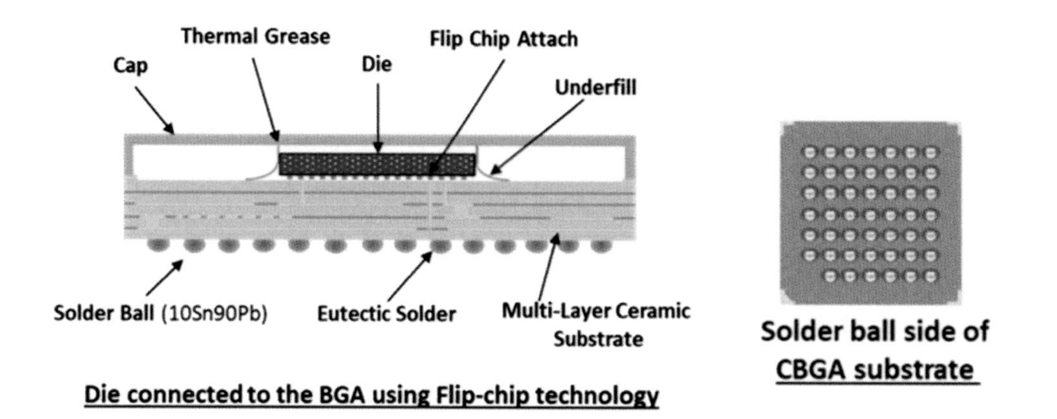

Figure 356. – CBGA Cross section using the flip chip tech. to connect the die to the CBGA

The advantages of Ceramic BGA on traditional BCB/FR-4 BGA is having good high-frequency performance and electrical performance, high thermal conductivity, chemical stability, excellent thermal stability, and other properties that organic substrates do not have. It is an ideal packaging material for the generation of large-scale integrated circuits and power electronic modules. Dicing CBGA is similar to dicing 96% and 99.6% Alumina. Same mounting methods and similar blade and dicing parameters.

18.30 - QFN - QUAD FLAT NO-LEAD PACKAGE

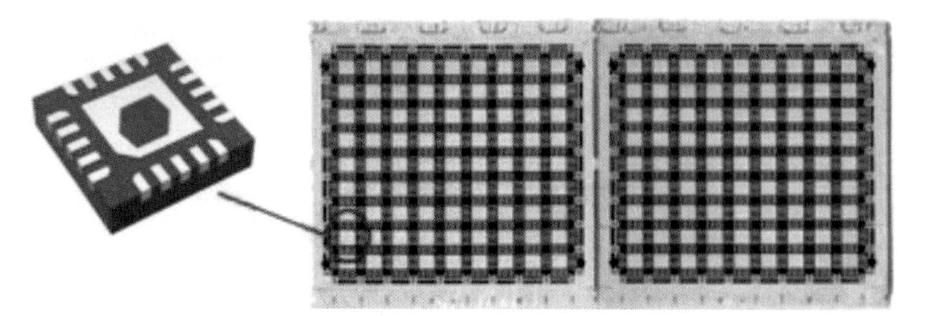

Figure 357. – QFN Quad Flat No-Lead substrate

QFN stands for quad flat no-lead package. It is a leadless package that comes in small size and offers moderate heat dissipation in PCBs. Like any other IC package, the function of a QFN package is to connect the silicon die of the IC to the circuit board.

QFN packages come with a die that is surrounded by a lead frame. The lead frame is made up of copper alloy C-194 with a hardness of ~ 135–145HV (1/2 hard) with a Ni-Pd-Au or Sn coating. The metalized terminal pads are located at the bottom surface. These terminal pads are present along the four edges of the bottom surface and provide electrical interconnections to any PCB. Following is a generic QFN substrate sketch. (see fig. 358).

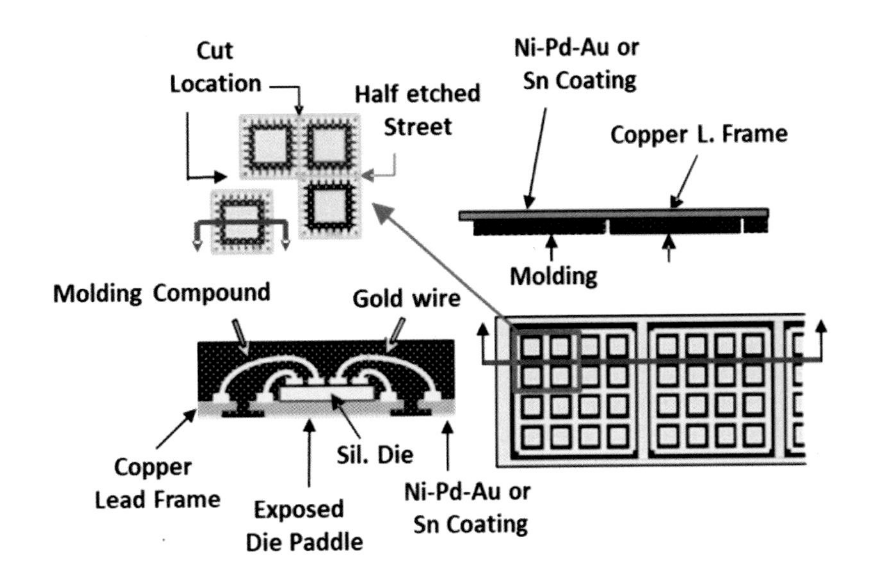

Figure 358. – A generic QFN substrate top and cross section view

There are three main family type of QFN substrates used in the marketplace:

- Standard HE (Half Etched) QFN: 0.8–1.2 mm Thick (Up to 0.2 mm copper lead frame)
- Thin QFN: 0.4–0.7 mm Thick (Up to 0.15 mm copper lead frame)
- Power QFN: 2–4 mm Thick (~ 0.5 mm copper lead frame)

The copper lead frame is soft and ductile and creates problems mainly copper burrs and copper smearing during the dicing process.

The process on the copper lead frame is a mold array process (MAP). The molding is a polymer compound reinforced by silica particles [SiO] in the range of 30–70 mic. in size. The molding compound is relatively brittle compared to the copper lead frame which tends to cause chipping during the dicing process. QFN presents an excellent example of a complex substrate composed of both ductile (copper) and brittle (plastic molding) materials. The combination of dicing a ductile material with a brittle material in the same dicing process creates real challenges. Fig. 359 demonstrates a generic deformation graph showing the difference between a brittle material and a ductile soft material.

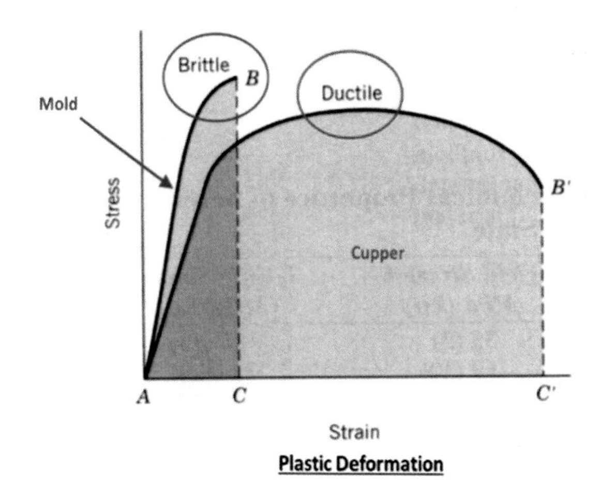

Figure 359. – Generic deformation graph

Different assembly houses are using different names for the QFN packaging. Following is a list of the different names:

- QFN – Quad flat no-lead
- MLF – Micro-lead frame
- MLPD – Micro-lead frame package dual
- MLPM – Micro-lead frame package micro
- MLPQ – Micro-lead frame package quad
- VQFN – Very thin quad flat no-lead
- DFN – Dual flat no-lead package

Typical die sizes of standard QFN are 3x3mm up to 10 x10mm and in some cases smaller down to 1x1 mm which are more common on thin QFN but also on standard QFN. Most blades used for dicing QFN are resinoid blades. The reason for using resinoid blades is the flexibility of making many matrix changes in order to deal with the quality issues related to the complex substrate materials. There are some QFN applications using Metal sintered blades however it is on the minority side.

Most blade diameters are 2" O.D. and in some cases 3" O.D. x thickness of .008"–.020". The diamond grit sizes range between 45 to 70mic. depending on the application spec. Feed rate is an open issue depending on specs, it can be between 10mm/sec up to 100mm/sec. Spindle rpm ranges between 20–30Krpm. The lower RPM is mainly used on

QFN substrates with Sn plating. Higher rpms are performing with more frictions resulting with higher temperatures that can melt the lead Sn plating and cause lead delamination (see fig. 360).

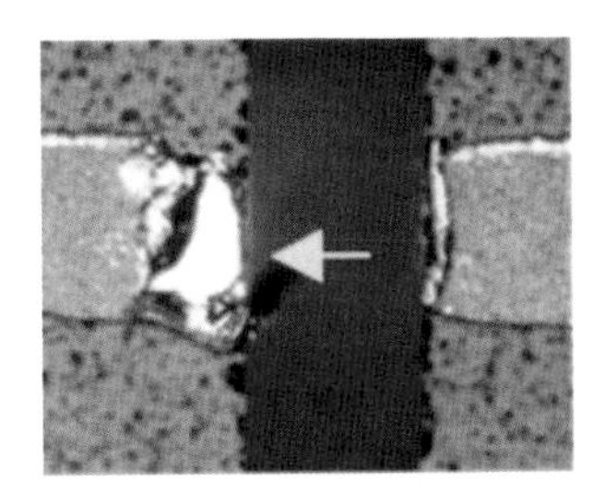

Melted Sn plating on lead

Figure 360. – Melted SN on the copper leads

18.31 - PQFN PACKAGE

PQFN (Power Quad Flat No-lead) is a QFN package type that is suitable for power applications. PQFN is an SMT semiconductor technology designed for PCB-mounted applications and is a highly efficient space saving package. The device sizes vary between 5x5 mm to 12x12 mm. The manufacturing process is similar to standard QFN, a copper lead frame mold array process (MAP). Normally PQFN substrates requires a higher torque spindle. 2" spindle can be used however in many PQFN applications, 4" spindle is a better choice using 4.5" and 4.6" resin blades x .008" to .020" thick blades and diamond grits of 45 mic. up to 105 mic. in extreme loading applications. Spindle speeds vary between 10–15 Krpm and feed rates are 1–5 mm/sec depending on quality and loading. Not so new but related at the beginning to PQFN and later done also on many standard QFN designs is the "Wettable"' flank design. The leads on the sides do have a small flat step to improve the soldering bond between the QFN package to the PCB. (see fig. 361).

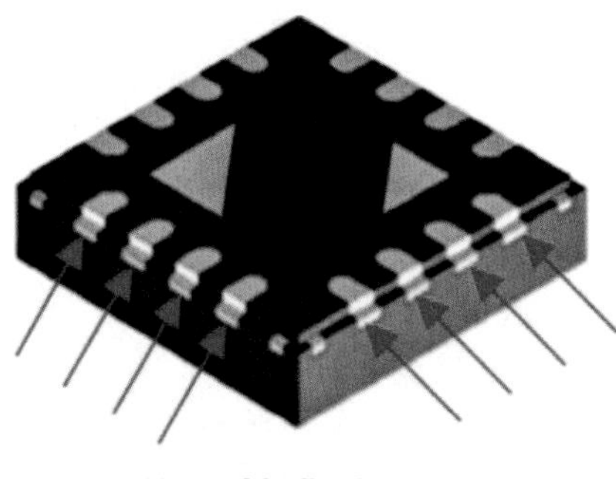

Figure 361. – New "Wettable" flank design

Wettable flanks (WF) are a modification to the exposed terminal /leads ends, which promote solder wetting and a better solder bond that is visible. Using a QFN package with a wettable flank enables optical inspection of the soldering. The Wettable flanks is actually a small step created by a first wide shallow cut and then a thinner full cut to separate the devices (see fig. 362).

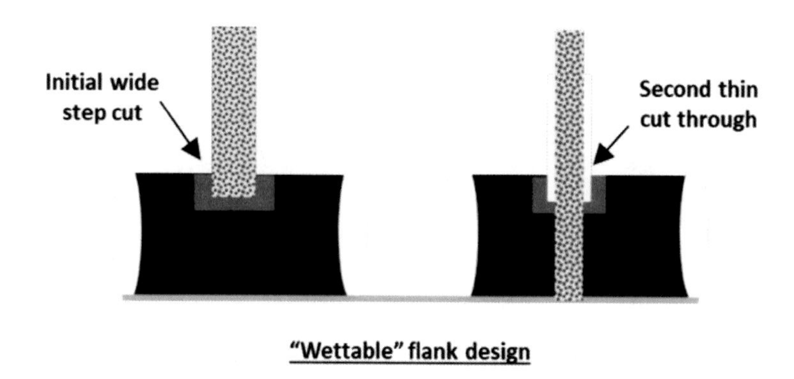

Figure 362. – Initial dicing full copper leads in 2 steps

Some customers are creating the sidestep during the half-etched process that will be discussed, however many customers prefer to do the wide shallow cut by dicing. The main advantage of the Wettable flank design is the improved soldering. (see fig. 363).

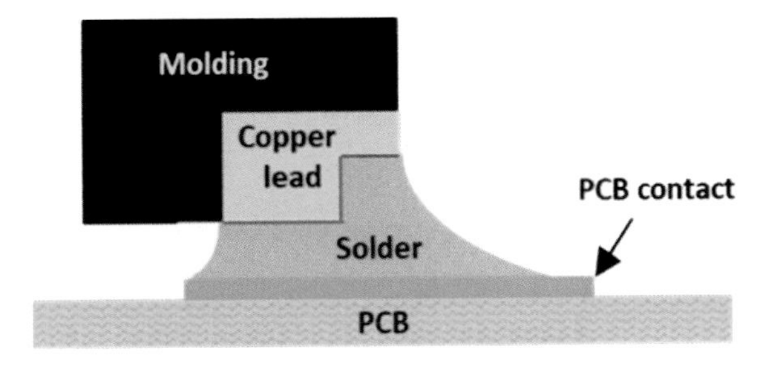

Figure 363. – Wettable flank design advantages

18.32 - DRESSING BLADES DURING DICING QFN

Dressing the blades is relatively easy mainly because resinoid blades do wear homogeneously on the edge while maintaining the blade thickness till the blade loses the exposure (see fig. 364).

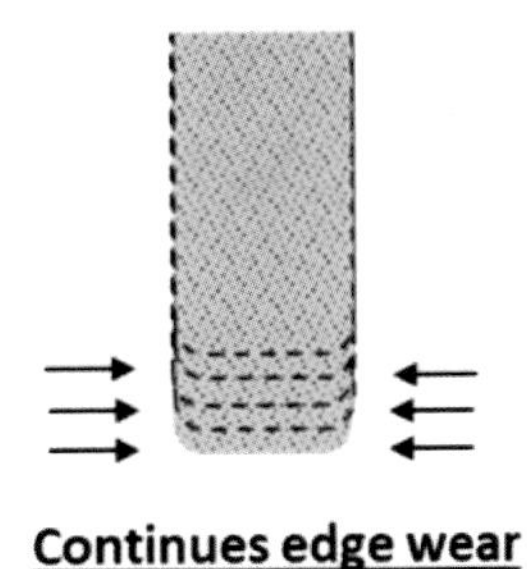

Figure 364. – Homogenous blade edge wear

Dressing QFN blades is done on production substrates using the override/Pre-cut process and can be done during dicing if needed.

COOLANT:

Coolant is a major factor in the QFN dicing process mainly on lead frames with Sn Coating. High temperatures created during the dicing process can easy melt the Sn plating on the leads (see fig. 360). One solution mentioned already was to slow down the spindle rpm, in most cases down to 20 Krpm on 2" O.D. blades. The other and probably the main solution to eliminate the Sn melting and to minimize chipping on the molding is to add an additive to the D.I. coolant. The additive is lowering the surface tension of the coolant helping the coolant to better penetrate to the kerf area and to better cool the blade edge that can reach high temperatures of a few hundred degrees C. Lowering the surface tension of the coolant is also helping to better wash away the residue from the dicing. Another parameter that is helping the dicing process is chilling the coolant down to 10°C – 12°C. All the major dicing manufacturers sale recycling and chilling units that can handle both the coolant additive and chilling the water. Most units can handle two dicing systems. Some large assembly houses using a large number of dicing systems are using a "mother" chilling unit that can handle a large number of saws.

SUBSTRATE'S GEOMETRY AND DESIGN PROBLEMS:

Following are a few issues just to understand the history of the initial QFN lead-frame design. The lead-frame design had a lot of metallization (copper + plating) in the dicing streets. This was loading drastically the blade, affecting quality and yields. A major improvement was to chemically etch the center of the lead-frame streets area to minimize the amount of copper that the blade is facing. Today all QFN lead-frames are lead etched which is a major improvement. (see fig. 365).

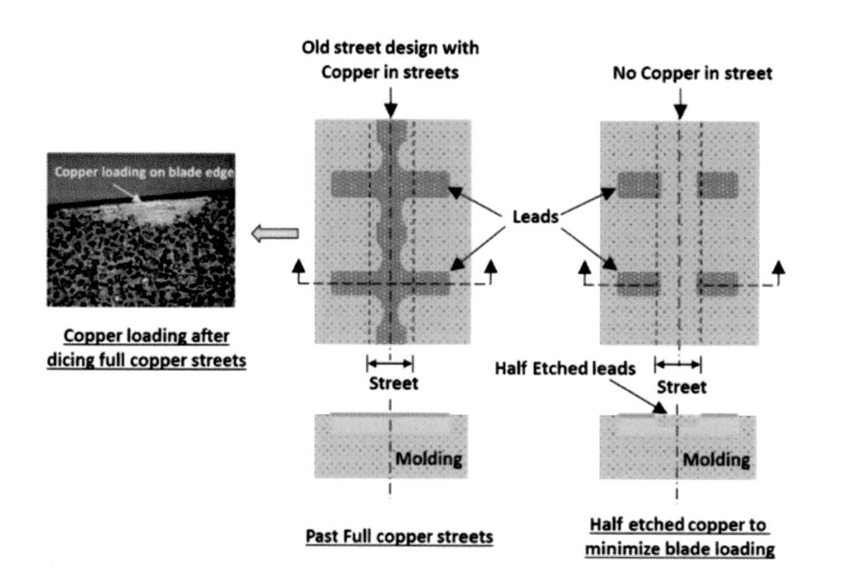

Figure 365. – Initial full copper leads and the newer improved half etched leads

One major problem in dicing QFN type substrates is poor flatness, in some cases causing mounting problems and cut accuracy problems on the Z axis.

The main reasons for the flatness problem are the polymer molding compound process. It is difficult to maintain a perfect stress-free molding. There are a few reasons for this problem, one is the Silica particle fillers, the size (30–70 mic.), the concentration and other molding process parameters like air bubbles and more. Regardless of the mounting method during dicing, flatness is a real problem affecting yields and costs.

A major problem related to flatness is dicing the wide shallow cut that was described already. A regular shallow cut can cause in some areas a way too shallow cut and in other areas a too deep cut. Both options can cause rejects on the lead's geometry.

On the saw that measure the uneven flatness for the Z movement to follow the up and down necessary corrections and to maintain the same cut depth throughout the substrate. The inspection device can be a mechanical height contact measuring device or a non-contact device. (see fig. 366).

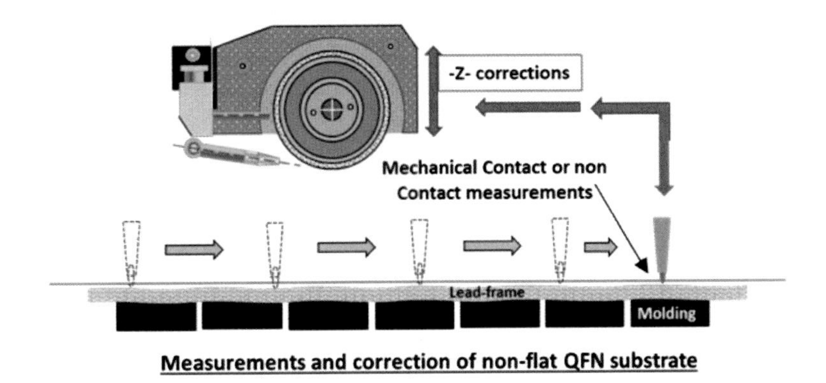

Figure 366. – Z – slide corrections on non flat substrates

Molding and copper lead frame openings is helping to minimize substrate stresses and improve flatness. In addition, opening in the lead frame minimizes blade loading as the blade has some releases and not a continuous blade to substrate contact. (see fig. 367).

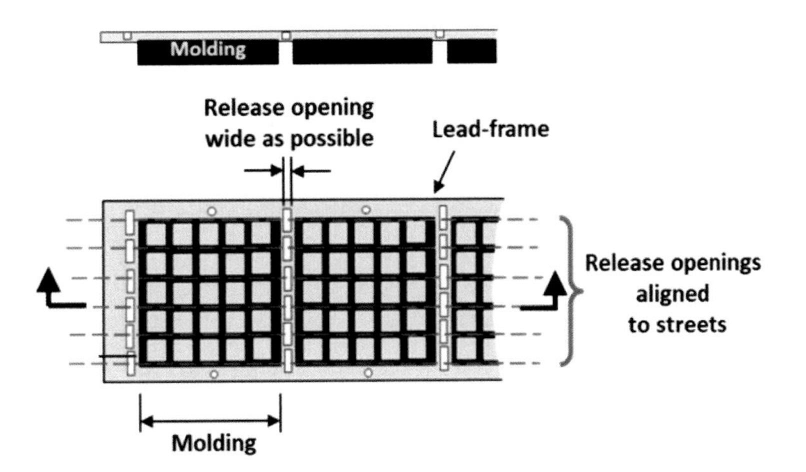

Figure 367. – Molding releases and metal lead frame releases helping to improve substrateflatness and minimizing loads.

A lot of customers are using a solid molding design in order to gain more real estate and some after improving their molding process to minimize stresses. (see fig. 368).

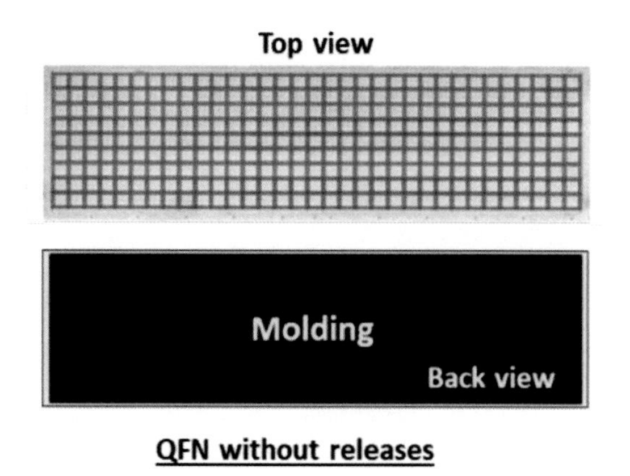

Figure 368. – Solid molding to gain more real estate

An important parameter that can improve cut quality and minimize blade breakage is the QFN substrate manufacturers. It is critical to minimize the lead frame extension outside the molding. A large, unsupported extension of the lead frame can cause blade vibration, poor cut quality and shorten blade life due to blade breakage. (see fig. 369).

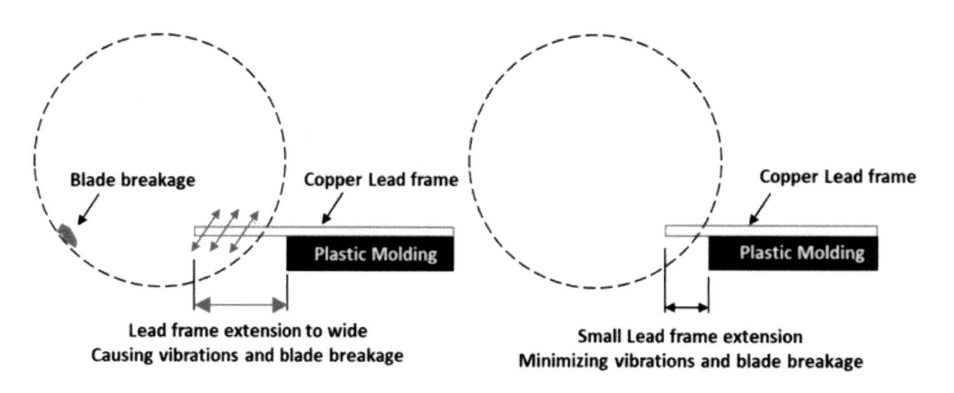

Figure 369. – Old and new improved shorter metal lead frame

18.33 – QFN – DICING SAWS AND SUBSTRATE MOUNTING

Tape mounting saws and tapeless mounting saws are used for dicing QFN substrates. It depends on mass production requirements and the QFN geometry. Small devices in the range of 1–2 mm square are difficult to dice on a tapeless chuck as the devices may move during dicing and are diced on UV tape. Mounting QFN substrates is in a way similar to mounting BGA and has similar mounting related problems. See the BGA section and fig. 370 of a generic tapeless mounting.

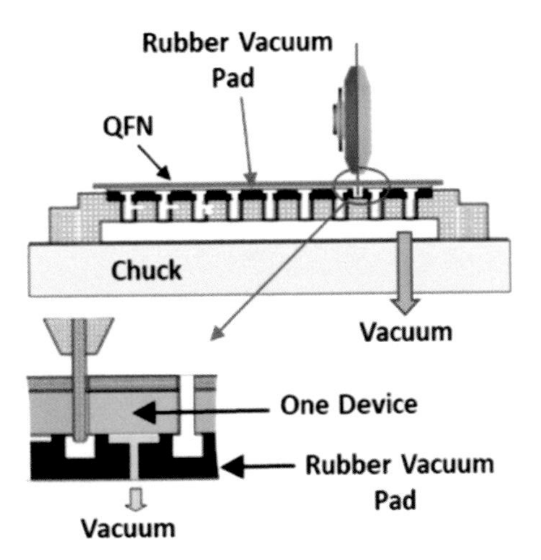

Figure 370. – Dicing QFN on tapeless chucks

The UV tape mounting has one advantage of dicing multiple substrates that can be aligned individually on the same UV tape mounting. (see fig. 371).

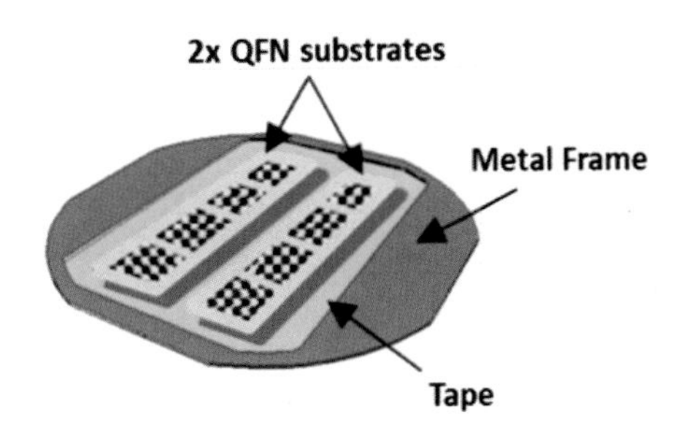

Figure 371. – Dicing multiple QFN substrates mounted on UV tape

DICING SEQUENCE:

Before getting into the QFN dicing quality specs, let's cover the dicing sequence of the tape mounting process and the tapeless mounting method.

The idea is to minimize device movements on both options and maintain the best yields. The tape mounting process is easier to maintain regarding device movements mainly on smaller devices of 1–2 mm square. (see fig. 372).

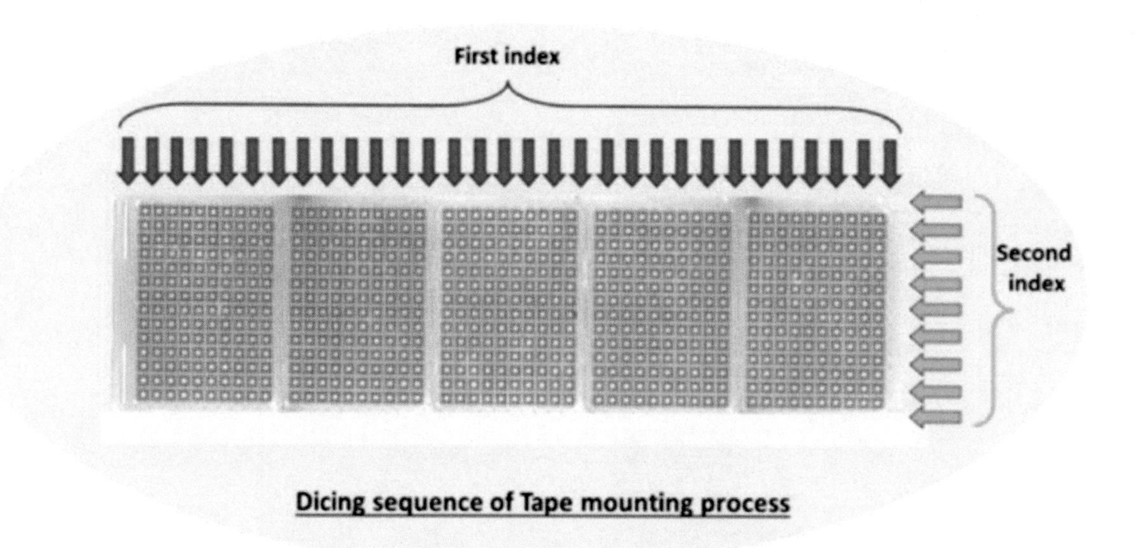

Figure 372. – Dicing sequence of QFN mounted on tape.

Tapeless mounting creates a lot of mounting problems mainly device movements, so the dicing sequence is way more important. The dicing sequence varies between customers. Fig. 373 demonstrates a common one that also makes sense.

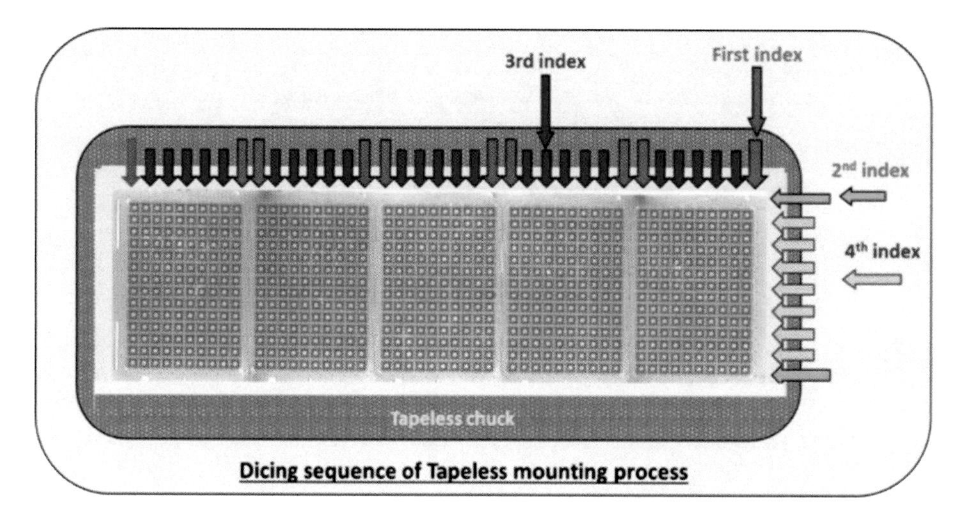

Figure 373. – Dicing sequence of QFN mounted on tapeless chuck

In some extreme cases with large unsupported lead frames outside the molding, the outer lead-frame area both on Y and X axis are first diced out.

18.34 - QFN DICING QUALITY SPECS

After reviewing some of the major QFN application parameters, it is time to review the different common cut quality specs. They vary between customers however they are on average common. As we are dealing with both soft and ductile materials and with harder and brittle materials, both need to be addressed.

MOLDING:

The main problem with the molding is chipping. The main reasons for chipping on the molding are: – Mold brittleness, silica filler particle size causing pullout, poor coolant, too aggressive dicing parameters and too large diamond grit in the blades. All the above reasons can be optimized. Fig. 374 shows a relatively moderate mold chipping.

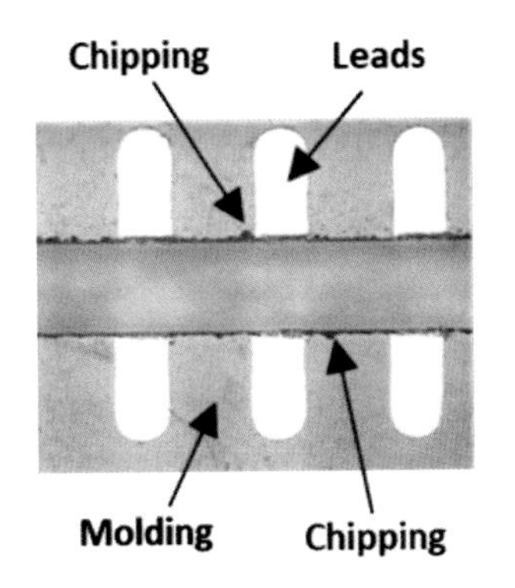

Figure 374. – QFN mold Chipping

Following are chipping specs: (see fig. 375).

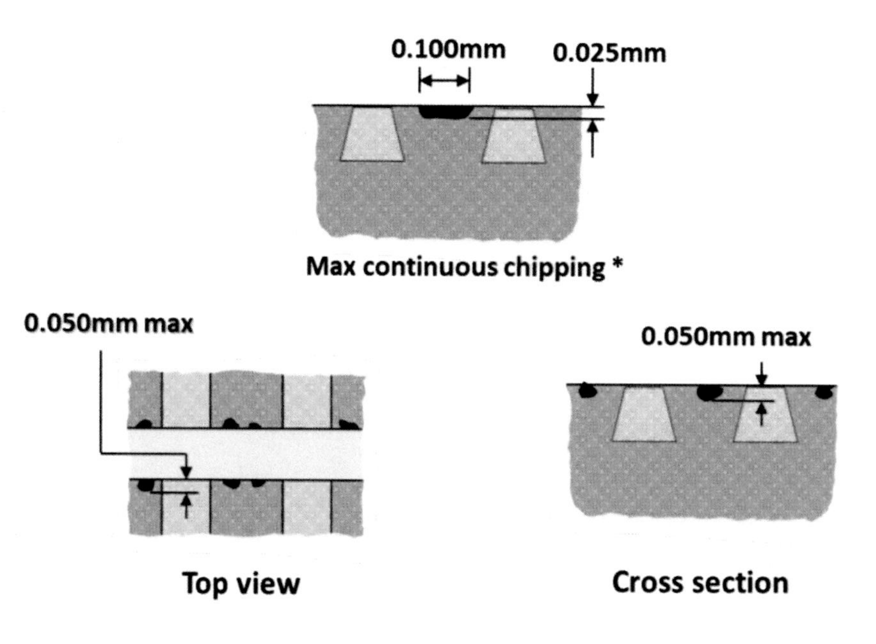

Figure 375. – QFN mold chipping spec

Copper burs are measured on 3 axes. (see fig. 376).

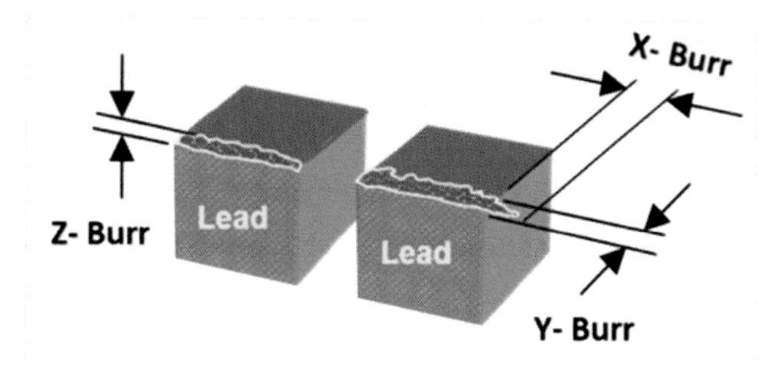

Figure 376. – Copper lead burrs

Burs specs. (see fig. 377).

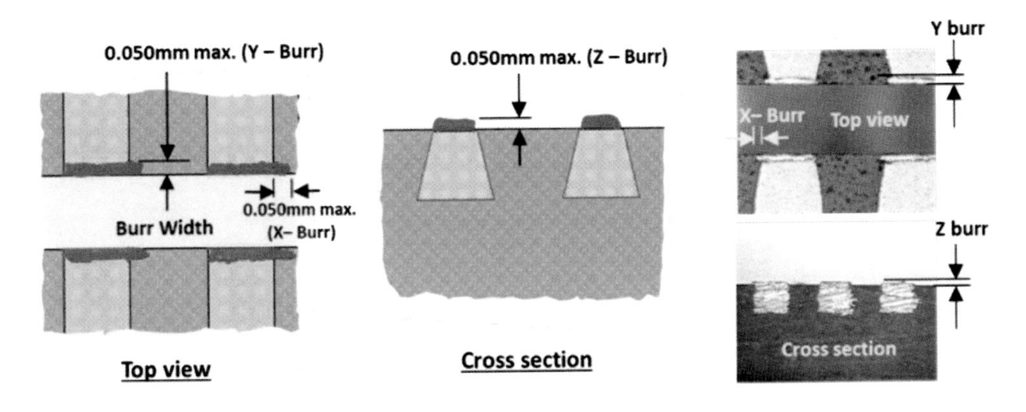

Figure 377. – Copper leads spec

Smearing spec. (see fig. 378).

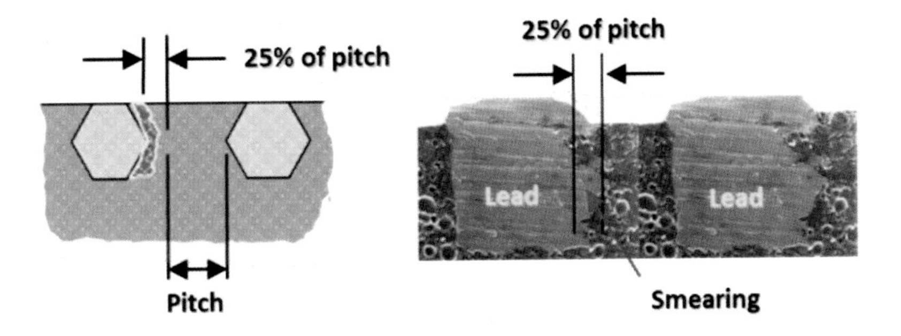

Figure 378. – Copper Lead smearing spec

18.35 - TIC / ALTIC - MAGNETIC SLIDER / HEAD

TIC/AlTiC stands for Aluminum-Oxide Titanium Carbide (magnetic recording heads)

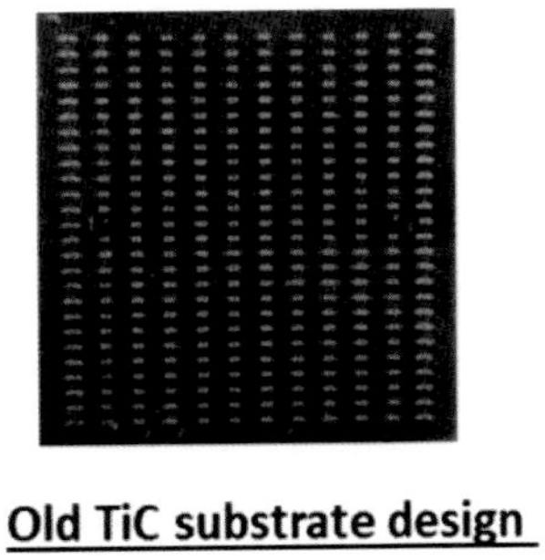

Old TiC substrate design

Figure 379. – Old standard TiC square substrate

The Slider/Head, in my opinion, is the most complicated, accurate, and complex dicing process that I know. It is an extremely interesting process with a lot of fine mechanical jigs, special blades, and unique process developments over the last 50 or more years. The initial head geometry design was quite large, but over the years it became really small and really challenging. The TiC material is hard and brittle, but in general, it is not too difficult to dice. The main challenges are meeting the very tough and demanding quality and accuracy specs. Let's talk in general about the hard disk drive (HDD) where the slider/head is a major part of the system. Fig. 380 shows a generic hard drive that is used almost in any computer/laptop.

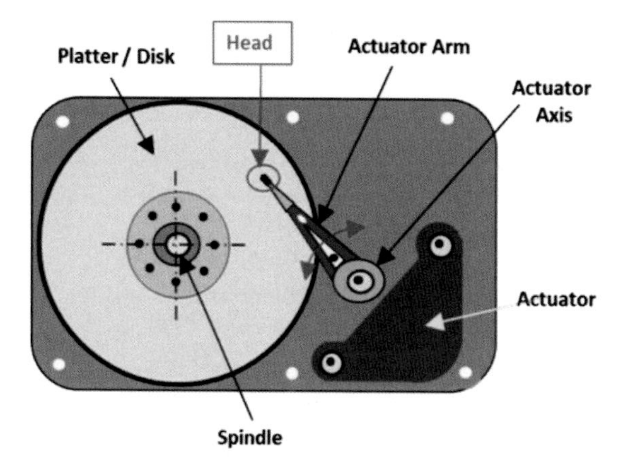

Figure 380. – Generic computer hard drive

A typical HDD (Hard Disc Drive) design consists of a spindle that holds a flat circular disk, called platter, which holds the recorded data. The platter is made from a non-magnetic material, usually aluminum alloy, glass, or ceramic. It is coated with a shallow layer of magnetic material typically 10–20nm in depth, with an outer layer of carbon for protection. The flat circular disk is an electro-mechanical data storage device that stores and retrieves digital data using magnetic storage. The disk drive uses magnetism to store information in a layer of magnetic material below the surface of the spinning disk. The information is stored in the magnetic layer by creating small magnetic domains in this material. In some HDD there are more than one disks for more information. Today standard consumer – grade HDD disks are rotating at speeds varying from 5,400 or 7,200 RPM. Information is written to and read from the disk as it rotates past devices called read-and-write heads (Slider) that are positioned to operate very close to the magnetic surface. When the disk rotates, an air cushion develops making the head float over the disk with a gap of 3 to 20-millionths of an inch above the disk. The head can be called an air-slide. The read-and-write head is used to detect and modify the magnetization of the material passing under it to read and write the data from the hard drive disk. The head is a small part of the disk drive, it is the most sophisticated part of the hard drive and has gone through several changes over the years. The head slider material of the present invention is a sintered body containing 100 parts by weight of alumina, 20 to 150 parts by weight of titanium carbide and silicon carbide in total, and 0.2 to 9 parts by weight of carbon. The following review is going through a few generic dicing processes involved in manufacturing a small head slider. Which is the heart of the hard drive system. Fig. 355 is a generic sketch of one head slider.

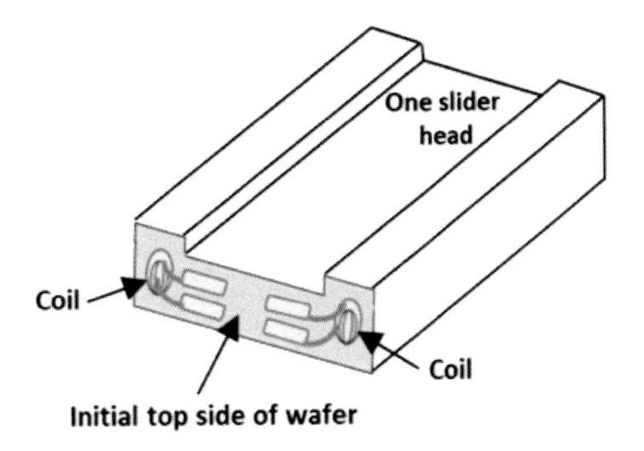

Figure 381. - Generic 3D head slider

Today's Head slider designs are way smaller compared to the designs 20–30 years ago. This also affects the substrate diameter and mainly the thickness of the substrate. The dicing process to manufacture the final head geometry includes a few steps: (The below process parameters are showing only general information and dimensions in order to understand in general the many steps involved in the manufacturing; however, they do change between the different manufacturers in the process steps/parameters and the spec numbers).

Substrate (Wafer) Fabrication (Fab type semiconductor process)

Substrate Mounting

External initial substrate separation/dicing the individual segment row groups

Second Substrate mounting [on some applications]

Row Slicing

Lapping the rows in reference to the coil

Aligning the rows on the mounting medias (a few options)

Head Parting

18.35.1 - TIC - WAFER FABRICATION (FAB TYPE SEMICONDUCTOR PROCESS):

This process is done in a fab type facility. It is similar to a semiconductor wafer process in a clean room facility. The process includes many accurate steps. (see fig. 382).

Figure 382. – Top side TiC substrate / wafer

The below are sketches of old and newer TIC substrate designs with their diameters. The larger diameters are designed in many segment groups, each segment with a few rows' indexes. The first initial TIC substrates were 4–5 mm thick however, the newer substrates are getting thinner and thinner than 1 mm. (see fig. 383).

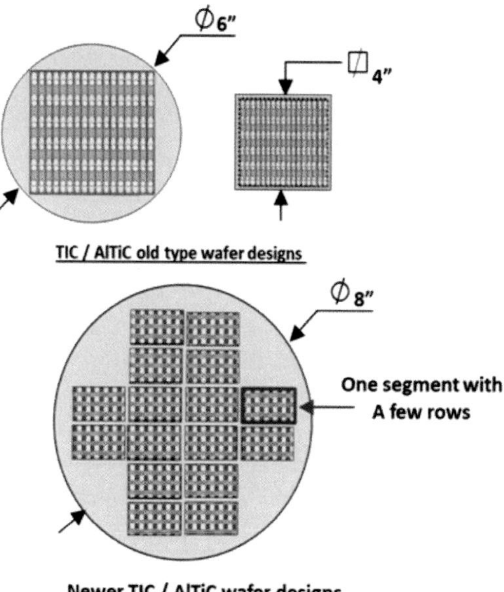

Figure 383. – Old and newer TiC substrates designs

18.35.2 - TIC - BLADES TYPE AND BLADE DRESSING/ O.D. GRINDING

Handling the blades prior to using them in production is done in what is called the wheel-room, a very important process prior to the manufacturing process.

Most blades used in the head manufacturing process are 4"– 4.6" O.D. nickel blades with some metal sintered blades with fine diamond grits depending on the process used. The diamond grits are 3–6 mic., 4–8 mic., 8–10 mic., 8–12 mic., 8–16 mic.. All blades are optimized on diamond concentration and nickel bond hardness per the specific process requirements. There are very strict requirements on blade flatness. This is important for cut perpendicularity and kerf width. There is a strict functional incoming blade inspection at the manufacturing house. All blades in most process applications are O.D. grinded on the same production flange that is used later to dice the actual production. The O.D. grinding has two purposes, one to get a perfect square edge and second to get a perfect runout of the blade O.D. to the spindle rotation. (see fig. 384)

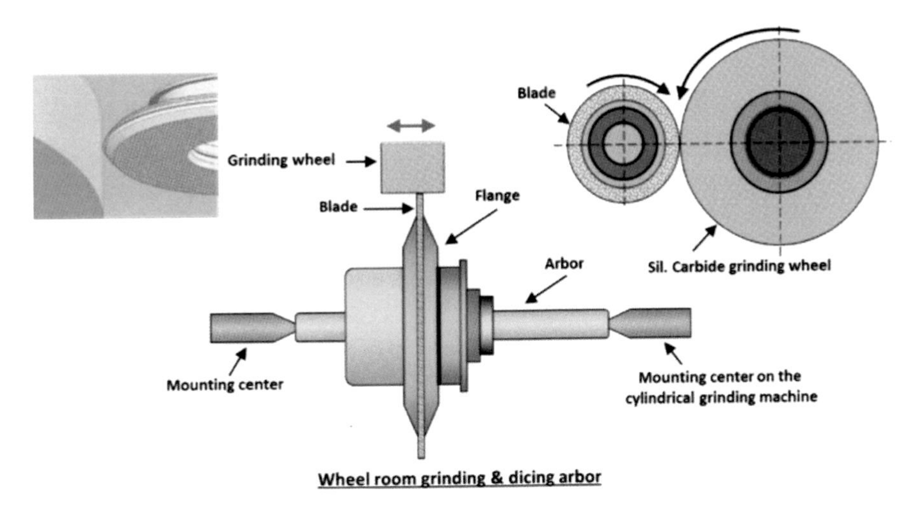

Figure 384. – Wheel room O.D. grinding set-up

The flange and blade are mounted on a very accurate hardened and ground mandrel with axial and side runout of 0.001 mm. (see fig. 385).

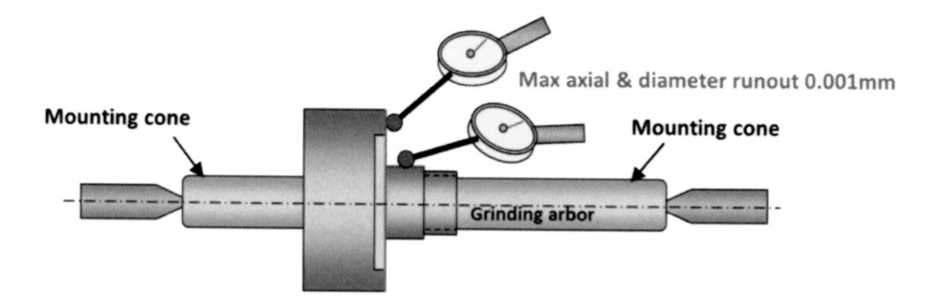

Figure 385. – Checking the grinding arbor accuracy

The accuracy of the grinder arbor mounting the flange is critical. Any small nick on the arbor mounting surfaces can deflect the flange/blade assembly and cause the blade to wobble during the grinding process. Correcting the arbor can easily be correct on the grinding machine. Fig. 386 shows the deflected flange causing the blade to wobble and correcting the arbor on the grinding machine.

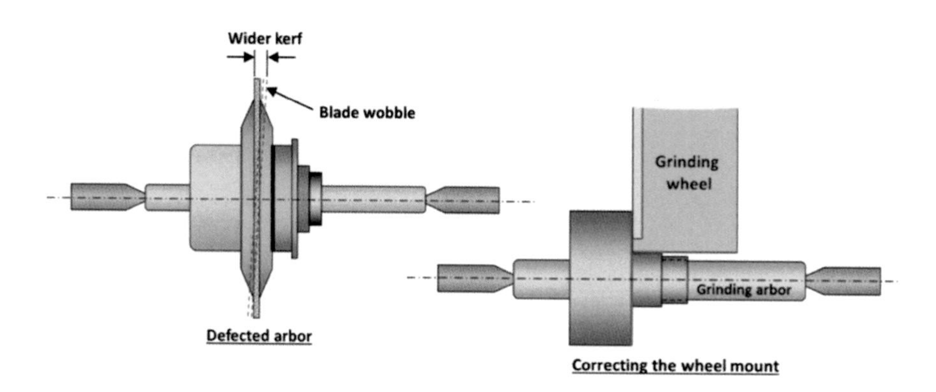

Figure 386. – Correcting the accuracy of the grinding arbore

The O.D. grinding of thin nickel blades is a slow process, mainly because of the difficulty of having a large and wide grinding wheel in contact with a very thin blade. This requires a very small infeed of the wheel and a very slow travel of the grinding wheel over the blade. The soft silicon carbide grinding wheel has a mesh size of about 400 mesh and is rotating close to 2000 rpm (depending on the grinding wheel O.D.). Most grinding wheel geometries are 16" O.D. x 1.5" wide. The arbor with the blade is rotating at about 100 rpm. The reason for the slow blade rpm is to minimize the wear on the grinding wheel and to create more wear on the blade. As we are dealing with relatively thin nickel-type blades with a high ratio of blade exposure to blade thickness, the infeed of the grinding wheel is about 0.001mm per pass of the blade over the grinding wheel. It is a very delicate process that takes about 30 minutes till the runout is completed and the blade edge is flat. Fig. 361 shows a cylindrical grinding machine during the grinding process.

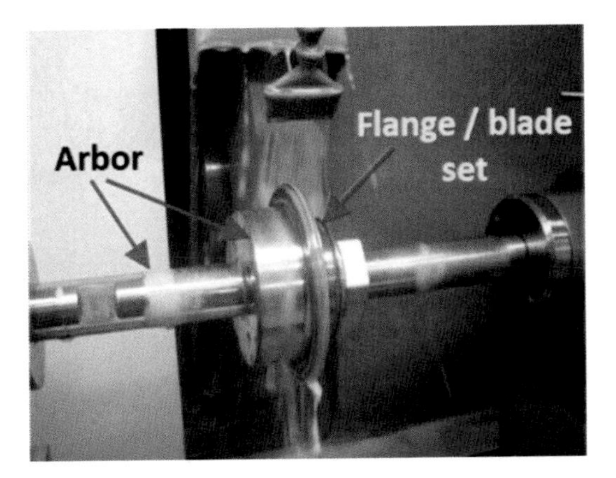

Figure 387. – Actual O.D. Grinding of a single blade

After the grinding process, the blade goes to an inspection process, which involves dicing a dummy TIC substrate similar to a production substrate in order to inspect the kerf size, cut perpendicularity and chipping size. This test starts with an override process up to production speed. If the blade does not meet the quality spec, a second dressing and quality dicing test is performed. If the second dressing does not meet the spec, the blade is rejected. Testing the blades is performed on a dedicated saw for testing and not in the production line. During the production row dicing, the loads are rising, and the blade needs to be dressed from time to time to minimize loads and meet the spec again. This will be discussed in the next steps. Fig. 388 demonstrates the cut quality of nickel blades after the grinding process and the tests on a dummy TIC substrate.

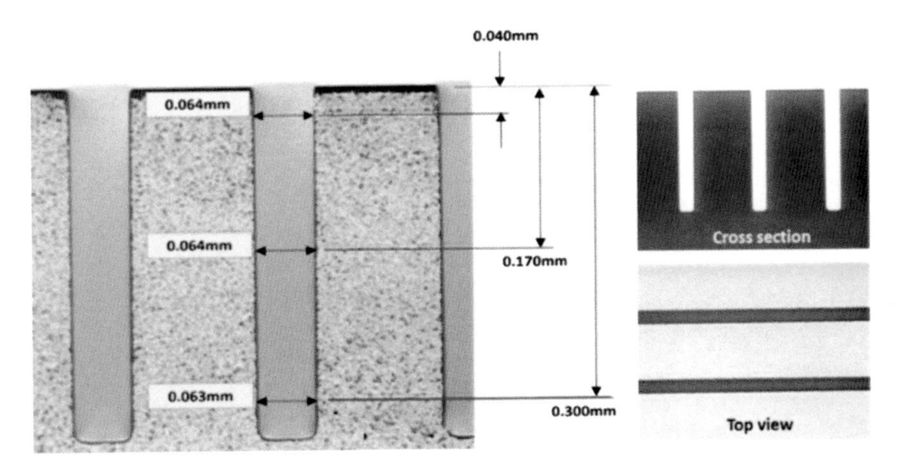

Figure 388. – Cut test accuracy after blade O.D. grinding

18.35.3 - TIC - WAFER MOUNTING

Below are two mounting methods of TIC substrates after the substrate fabrication. One is an old method and one of a newer method. (see fig. 389).

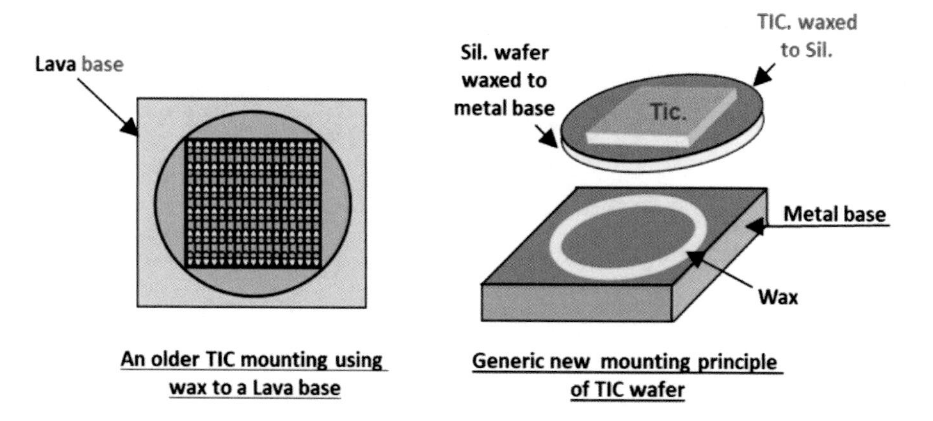

Figure 389. – Substrate mounting methods

18.35.4 - TIC - EXTERNAL INITIAL SUBSTRATE/ WAFER SEPARATION / DICING THE INDIVIDUAL SEGMENT ROW / GROUPS.

Below are a few options of wafer separations prior to the row slicing process. (see figs. 390, 391, and 392).

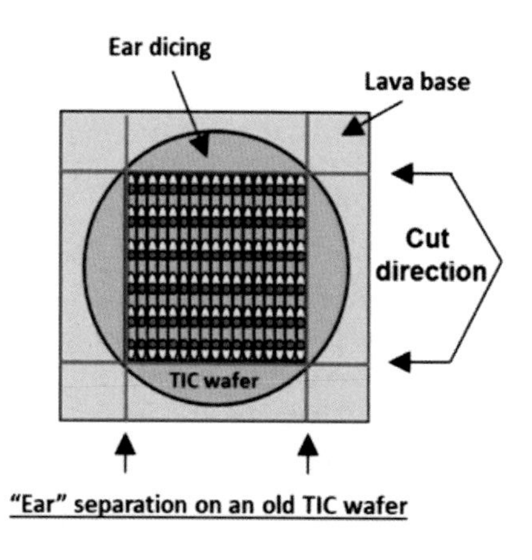

Figure 390. –

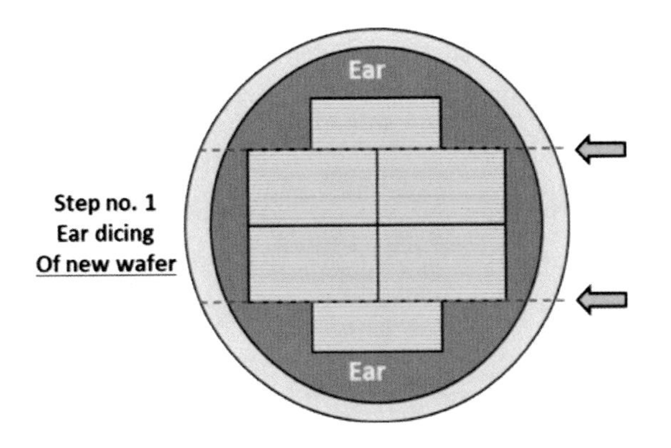

Figure 391.

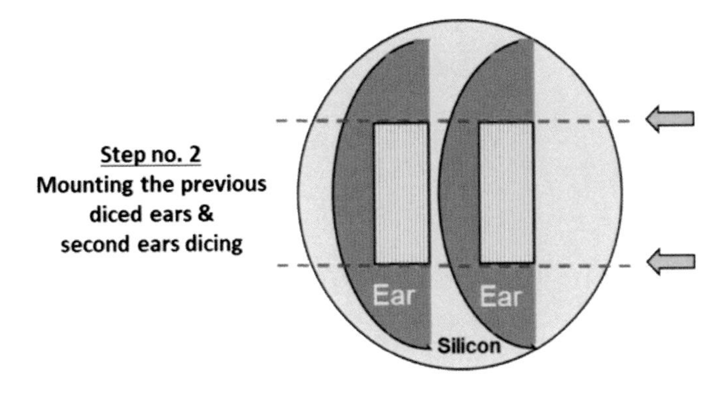

Figure 392.

In some cases, the outside of the segments is already one outside edge of a single row. The blades used for the row dice and the row segments vary in the industry, however they are in the following range:

Blade binder – Nickel

Blade O.D. – 4.0"–4.4" (3.80"–4.00")

Blade Thickness – 0.095 mm–0.100 mm

Diamond grits – 4–8 mic, 8–12 mic, 10–16 mic.

Cutting parameters:

Spindle rpm – 8–12 Krpm

Feed rate – 100 mm / min (+ –)

18.35.5 - TIC - COOLANT DESIGN

The coolant is a major factor, it is divided into two parameters. The cooling nozzle geometry is critical in order to minimize vibration of the thin blades. This was discussed already; a much larger diameter cooling nozzle is used in the range of 6–7 mm. A round nozzle is used in order not to deflect the water stream and in addition the coolant nozzle is a bit far from the blade, in order not to vibrate the blade. (see fig. 393).

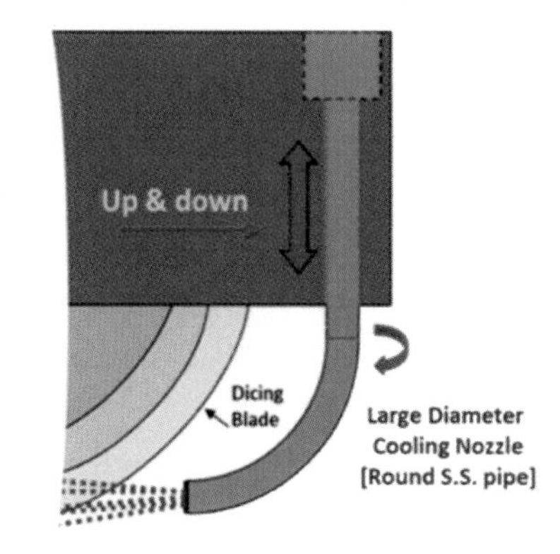

Figure 393. – Special large diameter cooling nozzle

The other important parameter regarding the coolant is adding additives to lower the surface tension in order to cool better the blade and the substrate and to better wash away the powder residue from the dicing. Some manufacturers are also chilling the coolant not necessarily to very low temperatures but to maintain an even coolant temperature during the dicing process and between all saws.

DYNAMIC BALANCING:

This process is done in the wheel room right after dressing the blade and achieving the perfect run-out of the blade edge to the spindle rotation. See in the dynamic balancing section. A typical dynamic vibration spec no. is < 0.001 cm /sec

Second substrate mounting [On some applications], lapping and row dicing:

Depending on the history of the process old and new, a second mounting is performed prior to some of the row slicing processes. Fig. 394 shows an old type TIC substrate.

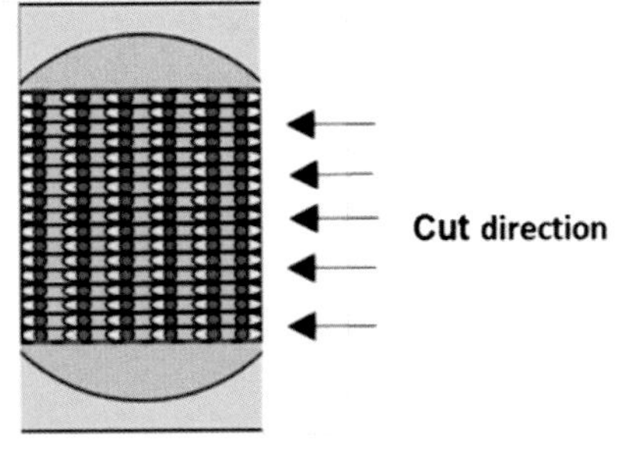

Figure 394 – Row dicing on substrate mounted on hard base

18.35.6 - ROW DICING

Dicing the rows on those "old days" required a lapping process after the row slicing to meet an accurate dimension between the row side and the coil. (see fig. 395).

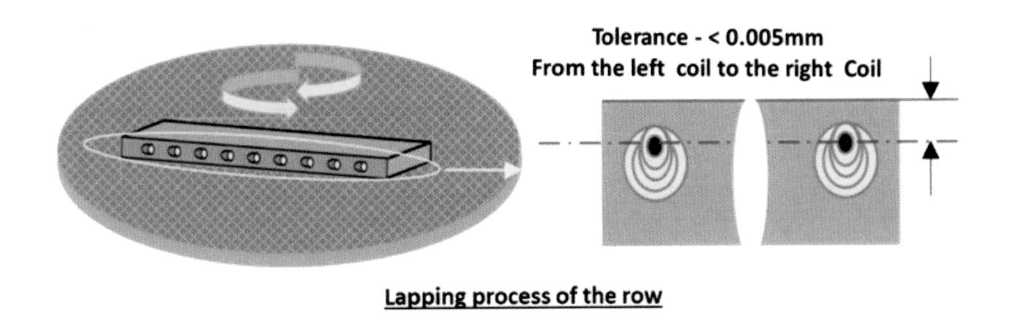

Figure 395. – Lapping the diced rows prior to parting

The newer processes are very different and involve some amazing engineering designs for mounting jigs. After singulating the segments out of the TIC substrate / wafers, each segment is glued and mounted to an accurate ceramic or carbide holder. Each segment includes a few rows to be diced in a unique process. The first initial method was to first lap the side of the side row to a flat surface and to the right dimension to the coils. The lapped side surface is then mounted to an accurate metal jig with a side vacuum that holds the row after it is singulated from the segment. This process is repeated till all rows are singulated. Fig. 396 is a top view of the row segment mounted to an accurate ceramic holder and facing the side vacuum jig.

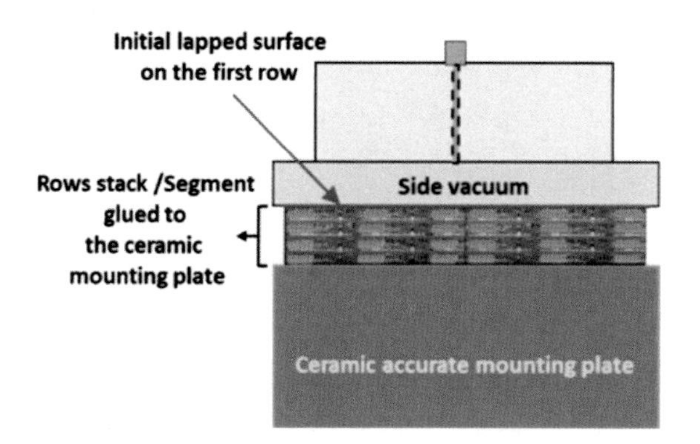

Figure 396 – Row dicing on side vacuum jig

Below is a more detailed view of the process. (see fig. 397).

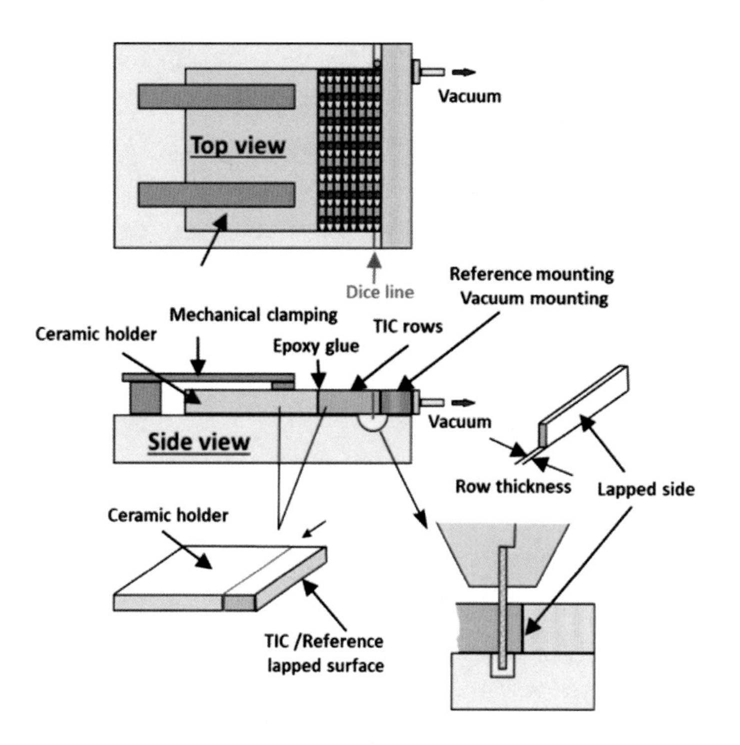

Figure 397. – Detailed view of row mounting and dicing

The above process was optimized and upgraded during the years to the degree that it became easier and more accurate to achieve the specs and the yields. A unique process set-up was developed with the help of the dicing saws suppliers. A combination of grinding the side front row is performed prior to the singulation process of the row in the same mounting set-up. This set-up is done with different similar designs among the different manufacturing locations. Fig. 398 demonstrates one generic set-up just to explain this unique process.

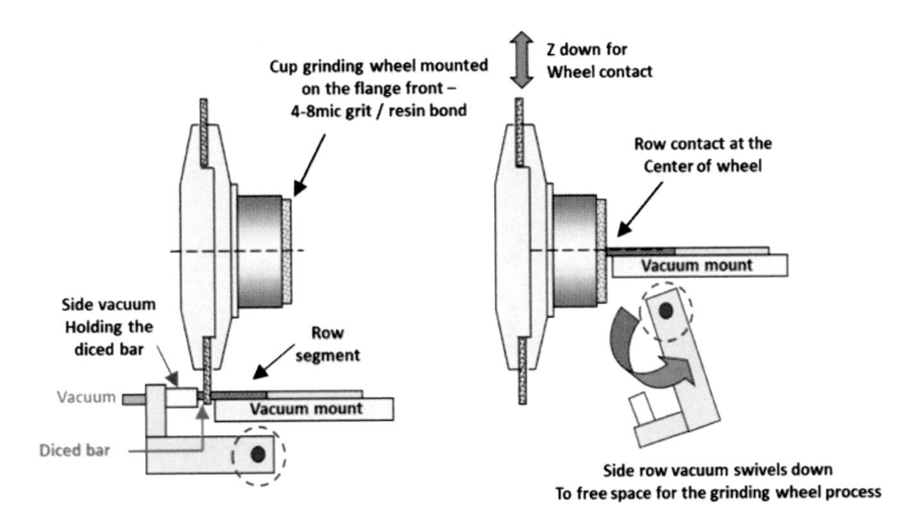

Figure 398. – Grinding and row slicing at the same set-up

Two passes are made with the grinding wheel at the center Height of the wheel.

After getting close with the grinding wheel to the side rod, the – X – slide is moving till there is a min. contact between the wheel and the row. About 0.005 mm is grinded from the row side. This is done in order to minimize chipping and to minimize the lapping cycle which is the next process done on each rod. After singulating the rows, they are lapped to meet the required dimensions from the row side surface to the front row geometry. In some cases, there is a point or line mark on the top row. The normal tolerance is <0.005 mm between the left side to the right side, in some cases even tighter. Fig. 399 shows a generic marking.

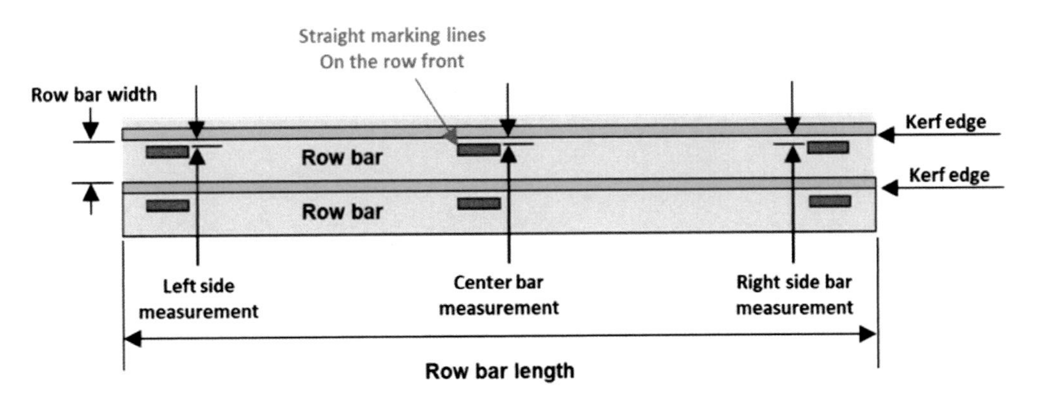

Figure 399. – Marking aligning marks on the wafers

The Row dicing process using relatively thin blades with fine diamond grits creates loading on the blade affecting the quality and accuracy. Fig. 400 shows a good edge quality and more chipping after blade loading.

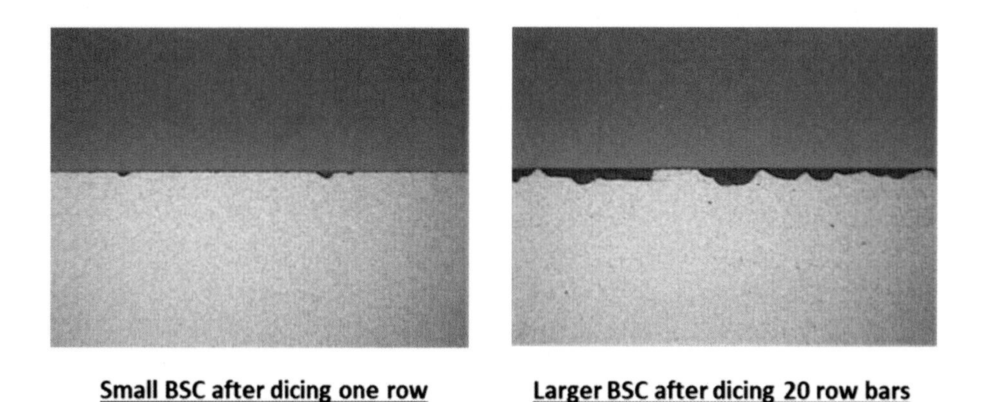

Figure 400. – Back side chipping samples after Row dicing

In order to minimize the loading and maintain the quality spec an online dressing is performed. Special mounting jigs are designed and used in mass production. The media used for the online dressing is aluminum oxide with a mesh size of 1000 and in some cases 2000. The blade is programmed to dice through the Aluminum oxide after every few cuts, and in some cases, even after every cut of a single TIC row bar dicing. The purpose of the online dressing is to have a continued diamonds well exposed while minimizing blade side wear. Fig. 401 shows a generic chuck design with the row mounting and the dressing media mounting.

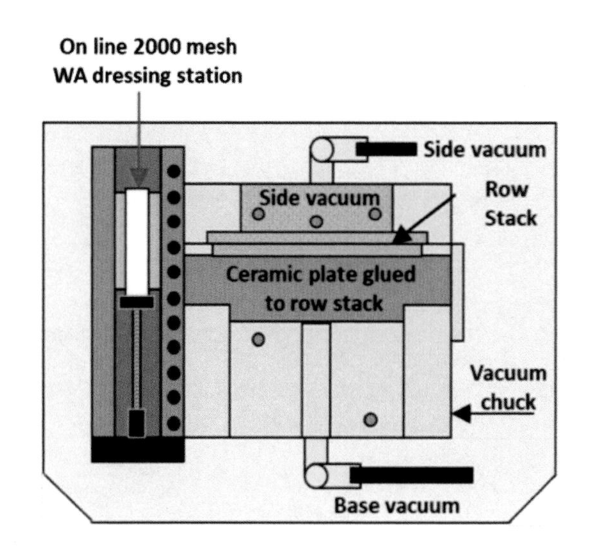

Figure 401. – Row mounting jig and dress station

The quality specs of the rows vary between the different head manufacturers. Below are a few spec no.'s just to get an idea:

Kerf size – + – 0.005mm from the nominal size.

Cut perpendicularity – < 0.005mm and in some <0.010mm. (see fig. 402).

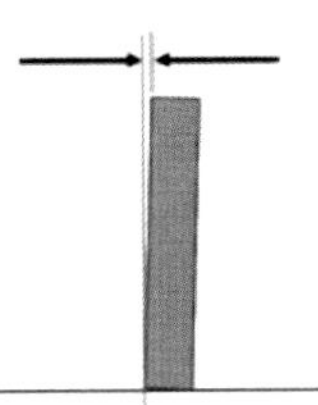

Figure 402. – Cut perpendicularity

Top and back side chipping – < 0.010mm (Top side chipping is normally less than 0.005mm)

Skew / Wheel walk: – < 0.005mm. The Skew and wheel walk phenomena is important in order to meet the tight row specs. (see fig. 403).

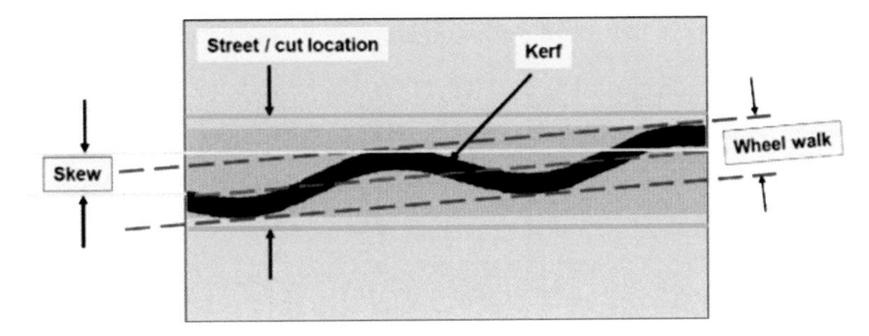

Figure 403. – Skew and wheel walk illustration

18.35.7 - PARTING THE ROWS.

The last process of the small heads is to separate them from the diced and lapped rows.

This process is called PARTING. The only way to separate the rows into individual heads in an economical process is to dice them in batches at the same time. This requires an initial very accurate mounting process to align the rows in order to meet the dimension spec of all heads after each cut.

A few aligning and mounting processes were developed, all of them are using sophisticated optical instruments for aligning. Below is a sketch showing the top view of the important aligning geometries that need to be aligned. (see fig. 404).

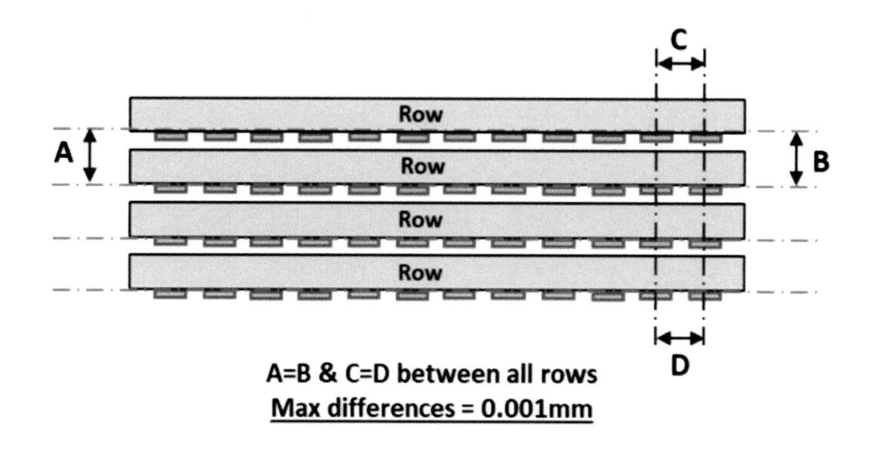

Figure 404. – Important aligning points

Mounting is the main difference among a few design options. Following are a few design options among others. (see fig. 405 and 406).

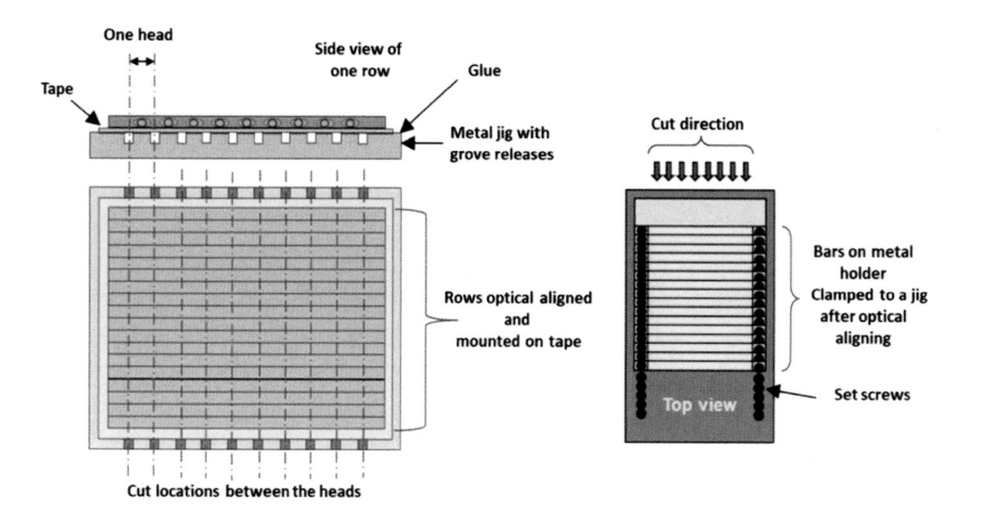

Figure 405. – Different mounting Jigs and aligning options

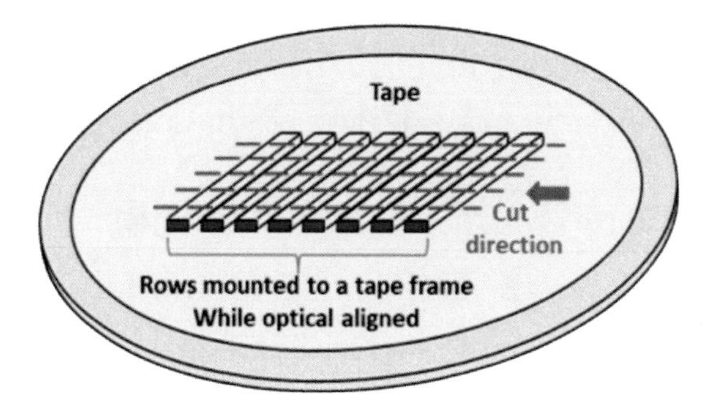

Figure 406. – Tape mounting and special aligning

Some manufacturing houses are using gang type saws for the parting process. This process requires another accurate setup to maintain, the blade spacing to meet the accuracy dimension of the head size and at the right location. Assembling a gang arbore requires special technical know-how of handling the spacers, the blades and special clamping torquing. Needless to say, this process also requires the row aligning process if the dicing is done using the batch row design. (see fig. 407).

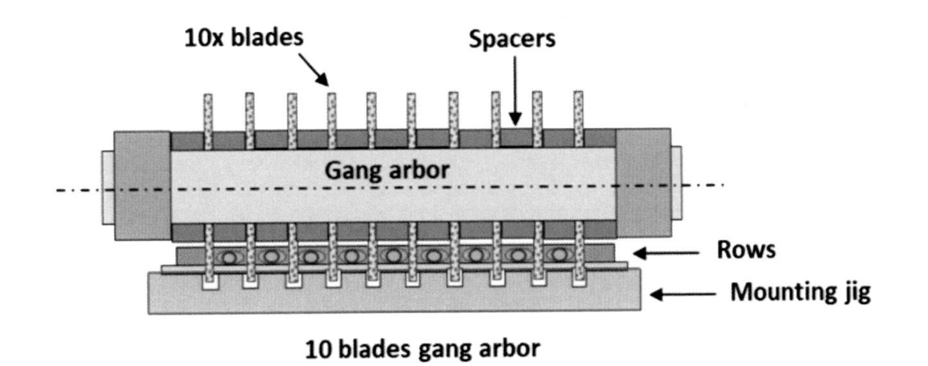

Figure 407. – Gang arbore dicing on special mounting jig

The gang process is saving time if the head geometry meets the spec. However, one disadvantage is when one blade is not meeting the spec or if it breaks, it is a production stopper; but in a production mode a replacement Arbore is ready to be mounted.

Blades used for the parting process are also 4.0–4.5" O.D. nickel blades but thinner in the range of 0.060 mm with smaller diamond grit sizes of 3–6 and 4–8 mic. The cutting parameters depend on the saw, the mounting method and final quality specs.

Spindle speed: 8–10 Krpm, Feed rate: 1–4 mm/sec., Kerf size tolerance – ±0.002 mm,

Top and bottom chipping: 0.005–0.010 mm depending on the head type. Cut perpendicularity: <0.005 mm. Bottom kerf radius is important when dicing on tape to eliminate any lip at the bottom side of the head. The max radius is in the range of 0.015 mm max on each side. A routine blade edge is performed by O.D. grinding or by dressing the blades on a nickel/diamond dress plate continuing with a fine dressing on silicon carbide dress media (see fig. 408).

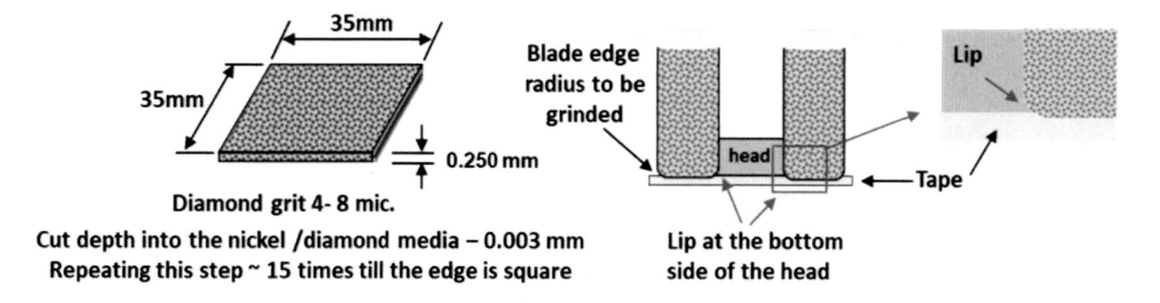

Figure 408. – Correcting the blade edge flatness on a nickel diamond dress media

Beside the relatively slow process parameters, mainly the feed rates to maintain good cut quality, a lot of other process parameters are critical and time consuming. The other main parameters to continuously control are: Mounting Jig accuracy, maintain coolant additive % and coolant temperatures, flange condition, blade loading, dynamical balancing and of course cut quality.

As mentioned, the head manufacturing process is one of the most complicated, most accurate and demanding applications.

CHAPTER 19

WEAR RATE CALCULATION

Before ending this blade and dicing technical review, one more discussion regarding measuring blade wear. Most customers are just measuring the blade wear per the cut length. Some are counting the no. of substrates they can cut using one blade. Following is a more detailed review that is mostly used in the grinding industry and is adapted by some high dicing production valium users.

The professional wear term is called "G Ratio" and is the relation between the amount of material (Substrate) removed and the amount of blade consumed, all in volume measured. The formula of the G ration is in Fig. No. 409.

$$\text{G Ratio} = \frac{\textbf{Volume of Substrate removed}}{\textbf{Volume of blade consumed}}$$

Figure 409. –

Lower wear rate & higher G ratio = Longer blade life

The blade wear calculation is simple: (see fig. 410).

$$\text{Wear Rate} = \frac{\textbf{µm of radial blade wear}}{\textbf{Meters of cut length}}$$

Example:

Total radial blade wear = 1.00mm

$$\text{Wear rate} = \frac{\text{1.00mm}}{\text{1000 meters}} = \underline{\textbf{0.001mm / meter}}$$

Total cut length = 1000 meters

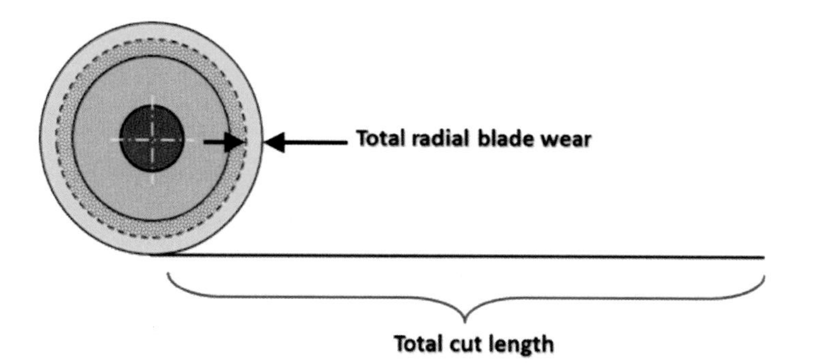

Fig. 410.-

Volume calculations for G Ratio calculations: (see fig. 411).

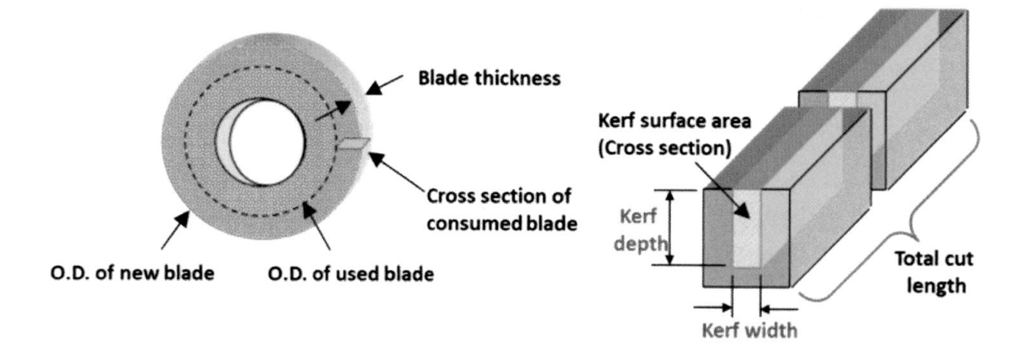

Figure 411.

Blade volume calculation including the blade I.D. = R^2 x π (3.24) x blade thickness

Blade wear volume calculation=

New blade volume – Used blade volume = Consumed blade volume

(Radius of new blade^2 x π (3.14) x blade thickness) – (Radius of used blade^2 x π (3.14) x blade thickness

Calculation of substrate volume removed:

Kerf surface area = kerf width x kerf depth

Total consumed substrata valium = Kerf surface area x total cut length in mm

Example of G Ratio calculations:

New blade diameter – 56mm / radius = 28mm

New blade thickness – 0.250mm

Cut depth – 0.300mm

Cut length – 1000meter.

Total wear rate – 1mm on the exposure / 1000meters = 0.001mm / meter (= 2mm on the diameter)

Blade diameter after dicing – 54mm / Radius = 27mm

Calculation of blade consumption (mm)

New blade volume:

$28^2 \times 3.14 \times 0.250 = 615.44^3$

Used blade volume

$27^2 \times 3.14 \times 0.250 = 572.26^3$

Used blade consumption:

$615.44^3 - 572.26^3 =$ **43.18^3**

Calculation of substrate volume removed:

1000 meters (1,000,000mm)

$0.250\text{mm} \times 0.300\text{mm} \times 1{,}000{,}000\text{mm} = 75{,}000^3$ (Total substrate removed)

$$\text{G Ratio} = \frac{\textbf{Volume of Substrate removed}}{\textbf{Volume of blade consumed}}$$

$$\text{G ratio} = \frac{\mathbf{75000}}{\mathbf{43.18}} = \underline{\mathbf{1736.91}}$$

CHAPTER 20

SHORT SUMMARIZE.

The idea of this technical review was to cover the blade side and the dicing side of the many different applications in the microelectronic marketplace. There are many good technical papers dealing with different specific applications, but I could not find an existing general, detailed review covering all the blade and dicing application aspects, which led me to write this technical review. All the information is based on personal dicing blade development that I was in charge of and personally involved with. The application side is also based on being in charge of the application development center, which involves many new applications. In addition, a lot of good information was gathered by meeting many customers and attending technical seminars I provided worldwide. I would like to mention that not all the applications were reviewed, just the major ones. Some of the dicing parameters and specs described may be different between manufacturers, but they are all close to what is used in the marketplace. The idea is to provide a good understanding of what the market is using. Some photos and sketches are taken from internet sites after getting a permit from the different vendors. All the vendor permit names are indicated in the reference section or below the photos. I would like to take this opportunity to thank them all for their great help. I hope I managed to highlight all the major blade and dicing process-related information. This technical review is meant to help young process engineers as well as experienced engineers and managers facing new applications. For specific questions, I can be reached via Linkedin.

REFERENCES

finishing.com

https://www.finishing.com/index.shtml

What Is Induction Heating?

http://www.gh-ia.com/induction_heating.html

Handbook of Thermoset Plastics – ScienceDirect

https://www.sciencedirect.com/book/9781455731077/handbook-of-thermoset-plastics

Xinxiang New Zuan Diamond Tools Co – Vitrified wheels

www.superabrasivetools.com

Identifying Oil Canning! An Auto Form Strategy – Forming World

https://formingworld.com/identifying-oil-canning/

Archimedes' Principle

https://courses.lumenlearning.com/physics/chapter/11-7-archimedes-principle/

KEYENCE CORPORATION – non contact thickness measurement

https://www.keyence.com/ss/products/measure/select/industry/semiconductor.jsp

Wafer ESD in dicing saws and the effect of the countermeasures

https://www.disco.co.jp/eg/solution//technical_review/doc/TR16-02_Wafer%20ESD%20in%20dicing%20saws%20and%20the%20effect%20of%20the%20countermea-sure_20160610.pdf

TURBULA® powder mixing

https://www.wab-group.com/en/mixing-technology/3d-shaker-mixer/product/turbula/

Colibri Air-Bearing Dicing Spindles

https://colibrispindles.com/colibri-air-bearing-dicing-spindles/

Design of bronze-bonded grinding wheel properties

https://daneshyari.com/article/preview/10672971.pdf

FARMAN MACHINERY IND. CO., LTD – Dressing machines

https://www.farman.com.tw/products.htm

Dicing technologies for SiC

https://www.disco.co.jp/eg/solution/technical_review/doc/TR16-05_Dicing%20technologies%20for%20SiC_20160610.pdf

Wafer Dicing Using Dry Etching on Standard Tapes and Frames

https://www.researchgate.net/publication/267327451_Wafer_Dicing_Using_Dry_Etching_on_Standard_Tapes_and_Frames

Research of diamond concentration in dicing blade effect by electroplating parameter – by ZZSM Engineering

https://iopscience.iop.org/article/10.1088/1757-899X/612/3/032049/pdf

Hyperion diamonds

https://www.hyperionmt.com/products/Abrasives/

Front book photo – Graphics from an ADT 2" dicing saw photo

Dicing of Fragile MEMS Structures

January 2008 Materials Research Society symposia proceedings. Materials Research Society 1139

https://www.researchgate.net/publication/231857036_Dicing_of_Fragile_MEMS_Structures

MEMS and MEMS Microfabrication By Sensera

https://sensera.com/2020/wp-content/uploads/2018/04/What-are-MEMS.pdf

Introduction to MEMS (Micro-electromechanical Systems)

http://www.primetechnologywatch.org.uk

Hamamatsu – Stealth Dicing™ technology

https://www.hamamatsu.com/eu/en/product/type/L9570/index.html

Dicing Opto – Electronic Components for the Communication Market

https://www.adt-co.com/userdata/SendFile.asp?DBID=1&LNGID=1&GID=710

Yole – Glass substrate for semiconductor applications 2020 www.yole.fr/Glass_Substrate_For_Semiconductor_Applications_Update_March2021.aspx

Swift Glass – Glass wafer fabrications

https://www.swiftglass.com/glass-wafers

Silicon on glass – handbook

https://www.mems-exchange.org/catalog/P3715/file/0655968ae8db/

SAW AND INTERDIGITALTRANSDUCERS

https://www.ijser.org/researchpaper/SAW-AND-INTERDIGITAL-TRANSDUCERS.pdf

Avoidance of Ceramic-Substrate-Based LED Chip Cracking

Induced by PCB Bending or Flexing

https://cree-led.com/media/documents/LED_Chip_Cracking.pdf

The function of silicone coating on the LED chip of SMT LED?

https://www.quora.com/What-is-the-function-of-silicone-coating-on-the-LED-chip-of-SMT-LED

Chemical compatibility of LEDs – Osram

file:///C:/Users/glevi/Downloads/Chemical%20Compatibility%20of%20LEDs%20(1).pdf

High and Low Temperature Cofired Multilayer Ceramics (HTCC and LTCC)

file:///C:/Users/glevi/Downloads/02_03SCHOTT_DB_HTCC_TTCC_RZ_E_2018_08_22%20(3).pdf

High Temperature Cofired Ceramic (HTCC) Package Design and Applications-AMETEK

https://www.ametek-ecp.com/-/media/ametek-ecp/v2/files/productdownloadabledocuments/datasheets-hermetic-packaging/hightemperaturecofiredceramicpackagedesignandapplicationsimaps2014.pdf

Public Material – und Design rules for Working Group Microsystems, LTCC and HTCC
https://www.ikts.fraunhofer.de/content/dam/ikts/downloads/electronics_and_microsystems/hybrid_microsystems/20170714_DesignGuide_LTCC-HTCC_public%20.pdf

Technology Overview LTCC – White paper by Plextek RF Integration

https://www.prfi.com/wp-content/uploads/2021/05/LTCC_technology_overview.pdf

A comprehensive overview on today's ceramic substrate technologies

https://www.researchgate.net/publication/224597543_A_comprehensive_overview_on_today's_ceramic_substrate_technologies

Dicing Through Hard and Brittle Materials in the Micro Electronic Industry

https://www.adt-co.com/ADT2014//userdata/SendFile.asp?DBID=1&LNGID=1&GID= 574

Development & prototyping of MEMS inkjet devices

https://www.engineeringsolutions.philips.com/app/uploads/2017/03/mems-inkjet-print-head-presentation.pdf

Si-MEMS Heads – What is the Fuss About

https://inkjetinsight.com/knowledge-base/si-mems-heads-what-is-the-fuss-about/

The ultrasound transducer & Piezoelectric Crystals

https://ecgwaves.com/topic/the-ultrasound-transmitter-probe/

Inkjet Printing Technique and Its Application in Organic Light Emitting Diodes

https://www.oldcitypublishing.com/wp-content/uploads/2017/04/DAIv2n3-4p339-358AC.pdf

Ball Grid Array (BGA): Features, Soldering Technique, and X-Ray Inspection

https://www.protoexpress.com/blog/bga-features-soldering-x-ray-inspection/

A Brief Introduction of BGA Package Types

https://www.pcbcart.com/article/content/introduction-of-bga-package-types.html

What Are QFN Packages?

https://www.protoexpress.com/blog/what-are-qfn-packages/

CONSIDERATIONS FOR MLP/QFN SUBSTRATE SINGULATION

https://www.adt-co.com/userdata/SendFile.asp?DBID=1&LNGID=1&GID=708

Assembly guidelines for PwrQFN (Power Quad Flat no-Lead) packages

https://www.nxp.com/docs/en/application-note/AN2467.pdf

Saw Singulation Characterization on High Profile Multi Chip Module Packages with Thick Leadframe

https://nazrulanuar.com/author/wp-content/uploads/2008/11/c3-0004pg298-302.pdf

HDD from Inside: Hard Drive Main Parts

https://hddscan.com/doc/HDD_from_inside.html

Anatomy of a Storage Drive: Hard Disk Drives

https://www.techspot.com/article/1984-anatomy-hard-drive/

Schenck balancing

https://schenck-rotec.com/homepage.html

JP – balancing machines

http://www.jp-balancing.com/

Hanmi Semiconductor Co., Ltd

http://www.hanmisemi.com/_2013dev/en/core2020/core_tech01.asp